WASTE WATER

WASTE WATER

Editor

Cullen Jeffries

scitus
academics

Waste Water

Edited by **Cullen Jeffries**

This Book is Printed on Acid-Free Paper. ♾

ISBN: 9781681179728

Library of Congress Control Number: 2017949111

SCITUS Academics LLC,

616, Corporate Way, Suite 2, 4766,

Valley Cottage, NY 10989

www.scitusacademics.com

Printed in United States of America.

Preface

Waters that are used for drinking, manufacturing, farming, and other purposes are degraded in quality as a result of the introduction of contaminating constituents. Organic wastes, suspended solids, bacteria, nitrates, and phosphates are pollutants that commonly must be removed. As mankind eventually adopted a more settled, non-nomadic way of life, people become increasingly involved in the technical aspects of water. To a large extent, the primary concerns in the beginning were utilization and improvement of existing water resources, together with protection against the hazards and potential harm associated with uncontrolled natural water. It was only toward the end of the nineteenth century that wastewater become an issue in science, technology, and legislation, specifically, its production and treatment, in terms of both municipal and industrial sources. Water supply and water quality degradation are global concerns. Many natural water bodies receive a varied range of waste water from point and/or non point sources. Water has been used in abundant quantities by chemical, petrochemical, petroleum refining and other process industries. However, in recent years, the increased cost of wastewater treatment to meet environmental requirements and the scarcity of less expensive industrial water have provided process industries with strong incentive to minimize the amount of water consumption and wastewater discharge. The major concern is to emphasize the importance of water reuse and a number of efforts have been made towards achieving the goal of extensive water reuse in various process industries. Hence, there is an increasing need for better tools to assess the effects of pollution sources and prevent the contamination of aquatic ecosystems. So, both optimized waste water treatment technologies and modern tools to assess the effects of pollution sources are necessary to prevent the contamination of aquatic ecosystems.

This volume covers a wide spectrum of issues related to waste water monitoring, the evaluation of waste water effect on different natural environments and the management of water resources. It deals with interdisciplinary topics concerning waste water treatment technologies, water quality monitoring and evaluation of waste water impact on natural environments. This book will be helpful for graduate students, environmental professionals and researchers of various disciplines related to waste water.

Table of Contents

CHAPTER 1

Pulp Mill Wastewater: Characteristics and Treatment

María Noel Cabrera*

[1] *Universidad de la República, Engineering School, Chemical Engineering Institute, Forest Process Engineering, Montevideo, Uruguay*

ABSTRACT

The production of chemical pulp in recent times is 180 million tons per year; while the production of eucalyptus pulp has increased intensively, especially in the southern hemisphere. The pulp and paper industry has long been considered a large consumer of natural resources (wood and water) and one of the largest sources of pollution to the environment (air, water courses and soil). Important efforts are being made to reduce the pollutant levels and water consumption of the industry. The wastewater composition, and therefore, the efficiency of effluent treatments and characteristics of the discharges to water are strongly dependent on the applied technology and raw materials. Despite a large body of literature on softwood-based wastewater, few studies have examined the characteristics of kraft eucalyptus bleaching effluents and their behaviour in the different biological treatments. The largest secondary treatment systems today use the activated sludge process. Sixty to seventy-five per cent of all the biological effluent treatment plants within the pulp and paper industry use this kind of treatment system. This chapter reviews the current pulping technologies at mills and compares the chemical composition and biological treatment of wastewater between softwood and hardwood bleached pulps.

Keywords: pulp mills, hardwood, softwood, kraft pulping, ECF-TCF bleaching

1. INTRODUCTION

The pulp and paper mill industry is an intensive consumer of water and natural resources (wood), discharging a variety of liquid, gaseous and solid wastes to the environment. Since the 1970s, a growing awareness of the effects of pulp and paper wastes in the ambience had prompted water and energy consumption levels and the loads of toxic compounds discharge to reduce. One of the most important implemented changes in this regard was made within the mill, wherein chlorine was completely substituted by, that is, chlorine dioxide as the bleaching chemical agent. Another major issue was the implementation of secondary biological treatments. The wastewater composition and hence the effluent treatment efficiencies and characteristics of the discharges are strongly dependent on the technology applied and the raw materials. In the last 25 years, however, the global distribution of pulp producers has significantly changed and so have the species of wood used. Eucalyptus pulp production, for example, is becoming a leader in the hardwood pulp market; Brazil went from being a pulp consumer to a world leader in hardwood pulp production, and since 2008, it has been the fourth largest pulp producer in the world.

2. WOOD PULP MARKET

Cellulose pulp is the main raw material in the production of different types of paper and paperboard. It is also used as the absorbent material in diapers and other sanitary products.

The global pulp market has changed intensely in recent years. A few decades ago, this industry was characterized as national character as a supply industry inputs for domestic production of paper and paperboard. Globalization has led to increased competitiveness in the international market, as new players have emerged both at the level of producers and consumers. Within the latter, the appearance of China and India have strongly modified cellulose demand worldwide [1].

Figure 1 graphically shows the evolution of world's production of wood pulp between 1979 and 2013 according to the data published by FAO [2–6].

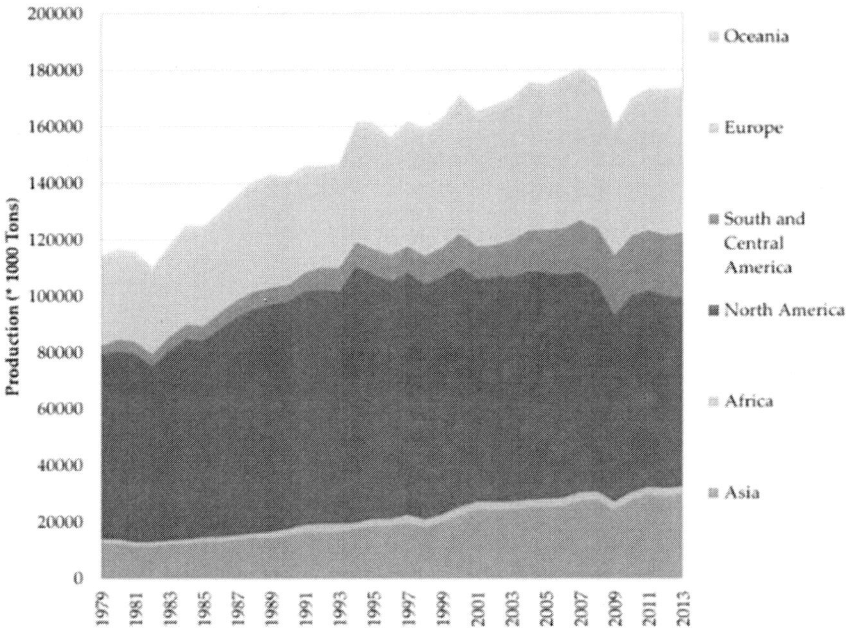

Figure 1. World pulp production 1979–2013. Data obtained from Ref. [2].

It is clearly shown that the world's wood pulp production increased to about 50% in this period, from 120 million tons in 1979 to nearly 180 million tons in 2013. Part of this growth can be explained by the explosive increase in production in non-traditional wood pulp producing regions such as Asia and South America. The main producing regions are still North America with 38% and Europe with 28%, even though in 2013, Asia produced 17% of the wood pulp and South America about 13% [6].

Wood pulp grades are categorized according to the pulping process, which can be classified as mechanical, semi-chemical and chemical pulps. In a mechanical process, logs or wood chips are mechanically grinded by abrasive action. In a chemical cooking process, a significant part of the wood components (mainly lignin) is chemically dissolved to obtain a solid compound with high cellulose fibre content. There are two main methods of chemical pulping: (1) sulphite pulping and (2) sulphate (kraft) pulping. The first process—sulphite cooking process—uses aqueous sulphur dioxide (SO_2) and a base of calcium, sodium, magnesium or ammonium. The kraft process uses a treatment comprising a mixture of sodium hydroxide and sodium sulphide, known as white liquor, at a high pressure and temperature. The semi-chemical pulping process combines chemical and mechanical methods, where wood chips are first softened or partially cooked with chemicals and then mechanically pulped [7].

Figures 2 and 3 illustrate the different kinds of pulp produced in 1979 and 2013.

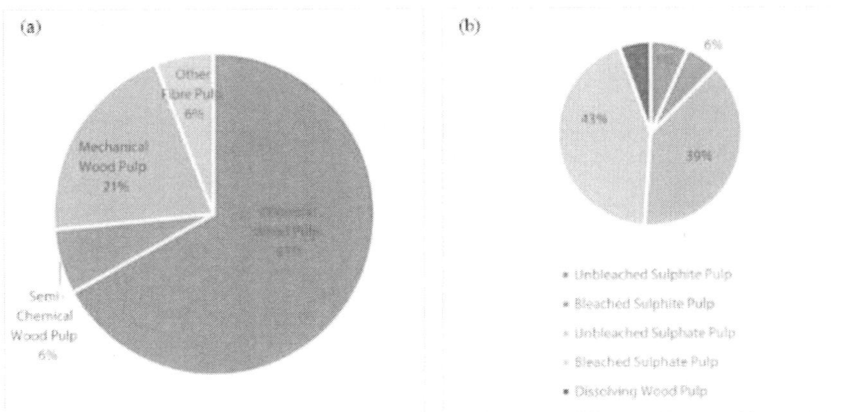

Figure 2. (a) World pulp production by type of pulp in 1979; (b) different kinds of chemical pulps produced in 1979 (Data from FAO [2]).

Figure 3. (a) World pulp production by type of pulp in 2013; (b) different kind of chemical pulps produced in 2013 (Data from FAO [6]).

The rise in wood pulp production is due to an increase in chemical pulp production, as the production of mechanical pulp has declined in the same period. Mechanical pulping has the advantage of converting up to 95% of dry weight wood into pulp, although considerable amounts of energy are required to do so. The pulp obtained produces a highly opaque paper with good printability, but the physical properties are inferior than chemical pulps and yellowing when exposed to light. Moreover, mechanical pulps are mainly produced from softwood [7].

There are significant changes in the production of chemical pulp. The use of sulphite cooking process in pulp production compared to kraft pulping technology decreased steadily, from 60% in 1925 to 20% in 1967 and 9.2% in 1979 to only 2.4% in 2013 [6, 8]. The superiority of kraft pulping process is explained by the following facts: (1) all wooden materials including low-quality wood can be used as raw material; (2) superior fibre strength of pulp compared to other chemical pulping methods; (3) more simple chemical and energy recovery process; (4) scale of economy of kraft methods prevents competition and (5) low environmental risks in modern mills [9].

A second classification considers the type of wood used by distinguishing softwood or long fibre (produced mainly from pine and spruce) from hardwood or short fibre (produced from eucalyptus, birch, poplar, etc.) [10]. A gradual move from softwood to hardwood can be observed. In 2013, 56% of bleached kraft pulp was produced with long-fibre wood (softwood), while the remaining 44% was produced with short-fibre wood (hardwood) (according to data from Ref. [6]). In 1980, the production capacity of bleached kraft pulp corresponded to 63% of softwood pulp. The entry into the market of non-traditional producing countries such as Brazil, Indonesia, Spain and Portugal, significantly increased the production of hardwood pulp. Eucalyptus bleached pulp production is rapidly increasing (from 8 million tons in 2003 to nearly 15 million in 2015), and eucalyptus wood is thus considered to be the most important raw material of hardwood bleached market pulp in the world [11].

As kraft pulping is by far the most common process used these days, this chapter will focus in the wastewaters generated in this process.

3. MAIN PROCESSES DESCRIPTION

3.1. Mechanical pulping

The oldest method of mechanical pulping is the groundwood process. In this process, round logs are forced against a rotating pulp stone (revolving at peripheral speeds of 1000–1200 m/min), under specified conditions of pressure and temperature. Atmospheric grinding, pressure grinding and thermo-grinding could be done according to the applied temperature and pressure. In all of them, the temperature levels obtained from the heat applied or from rubbing the logs on the stone soften and break down the fibres structure; and cracks the fibres from the wood matrix [7, 8].

Another common method is the refiner mechanical pulping (RMP). The wood chips are pulled between two rotating disks. Among them, thermomechanical pulping operates like RMP, but under higher temperature and pressure. The high temperature and pressure levels soften the lignin even more than frictional heat, making fibres separation easier. Thermomechanical pulp is stronger than refined mechanical pulp, and still retains the high-yield and cost-effectiveness of mechanical pulps [7].

3.2. Chemical pulping

3.2.1. Sulphite pulping

Sulphite process is very versatile, and covers the entire pH range, achieving high fibre flexibility in pulp yields and properties. The cooking process involves the use of aqueous sulphur dioxide (SO_2) and a base: calcium, sodium, magnesium or ammonium. Sulphite pulping was developed in the second half of the nineteenth century and for several decades, the calcium acid sulphite process was the most common method. However, since 1950, the utilization of bases other than calcium has been a major development. The specific base used will determine the process's chemical and energy recovery system and water use. The use of the relatively cheap calcium base has become obsolete because the cooking chemicals cannot be recovered. Magnesium and sodium bases allow chemical recovery, and magnesium bases are currently the dominant choice in sulphite pulping process [7, 12].

3.2.2. Kraft pulping

In kraft pulping, white liquor, containing mainly active chemicals—sodium hydroxide and sodium sulphide—is used for cooking the chips at a high temperature (150–170°C) and pressure. Approximately, half of the wood composition degrades and dissolves during cooking. The spent cooking liquor (black liquor) contains reaction products of lignin and hemicelluloses, and is concentrated and burned in a recovery boiler that recovers the cooking chemicals and generates energy. The smelt is dissolved into water to form green liquor (mostly sodium carbonate and sodium sulphide), which then reacts with lime to convert the sodium carbonate into sodium hydroxide regenerating the white liquor. After cooking and washing, a brown pulp (brown stock pulp) is obtained. Printing, writing and tissue papers require the pulp to be bleached which removes the excess lignin and chromophores to produce a "white" pulp.

4. BACKGROUND OF PULP MILL EFFLUENTS: ENVIRONMENTAL FATE AND EFFECTS

The pulp and paper industry consumes enormous amounts of water and natural resources and is also one of the largest effluents generators. Before the 1970s, wastewaters from the pulp and paper mills were normally discharged directly to the rivers or lakes, without any treatment or even a rough primary treatment. The high organic loads and solid content in the effluents affected the aquatic ecosystem in several ways such as localized damage to the benthic community, oxygen depletion in large areas and numerous changes in fish reproduction and physiology. In the 1980s, studies in Scandinavia, along the Baltic Coast and the Gulf of Bothnia, showed alterations in fish reproduction and increase of diseases and parasites [13, 14].

Studies conducted in USA and Canada in the beginning of the 1990s, under the Environmental Effects Monitoring (EEM) program [15, 16], revealed delayed sexual maturity, smaller gonads, changes in fish reproduction and depression in secondary sexual characteristics in species living downstream of pulp and paper mills discharges.

From the end of the 1970s until now, the main concern regarding effluents is the formation of chlorinated compounds in bleaching plants. In 1985, 2,3,7,8-tetrachlorodibenzo-p-dioxin (TCDD) was discovered in the pulp mill effluents, which led to a general concern over the formation of chlorinated organic matter in chlorine bleaching. Consequently, the use of chlorine in the bleach plants gradually decreased until it was completely substituted with chlorine dioxide. In many countries, the environmental control authorities set strict restrictions on the discharges of chlorinated organics, measured as adsorbable organic halogen (AOX), into the aquatic environment.

In 1992, the Swedish Environmental Protection Agency limited organochlorines emissions to 1.5 kg AOX/t of pulp and in 1995, Finland's official limit was set at 1.4 kg AOX/t of pulp [14].

Several authors reported that with the replacement of chlorine with chlorine dioxide, the effluent quality improved in AOX levels and the elimination of detectable amount of dioxins, polychlorinated compounds and chloroform [12, 13, 17].

The European Integrated Pollution and Prevention Control [12] has created reference documents (BREF) that set the Best Available Techniques (BAT) for several industrial sectors. The pulp and paper industry has a very defined set of operations to be especially applied in the new mills. Similarly, the International Finance Corporation [18] among others has defined directives that could be required to give financial support for the construction of new mills. For kraft pulp, the most important guidelines are listed in Table 1.

Table 1. Main BAT guidelines from IFC [18] and/or IPPC Bureau [12] regarding wastewater load minimization in bleached kraft pulp mills.

Dry debarking of wood
Extended modified cooking to a low kappa number (batch or continuous)
Systems for collection and recycling of temporary and accidental discharges from process water spills
Closed screening
Efficient washing of the pulp ahead of the bleaching
Oxygen delignification ahead of the bleach plant
Elemental chlorine free (ECF) or total chlorine free (TCF) bleaching
Removal of hexenuronic acids by mild hydrolysis at the beginning of the bleaching process, for hardwood pulp, especially eucalyptus
Partial closure of the bleach plant combined with increased evaporation
Sufficient and balanced volumes of pulp storage, broke storage and white water storage tanks to avoid or reduce process water discharges
Recycling of wastewater, with or without simultaneous recovery of fibres
Separation of contaminated and non-contaminated (clean) wastewaters
Biological secondary wastewater treatment

5. MECHANICAL PULPING: WASTEWATER CHARACTERISTICS

Figure 4 shows a block diagram of the main part of the mechanical pulp production indicating the sources of emissions to the water from a pulp mill.

Table 2 shows the specific water consumption and loads before wastewater treatment from the mechanical pulping [12].

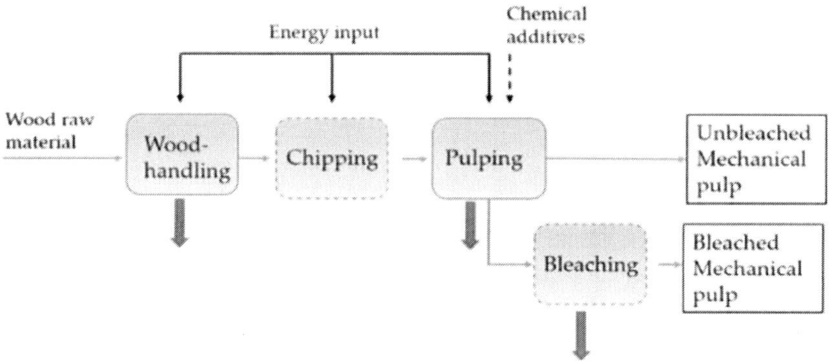

FIGURE 4. Main unit operations of the mechanical pulping. Light brown arrows indicate wastewater sources.

Table 2. Specific water consumption, organic and nutrient loads before wastewater treatment from the mechanical pulping. Data taken from Ref. [12].

Pulping process	BOD_5 (kg/ADt)	COD (kg/ADt)	Nitrogen (kg/ADt)	Phosphorous (kg/ADt)
GW	8.5–10	20–30	80–100	20–25
PGW	10–13	30–50	90–110	20–30
RMP	10–15	40–60	90–110	20–30
TMP	13–22	50–80	100–130	30–40

[i] - BOD5: Biochemical Oxygen Demand; COD: Chemical Oxygen Demand; GW: groundwood pulping; PGW: pressurized groundwood pulping; RMP: refined mechanical pulping; TMP: thermomechanical pulping; ADt: air dry tone (10% water and 90% oven-dry pulp).

6. KRAFT PULPING: WASTEWATER CHARACTERISTICS

6.1. Process description and emissions to water

A kraft pulp mill can be divided into four main parts: (1) raw material handling; (2) pulping line with an almost closed chemical and energy recovery system; (3) bleaching with an open water system and (4) the external wastewater treatment system. **Figure 5** shows the emissions sources to water from a kraft pulp mill.

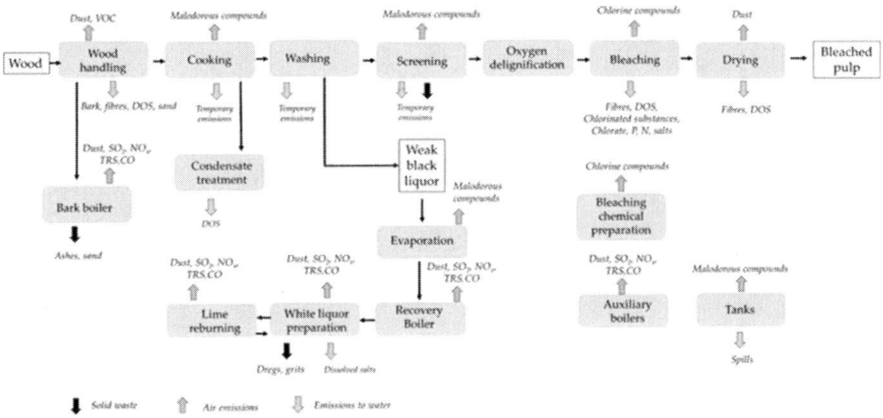

Figure 5. Main unit operations of kraft pulping. DOS: dissolved organic substances. Adapted from Ref. [12].

Table 3. Sources of effluents and effluents loads from kraft pulp mill [12, 19].

Department	Flow	TSS	BOD	AOX	COD	P	N
Debarking	2.5	4	2	0	5	20	0.2
Washing and screening	0.5	3	1	0	2	1	0.015
Bleaching	31	2	10	1.2	35	47	0.075
Condensates	1	0	1	0	3	0	0
Others	3	4	4	0	10	7	0.002
Total	38	13	18	1.2	55	75	300

[i] - Flow in m^3/Adt, TSS, BOD, AOX, COD and Nitrogen in kg/ADt. Phosphorous in g/ADt.

Table 3 shows the typical figures for the parameters in different sectors of a kraft pulp mill.

Data on current discharges to water (after wastewater treatment) expressed as loads based on available data from kraft pulp mills within the European Union are given in Table 4. Figure 6 presents a comparison of the discharges to water of different existing mills with the performance of the new mills in South America that are processing eucalyptus wood and applying the Best Available Techniques (according to the European IPPC Bureau [12] and the IFC Guidelines [18]).

Table 4. Reported annual average discharges from kraft pulp mills within the EU [12].

	Flow	COD	AOX	TSS	Total P[1]	Total nitrogen
Unbleached pulp	14–82	1.2–23	–	0.02–3	0–0.05	0.01–1.0
Bleached pulp	20–94	5–20[2] 7.5–42[3]	0–0.3	0.015–7	0.003–0.11	0.01–0.6

Flow in m^3/ADt, COD, BOD_5, AOX, TSS, nitrogen and phosphorous in kg/ADt.

[1] Eucalyptus strands contain higher levels of phosphorus compared to other forest species used for pulp production. The average level discharged with the effluent is up to 0.12 kg total-P/ADt.

[2] Emissions from eucalyptus pulp mills.

[3] Emissions from other hardwood (no eucalyptus) and softwood.

6.2. Bleaching effluents

Up to 85% of the total effluent volume is generated in the bleaching stage. Therefore, this part of the mill is broadly studied in order to minimize the effluent organic loads (especially the organochlorines loads) without impacting the pulp yield and brightness. Effluent loadings depend on the production process and the raw materials. The degree of delignification of the unbleached pulp, the bleaching process, the washing loss, type of wood, final brightness desired, chemical and water consumption and the degree of plant closure are important indicators of wastewater characteristics [12, 19]. To this end, kappa number is an important mill control parameter. The kappa number quantifies by a redox reaction to the amount of lignin (or the delignification degree) still in the pulp. The higher the kappa number, the higher the lignin content in the pulp. The low lignin amounts to be removed during bleaching, decreases the utilization of bleaching chemicals, which consequently reduces the load to the wastewater treatment. However, if the kappa number were to decrease too much during the cooking then the pulp yield and physical properties will be considerably low [10]. Table 5 provides performance data of the different processes [12].

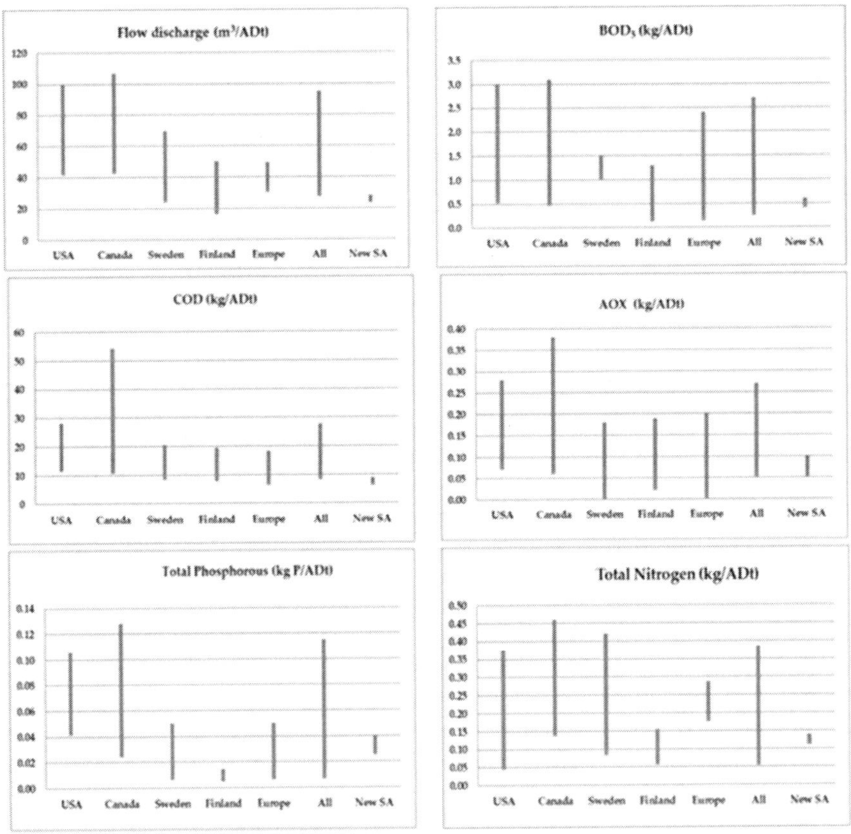

Figure 6. South America new mills performance compared with mills in North America and Europe (Data from EKONO and author personal sources). The vertical bars depicted in the graphs correspond to the 10th percentile to 90th percentile range. The column "All" corresponds to the average of the values reported.

Table 5. Kappa number currently achieved with different delignification technologies and comparison of the calculated effluent COD without considering the washing losses [12].

Delignification technologies	Kappa for hardwood	Kappa for softwood	Calculated COD load (kg/t) from the bleach plant	
			Hardwood	Softwood
Conventional cooking	14–22	30–35	28–44	60–70
Conventional cooking + oxygen delignification	13–15	18–20	26–30	36–40
Extended/modified cooking	14–16	18–22	28–32	36–44
Extended cooking + oxygen delignification	8–10	8–12	16–20	16–24

6.2.1. Hardwood and softwood bleaching effluents

The effluents from kraft pulp bleaching constitute varying quantities of organic and inorganic substances. The organic typically represents one-third of the dissolved material while the inorganics comprise two-thirds. The solid matter includes mainly fibres, pieces of fibres and the additives used in bleaching. The dissolved organic matter is composed of various species derived from the raw material and formed in the pulping and bleaching process (residual lignin, hemicelluloses and extractives) [19].

Wood material impact on the values of the effluents parameters can be assessed by comparing the figures for bleaching effluents derived from softwood and hardwood pulp. The former has higher COD and colour content than those of hardwood pulp. The compounds responsible for colour are lignin fragments of high molecular weight (HMW), which represents low biodegradability in the biological treatment [20]. Research has compared effluents from softwood and eucalyptus pulps [13, 20, 21] through AOX, COD, BOD_5 and colour behaviour of the different kinds of pulp production (conventional bleached pulps and oxygen delignified bleached pulps). According to the findings, softwood and eucalyptus effluents have the same trend in AOX levels. For both conventional pulps, the AOX levels were higher than the corresponding oxygen delignified pulps. Furthermore, as it mentioned earlier, the total COD levels are dependent on the initial kappa numbers. The COD compositions of eucalyptus and softwood effluents are significantly different, where the effluents from the eucalyptus pulps are more biodegradable. The compounds forming the kappa number in softwood and hardwood (especially eucalyptus) differ as well: in softwood, the kappa number mainly representative of lignin, whereas in eucalyptus, the hexenuronic acids (HexA) are a large contributor [22, 23]. In this regard, the most common way to remove the hexenuronic acids is in the early bleaching stages through hot acid hydrolysis (A) and hot chlorine dioxide bleaching (D_H) technologies [11, 22, 24].

6.2.2. Chemical composition of the wastewater

The two main types of bleaching methods in use are elemental chlorine free (ECF), when no molecular or gaseous chlorine is dosed in the bleaching, and totally chlorine free (TCF) bleaching [12]. ECF is dominating the bleached chemical pulp market. In 2012, ECF pulp production reached approximately 93% of bleached kraft pulp's world market share. TCF production has declined a little over the last 10 years [25].

Owing to the differences between both the bleaching technologies and chemical composition of the bleaching effluents, it is necessary to study in order to predict and understand the environmental impact associated, and consequently to develop the most suitable treatment that decreases effluent loads and toxicity. A significant number of studies pertaining to the chemical composition of bleaching effluents have been published. Several authors have worked in identifying the chemical compounds in filtrates. More than 500 organic compounds have been identified in bleaching effluents so far. Most compounds identified in bleaching effluents are derived from lignin or other wood components, such as extractives or carbohydrates [26].

The most important difference, when comparing softwood effluents with the eucalyptus effluents, is the higher lignin content in the former and the hexenuronic acid content in the latter [20].

Lignin degradation products were commonly considered as the major precursors of chlorinated compounds. However, the presence of monochlorinated compounds derived from glucuronxylans were identified to be the major components of chlorine dioxide bleaching filtrates of eucalyptus kraft pulps [27, 28].

Other important compounds found in the effluents are wood-derived components: resin acids, fatty acids, phytosterols and retene. Lipophilic hardwood extractives consist of a complex mixture of compounds such as sterols, long chain aliphatic acids and alcohols, waxes, glycerides and sterol esters. If high amounts of these compounds are found in kraft mill effluent, their origin is frequently the spills of black liquor and soap or black liquor transported with the pulp [14, 29].

6.2.3. Molecular weight distributions

Several authors [14, 30, 31] have worked in determining the molecular weight distribution of the components in the effluents. The importance of determining the molecular weight distribution comes from the fact that significant removal in the biological treatment system is achieved from the low molecular weight (LMW) material. Evidence of this is the increment in the proportion of organic compounds with high molecular weight after biological treatment. Improvements in the removal of high molecular weight material would lead to greater efficiency and improve the effluent quality. Traditionally, the separation between low molecular weight (LMW) and high molecular weight (HMW) is done at 1000 Da. Bleach kraft mill effluents have an extended molecular weight distribution; from diverse kinds of monomeric compounds to large and complex molecules with molecular weights between 10,000 and 30,000 g/mol. The molecular weight distribution depends on the raw material and the bleaching process used. For example, the average molecular weight of organic matter in hardwood kraft pulp effluents is lower than the corresponding softwood effluents [14].

The molecular weight fractions in the bleaching filtrates of oxygen delignified eucalyptus pulps were studied. The HMW fraction contributed to approximately 40% of the total effluent load of COD both in softwood and hardwood ECF bleached pulps production, and about 30–40% to TCF bleached pulps effluents [30, 31]. Additionally, the most remarkable differences between softwood- and hardwood-derived effluents are in the aromatic region. The aromatic lignin-derived structures such as syringyl and guaiacyl units are not important structural elements in HMW effluent materials from ECF bleaching of oxygen delignified hardwood kraft pulps, but are important in softwood HMW effluents [31, 32]. Similarly, the results show that all HMW effluents contained carbohydrates. The carbohydrates found in the examined HMW could have had oligosaccharides, polysaccharides or both present in the effluent, either in dissolved or colloidal form. As can be expected, the HMW hardwood kraft pulps fraction contained more carbohydrates (mainly xylan) than the corresponding samples from softwood kraft pulps. Concerning the presence of carboxylic acids, the HMW samples showed high levels of these groups. They were formed due to the oxidation of lignin structures in the bleaching process [30–32].

Regarding the low molecular weight (LMW) compounds, it can be broadly classified into three main classes: acids, phenolic compounds and neutral compounds. The phenolic compounds and some of the acids are degradation products from lignin, while the resin acids, fatty acids, terpenes and sterols are residues of extractives presents in the raw material [14].

6.2.4. Ecf and tcf wastewaters treatability

The biological treatment of the effluents from ECF and TCF is almost the same. There is a slight difference in the organic matter constitution among these bleaching effluents, but it is less than other parameters such as raw materials, effluents from the unbleached line, than the bleaching effluent itself.

TCF eucalyptus pulp produced an effluent with 3.5 times the BOD and twice the COD than ECF eucalyptus pulp effluent [30]. Similarly, TCF bleaching effluent had approximately twice the COD in softwood than the ECF effluents [33]. The larger amounts of COD and TOC in the TCF effluents can be explained because the bleaching reagents used in the TCF sequences (O_3, H_2O_2) are less selective towards residual lignin than the ClO_2 use in the ECF sequences. Bleaching of pulps with ozone is known to produce aldehyde and keto groups on carbohydrates, which are highly susceptible to oxidative degradation under alkaline conditions. An alkaline peroxide stage is used to further bleach ozone-treated pulps, resulting in an oxidative degradation of these carbohydrates and thus contributing to higher COD and TOC values in the TCF effluents. Moreover, the hardwood TCF effluents contained more carbohydrates (mainly xylan) than the ECF effluents. An explanation of these differences was that the process conditions in P-stage (long retention time under alkaline conditions) may favour dissolution of xylan from the pulp [30, 33].

However, while TCF effluent contains more dissolved organic matter, it is less coloured than ECF effluent, mainly because of the action of residual reagents (i.e. H_2O_2) in the TCF effluent. Normal values of colour at 525 nm in TCF effluents are 300 and 1300 C.U. in ECF effluents [31].

7. KRAFT PULPING: WASTEWATER TREATMENT

The typical pulp mill wastewater treatment should include primary treatment (neutralization, screening or sedimentation), principally to remove suspended solids, and biological/secondary treatment. The secondary treatment is mainly done to diminish the organic matter, which is removed by biological degradation, and is particularly useful for the removal of low molecular mass organic matter with a molecular weight of 800 Da or less. Some mills have tertiary treatment to further reduce toxicity, suspended solids, organics or colour [12, 13, 34].

Secondary biological treatment is applied in most types of pulp and paper mills. The most usual methods are activated sludge and aerated lagoons. Some variations of these systems include the use of filters and sequences reactors—Mobil Bed Bioreactor (MBBR) and Membrane Bioreactors (MBR).

Sometimes anaerobic treatment is used followed by an aerobic biological stage [12, 18].

Aerated ponds and activated sludge methods are the most common treatment systems in pulp and paper industry. In an aerated pond, wastewater is treated through a combination of physical, biological and chemical processes. They have large residence times between 3 and 20 days, and consequently a large volume. They work with low microorganism concentration (low solids concentration) about 100–300 mg/L. These ponds use aeration devices to add oxygen to the wastewater (normally surface turbine aerators or bottom aerators) and mix the contents of the pond, thereby enhancing the microbial activity. However, due to low efficiency levels and the large surface required, the use of aerated lagoons has drastically diminished [12, 13, 34].

The largest secondary treatment system is activated sludge (60–75% of all the biological effluent treatment plants in pulp and paper industry use activated sludge systems); even in new plants. The advantages of the aerated activated sludge systems compared to the aerobic ponds are that they achieve high removal efficiencies, the process can be well controlled, requires less surface and the microorganisms are adapted to the receiving wastewater. The disadvantages are the high construction and operation costs (especially the energy cost of the aeration systems), the high rate of sludge production and the loss of efficiency due to bulking problems, and consequently, the need to add nutrients to avoid this problem. Sludge handling and nutrient dosage are additional to the energy cost, which is the major component contributing to the operational cost of the biological treatment of process effluents within the pulp and paper industry [34].

7.1. Characteristics of activated sludge treatment

Two main units of the activated sludge plant are the aeration basin and the sedimentation basin. In the aeration basin, the effluent is treated with a culture of microorganisms (the activated sludge), which is present in a high concentration. Figure 7 shows a diagram of a pulp mill treatment with the activated sludge system. Activated sludge plants at kraft pulp mills have a retention time of about 15–48 h. The solids concentration in the activated sludge systems is typically 2000–6000 mg/L. The hydraulic residence time is 4–8 h for a conventional system and the cellular residence time (sludge age) is normally 5–15 days. Normal loads are between 0.05 and 0.1 kg BOD/kg sludge for extended aeration and 0.1–0.3 kg BOD/kg sludge for low load process. The common operating temperature is about 35–37°C and the dissolved oxygen (DO) concentration is 1.5–2.0 ppm. The nutrients concentration in relation to the organic matter is important in effluent treatment. Effluents from the wood processing industry generally have a

BOD:N:P ratio of 100:(1–2):(0.15–0.3) and the addition of supplemental nutrients is normally required [13, 34].

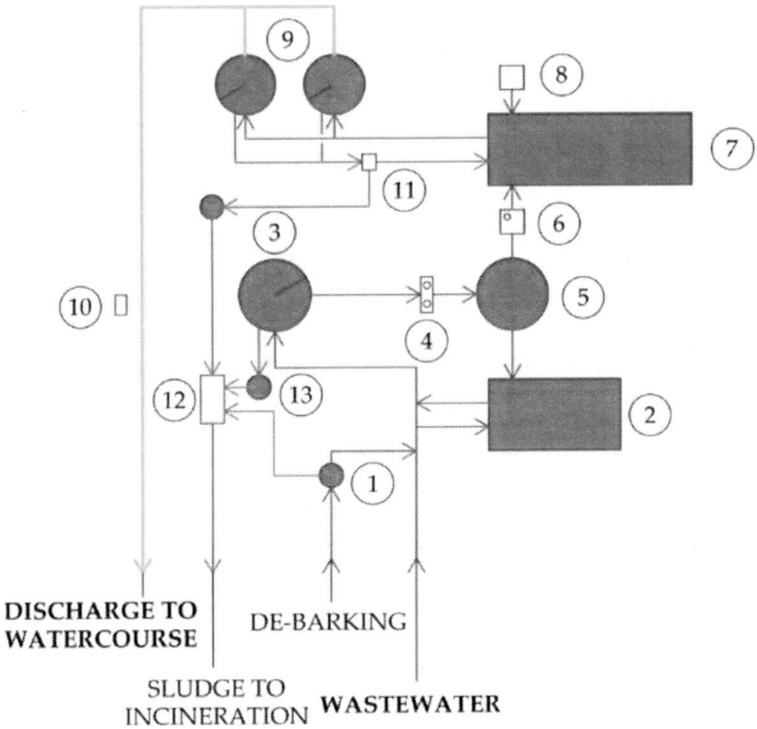

Figure 7. Diagram of a pulp mill treatment plant with activated sludge as biological treatment.

The removal efficiencies reached vary according to the wastewater residence time and the operating conditions. Normal efficiencies figures are between 85 and 98% BOD_5 removal and 60–85% for COD removal. For AOX, the reduction is about 40–65%, 40–85% for phosphorus and 20–50% for nitrogen. The overall efficiency of TSS removal using primary and secondary treatment is about 85–90% [12].

7.2. Aerobic treatability of the different effluent fractions

The COD of treated effluent represents how effective a treatment technology is in its ability to remove the total organic material present in the influent. BOD measurements by themselves do not quantify the non-biodegradable or slowly biodegradable organic portion of the effluent. Moreover, studies seem to indicate that the residual colour in pulp mill effluents could be linked to the recalcitrant COD [35].

Recalcitrant organic matter is supposed to be partly responsible for long-term toxicity in receiving waters [21]. As discussed earlier, it is widely reported that the residual recalcitrant organic matter is composed predominantly by high molecular weight components, which are not metabolized due to its size. However, the contributions of high and low molecular weight fractions in bio-treated effluents are dissimilar [36]. In the LMW fraction, a large-scale removal of the chlorinated phenolic compounds, chlorinated resin acids and sterols occurs. In the HMW fraction, the carbohydrates are strongly affected; however, other compounds such as oxidized lignin were less affected [30].

Some findings are possible by comparing the high molecular weight (HMW) and low molecular weight (LMW) fractions of the acidic and alkaline filtrates post biological treatment [32]. In the alkaline filtrate, the COD and TOC in the HMW fraction increased after treatment. The same behaviour was observed with the AOX and lignin content in the acidic filtrate. This is attributable to the formation of soluble bacterial products or to the adsorption of the LMW into HMW matter [32, 35]. In the LMW filtrates, the COD/TOC decreased after biological treatment, as a result of the large removal of highly oxidized organic carbon. The colour increased in the HMW fractions of acid and alkaline filtrates. The biological treatment often leads to increased colour in ECF bleaching effluents due to the creation of new chromophores in the HMW fractions [13, 32].

7.3. Bulking problems in the activated sludge systems

Two critical operational aspects of an activated sludge plant are maintaining proper control of the dissolved oxygen (DO) concentration in the aeration tank and preserving a good settling sludge. Reduced settleablility results in poor plant performance, as it is difficult to maintain a low concentration of suspended solids in the plant effluent [13, 34]. Activated sludge plants that treat pulp and paper mill wastewaters seem to be particularly prone to this. There are several reasons for poor separation properties, such as filamentous bulking sludge, bulking due to excessive extracellular polymeric substances (EPS), production or formation of small flocs and dispersed biomass [37, 38]. In pulp mill wastewater, bulking is often due to the presence of filamentous bacteria. Common conditions that favour bulking are working at feeding loads ratios out of normal range, deficiencies in nitrogen and phosphorous species or in the level of DO [13]. In kinetic terms, the floc forming microorganisms have a competitive advantage at lower substrate concentrations because that allows the compounds to utilize oxygen and nutrients more efficiently than the not floc forming microorganisms [37].

The presence of filamentous bacteria was examined for two years in 15 French pulp, paper and board mills wastewater. The study of 25 bulking cases attributed the source in 10 cases to be COD hydraulic overloads, in 8 cases to deficient aeration and in 5 cases to nutrient deficiency [39].

8. PARTIAL CLOSURE IN WATER CIRCUITS

The current market and environmental demands facing pulp and paper mills are the increased closure of the plant circuits and a further reduction or elimination of the wastes produced. The concept of a closed loop mill aims to eliminate discharges to the aquatic environment, recycle and reuse all possible solid and liquid process wastes, and reduce air emissions to the lowest possible quantity and toxicity. However, until today, no kraft mills are operating with complete closure and complete reutilization of the effluents. The most important problem experienced in mills that try to operate for long periods with zero discharge was corrosion caused by chlorides in a number of positions. Nevertheless, great progress has been made in minimizing impacts associated with pulp mill effluents. Water circulation closure methods include dry debarking, effective liquor spilling control, closed screening and washing, condensate stripping and other methods to minimize the loss of wood-derived organic matter. Extended and oxygen delignification can significantly reduce bleach plant effluent loads from kraft pulp mills. The bleach plant is the most important source of effluent within a pulp mill and the chlorinated effluents are more complicated to reutilize within the mill. For this reason, an important trend in bleaching development is to reduce volumes and decrease the effluent loads, especially of chlorinated compounds [40–42].

Up to now, a complete water closed circulation is not available; nevertheless, a partial closure of the water circuits is possible. This can be done segregating the acid and alkaline effluent streams and recirculating the liquids countercurrently from the last bleach stage through the sequence to the brown stock washer. The alkaline effluent could be used for washing the pulp in the unbleached part of the process.

9. CONCLUSIONS

The pulp and paper industry has been considered a large consumer of wood, energy and water, and an important contributor of pollutant discharges to the environment (air, water courses and soil). However, the last decades have seen a lot of effort in creating solutions such as generating less pollutant wastewaters and reducing the amount and load of the emissions to the environment. The implementation of several measures like the dry debarking of wood, the introduction of extended cooking and oxygen delignification, the reuse of condensates, improvements in washing efficiency and especially the total substitution of chlorine, has brought a significant reduction in effluent flows and in the chlorinated and organic loads generated within the mill. In addition, the introduction of end-of-pipe secondary, and even tertiary, treatments have reduced large amounts of pollutant loads to the environment. However, the need for tertiary treatment is not yet well proven; while it purifies the effluent, the energy costs are high and even forms sludge.

Effluent characteristics are dependent on the production process and the raw materials. ECF eucalyptus pulp production is increasing appreciably but not much information on its effluents is available. The main difference between softwood and eucalyptus pulps is in the kappa number: the kappa number is mainly formed by lignin content in softwood pulp, and the Hexenuronic acids are important contributors to kappa number in eucalyptus pulp. Hence, the bleaching conditions for eucalyptus are less severe and consequently the effluents characteristics are different. Eucalyptus bleaching effluents have lower COD, AOX and colour content and higher biodegradability than the softwood effluents.

The environmental impact of effluent loads and the appropriate treatment can be determined by studying the chemical composition and molecular weight distribution of the bleaching effluents. The HMW in hardwood bleaching wastewaters constituted an important but not prevailing fraction of the wastewater composition (30–65% of the total). The hardwood HMW fraction is mainly composed of non-aromatic structural compounds.

Aerated activated sludge is the most common treatment system in pulp mills. BOD_5 removals of 85–98% and COD removals of 60–85% are normally achieved with these systems. For AOX, the reduction is about 40–65%, 40–85% for phosphorus and 20–50% for nitrogen. Bulking problems are common in these systems mainly due to nitrogen deficiencies and phosphorous concentration or the level of DO.

Nowadays, plants that apply the best available technologies have their emissions controlled and present minimum environmental impact at the receiving waters.

The new developments are in the way to close even more the internal circuits in the plant, to reduce the flow discharged. Membrane technologies and similar technologies may be key in this regard.

REFERENCES

1. Diesen M. Economics of the pulp and paper industry. 1st Ed. Helsinki (Finland): Paperi ja Puu Oy; 2007, 186 p.
2. FAO. Forest products yearbook, 1979–1990 [Internet]. Rome (Italy); 1992, p. 392. Available from: http://www.fao.org/forestry/statistics/80570/en/ [Accessed 2009 Oct 8]
3. FAO. Forest products yearbook, 1991–1995 [Internet]. Rome (Italy); 1996. Available from: http://www.fao.org/forestry/statistics/80570/en/ [Accessed 2009 Oct 15]
4. FAO. Forest products yearbook, 1996–2001 [Internet]. Rome (Italy); 2003. Available from: http://www.fao.org/forestry/statistics/80570/en/ [Accessed 2009 Sep 8]
5. FAO. Forest products yearbook, 2002–2006 [Internet]. Rome (Italy); 2008. Available from: http://www.fao.org/forestry/statistics/80570/en/ [Accessed 2010 Oct 15]
6. FAO. Forest products yearbook, 2009–2013 [Internet]. Rome (Italy); 2015. Available from: http://www.fao.org/forestry/statistics/80570/en/ [Accessed 2016 Jul 11]
7. Smook GA. Handbook for pulp & paper technologists. 3rd ed. Tappi Press; Vancouver-B.C.; 2003, 425 p.
8. Sixta H, editor. Handbook of pulp. 1st Ed. Weinheim: WILEY-VCH Verlag GmbH &Co. KGaA; 2006, 1369 p. doi:10.1002/9783527619887.
9. Stenius P. Forest products chemistry. 1st Ed. Gullichsen J, Paulapuro H, editors. Fapet Oy; Helsinki; 2000, 348 p. i
10. Gullichsen J. Fiber line operations. In: Gullichsen J, Fogelholm CJ, editors. Chemical pulping. Fapet Oy; Helsinki; 1999, p. 693.
11. Colodette JL, Gomes CM, Rabelo M. Progess in eucalyptus kraft pulp bleaching. In: 2nd International Colloquium on Eucalyptus Pulp (2ICEP); Concepcion-Chile; 2005, pp. 1–18. Available from: http://www.eucalyptus.com.br/icep02/jorge_colodette.pdf. [Accessed 2009 Oct 15]
12. European Commission, Joint Research Centre, Institute for Prospective Technological Studies. Integrated Pollution and Prevention Control. Best Available Techniques (BAT) reference document for the production of pulp, paper and board. 2015, Available from: http://eippcb.jrc.ec.europa.eu/reference/BREF/PP_revised_BREF_2015.pdf. [Accessed 2016.01.20]
13. Springer A. Industrial environmental control: Pulp and paper industry. 3rd Ed. Atlanta, GA (USA): TAPPI Press; 2000.
14. Lehtinen K. Relationship of the technical development of pulping and bleaching to effluent quality and aquatic toxicity. In: Borton DL, Hall T, Fisher R, Thomas J, editors. Pulp & paper mill effluent environmental fate & effects. Destech Publications Inc.; Lancaster-PA; 2004.
15. Munkittrick KR, Servos MR, Carey JH, Van Der Kraak GJ. Environmental impacts of pulp and paper wastewater: Evidence for a reduction in environmental effects at North American pulp mills since 1992. Water Sci Technol. 1997;35(2–3):329–38.
16. Thompson G, Swain J, Kay M, Forster CF. The treatment of pulp and paper mill effluent: A review. Bioresour Technol. 2001;77(3):275–86.
17. Rodgers JH, Thomas JF. Evaluations of the fate and effects of pulp and paper mill effluents from a watershed multistressor perspective: Progree to date and future opportunities. In: Borton DL, Hall T, Fisher R, Thomas J, editors. Pulp & paper mill effluent environmental fate & effects. DEStech Publications Inc.;. Lancaster- PA; 2004
18. International Finance Corporation (IFC). Environmental, health, and safety guidelines pulp and paper mills. 2007. Available from: http://www.ifc.org/wps/wcm/ connect/ 6f13e78048855398afb4ff6a6515bb18/Final%2B-%2BPulp%2Band%2BPaper% 2BMills.pdf?MOD=AJPERES [Accessed 2010 April 5]
19. Dahl O. Process modifications to reduce effluent loads. In: Dahl O, editor. Environmental management and control. 1st Ed. Paperi ja Puu Oy; Helsinki; 2008, pp. 70–84. ISBN: 978-952- 5216-30-1.

20. Herstad Svärd S, Basta J, Wäne G, Jour P. Comparative characterization of the bleach plant effluents from modern ECF bleaching—eucalyptus and softwood pulps. In: 30th pulp & paper annual meeting, ABTCP. 3–7 November 1997, Sâo Paulo, Brazil; pp. 49–62.
21. Yousefian S, Reeve D. Classes of compounds responsible for COD and colour in bleached kraft mill effluents. In: International environemental conference & exhibit: Setting the environmental course for the 21st century. 6–10 May 2000, Denver, CO (USA); pp. 657–65.
22. Teleman A, Hausalo T, Tenkanen M, Vuorinen T. Identification of the acidic degradation products of hexenuronic acid and characterisation of hexenuronic acid-substituted xylooligosaccharides by NMR spectroscopy. Carbohydr Res. 1996;280(2):197–208. doi:10.1016/0008-6215(95)00309-6.
23. Ventorim G, Oliveira KD, Colodette JL, Da Costa MM. Effect of pulp kappa number, lignin and hexenuronic acid contents on oxygen delignification performance. Sci For Sci. 2006;71:87–97.
24. Ragnar M, Lindström ME. A comparison of emerging technologies. Pap ja Puu. 2004; 86(1):39–44.
25. Alliance for Environmental Technology. Trends in world bleached chemical pulp production: 1990–2012. AET Reports. 2013. Available from: http://www.aet.org/science_of_ecf/eco_risk/2013_pulp.html [Accessed 2016 Aug 10].
26. Mc Kague B, Carlberg G. Effluent characteristics and composition. In: Dence C, Reeve D, editors. Pulp bleaching-principles and practices. TAPPI Press; Atlanta- GA; 1996, pp. 751–65. ISBN:0-89852-063-0.
27. Freire CSR, Silvestre AJD, Pascoal Neto C. Carbohydrate-derived chlorinated compounds in ECF bleaching of hardwood pulps: Formation, degradation, and contribution to AOX in a bleached kraft pulp mill. Environ Sci Technol. 2003;37(4):811–4. doi:10. 1021/es0200847.
28. Freire CSR, Silvestre AJD, Neto CP, Silva AMS, Evtuguin DV, Cavaleiro JAS. Easily degradable chlorinated compounds derived from glucuronoxylan in filtrates from chlorine dioxide bleaching of eucalyptus globulus kraft pulp. Holzforschung. 2003; 57(1):81–7. doi:10.1515/HF.2003.013.
29. Freire CSR, Silvestre AJD, Pascoal Neto C. Oxidized derivatives of lipophilic extractives formed during hardwood kraft pulp bleaching. Holzforschung. 2003;57(5):503–12. doi:10.1515/HF.2003.075.
30. Dahlman O, Reimann A, Stromberg L, Mörck R. High molecular weight effluent materials from modern ECF and TCF bleaching. Tappi J. 1995;78(12):99–109. ISBN: 978-952-5216-30-1.
31. Mounteer AH, Passos FML, Borges AC, Silva DO. Detecting structural and functional differences in activated sludge bacterial communities originating from laboratory treatment of elementally and totally chlorine-free bleaching effluents. Can J Microbiol. 2002;48(3):245–55.
32. Souza L, Mounteer AH, Silva C., Dalvi L. A study on biological removal of recalcitrant organic matter in eucalyptus kraft pulp ECF bleaching filtrates. In: Tappi International Environmental Conference, CD-ROM. Atlanta, GA (USA): Tappi Press; 3–5 May 2003, pp. 1–8.
33. Cates D, Eggert C, Yang J, Eriksson K. Comparison of effluents from TCF and ECF bleaching of kraft pulps. Tappi J. 1995;78(12):93–8.
34. Hynninen P. Effluent treatment. In: Dahl O, editor. Environmental management and control. 1st Ed. Paperi ja Puu Oy; Helsinki; 2008. pp. 86–116.
35. Konduru RR, Liss SN, Allen DG. Recalcitrant organics emerging from biological treatment of kraft mill effluents. Water Qual Res J Canada. 2001;36(4):737–57. doi:10.1139/w02-006.
36. Jokela JK, Laine M, Ek M, Salkinoja-Salonen M. Effect of biological treatment on halogenated organics in bleached kraft pulp mill effluents studied by molecular weight distribution analysis. Environ Sci Technol. 1993;27(3):547–57. Available from: http://pubs.acs.org/doi/abs/10.1021/es00040a014

37. Andreasen K, Agertved J, Petersen JO, Skaarup H. Improvement of sludge settleability in activated sludge plants treating effluent from pulp and paper industries. Water Sci Technol. 1999;40(11–12):215–21.
38. Tsang Y, Chua H, Sin S, Tam C. A novel technology for bulking control in biological wastewater treatment plant for pulp and paper making industry. Biochem Eng J. 2006;32:127–34. doi:10.1016/j.bej.2006.08.014.
39. Fourest E, Craperi D, Deschamps-Roupert C, Pisicchio JL, Lenon G. Occurence and control of filamentous bulking in aerated wastewater treatment plants of the French paper industry. Water Sci Technol. 2004;50(3):29–37.
40. Axegård P, Carey J, Folke J, Gleadow P, Gullichsen J, Pryke D, et al. Minimum impact mills : Issues and challenges. In: Tappi minimum effluent mills symposium. San Francisco, CA (USA): TAPPI Press; 1997. Available from: http://infohouse.p2ric.org/ref/18/17842.pdf. [Accessed 2008.10.15]
41. Fiskari J, Toikkanen L, Gullichsen J. Experiments on bleaching effluent minimization. In: TAPPI pulping/process and product quality conference. Boston, MA (USA); 5–8 November 2000, pp. 771–8.
42. Stratton SC, Gleadow PL, Johnson AP. Pulp mill process closure: A review of global technology developments and mill experiences in the 1990s. Water Sci Technol. 2004;50(3):183–94.

CHAPTER 2

Technologies Used in the Wastewater Treatment for Nutrient Removal

Margarida Marchetto*

Department of the Sanitary and Environmental Engineering, Federal University of Mato Grosso, Brazil

ABSTRACT

In this paper will be showed some techniques employed effectively in wastewater treatment for nutrient removal. The results concerning biological phosphorus and nitrogen removal, from an anaerobic expanded bed reactor (AEBR), treating domestic wastewater, in an Intermittent Aeration Reactor (IAR) followed by dissolved air flotation (DAF); aiming to remove chemical oxygen demand (COD), nitrogen and phosphorus on the same unit. The DAF was used for separation of suspended solids. Conditions to obtain nitrification were assessed, denitrification and phosphorus removal in a reactor operated with alternate aerobic and anaerobic periods. The intermittent aeration system was operated with hydraulic detention time (θh) of 8 hour and 6 hours. The system with intermittent aeration (θh:6hr) showed stable results and presented average COD removal of 92% (90 to 94%) and average phosphorus removal of 90% (82 to 96%). For hour detention time, the average orthophosphate (PO_4^{-3}-P) removal was: 84% (60 to 94%) for raw samples and 94% (60 to 98%) for filtered samples (filter: 0.45 μm). Neither the use of the external carbon source nor pH correction was necessary.

Keywords: Biological phosphorus removal; Intermittent aeration activated sludge reactor; Dissolved air flotation

1. INTRODUCTION

In many developing countries, water scarcity and pollution pose a critical challenge in the area of environmental management. In urban areas, it is becoming difficult for the authorities to manage this question. The possibilities of wastewater reuse in agriculture, industry, urban uses, and environment, including groundwater recharge enhancement are discussed worldwide.

Water use has more than tripled globally lately. According to UNEP [1] these problems may be attributed to many factors as increasing population. Inadequate water management is accelerating the depletion of surface water and groundwater resources. Water quality has been degraded by domestic and industrial pollution sources as well as non-point sources. In some places, water is withdrawn from the water resources, which become polluted owing to a lack of sanitation infrastructure and services. Population growth in urban areas is of particular concern for developing countries and water issues are expected to grow.

The capacity building policy-making, institutional strengthening, financial mechanisms, and awareness rising and stakeholder participation are vital to implement these strategies for wastewater reuse.

In sanitary sewer, the phosphorus appears, mainly, as organic phosphorus, polyphosphate and orthophosphate [2]. The organic phosphorus comes from human and animal excretions, as well as from food remains. When organics passes through biological decomposition, they generate orthophosphates. In addition, polyphosphates may also derive from detergents [2]. That fraction has been suffering expressive growth due to increasing consumption of detergents.

The biological phosphorus removal (removal in "excess") is obtained by the selection of bacteria capable to store polyphosphate. The selection is made to expose the bacteria alternately to the anaerobic and aerobic conditions [3]. Under anaerobic condition, bacteria that accumulate phosphorus in excess, use the derived energy of the polyphosphates hydrolyses, to fix organic substratum, that are stored in the form poli-b-hidroxibutirato (PHB) or poli-b-hidroxivalerato (PHV) [3].

The removal of the phosphorus is more efficient in alternate conditions of aerobic and anaerobic [4,5]. Also it was verified, that the phosphorus removal is associated, among other factors, to the contact of the sludge recirculation with the raw affluent, in the initial zone predominantly anaerobic [6].

The technology of reactors with intermittent aeration, in continuous hydraulic permanent flow, constitutes an alternative to the conventional activated sludge. Specific conditions for aerobic and anaerobic activities are provided by these reactors [7].

Choi et al. [8] used synthetic wastewater to evaluate phosphorus and organic matter removal, in a single reactor, combining aerobic and anaerobic conditions. The removal efficiency was about 92%, for a total-P, 90% for PO_4^{-3}-P, while a great difference of phosphorus removal was observed with relationship to N/P ratio. Rittmann and Langeland [9] and Campos [6] observed that combination of nitrification and denitrification processes in same treatment system might result in an increase of phosphate removal efficiency.

For Kerrn-Jespersen et al. [10] the bacteria that store polyphosphate can be classified in two groups: a group which is capable to use oxygen or nitrate as electrons receiver, and another which is capable to just use oxygen.

In agreement with the mentioned authors, the differences observed in the phosphorus removal rates, under aerobic and anoxic condition, are due to the fact that, in aerobic condition, both groups of bacteria are capable to absorb phosphorus from the medium. In anoxic condition, the bacteria that can use nitrate as electrons receiver has capacity to store phosphorus, what explains the phosphorus removal, in aerobic condition, it is larger than the removal under anoxic condition.

2. METHODS

There are different technologies and methods for wastewater nutrient removal (**Table 1**). The author has developed a fully operational laboratory scale one system (**Figure 1**) [7], being the first one an Anaerobic Expanded Bed Reactor (AEBR), built in steel carbon, with 14.9 m of total height and useful volume of 32.0 m^3. The reactor was composed of two regions: reaction, and decantation. The reaction region presented cylindrical format with 12.0 m height, 1.5 m of base and 2.5 m of superior border diameter (operated with average θ_h of 3.2 hours) [11,12]. The second unit is an Intermittent Aeration Reactor (IAR) - with controlled dissolved oxygen (DO) concentration - seeking to the nitrification/denitrification, and subsequently, a unit of DAF. Both IAR and DAF units were in bench scale with continuous flow, being fed for only part of the of expanded bed reactor effluent, (range 11.5 to 15.5 L/hr).

Table 1: Wastewater Treatment Process.

	Preliminary	Primary	Secondary	Tertiary and Advanced
Purpose	Removal of large solids and grit particles	Removal of suspended solid	Biological treatment and removal of common biodegradable organic pollutant	Removal of specific pollutants, such as nitrogen orphosphorous, color, odor, etc.
Sample Technologies	Screening, Settling	Screening, Sedimentation	Percolating/trickling filter, activated sludge Anaerobic treatment Waste stabilization ponds (oxidation ponds)	Sand filtration, Membrane bioreactor, Reverse osmosis, Ozone treatment Chemical coagulation Activated carbon. wetlands

(Asano and Levine [18] and Campos et al. [19])

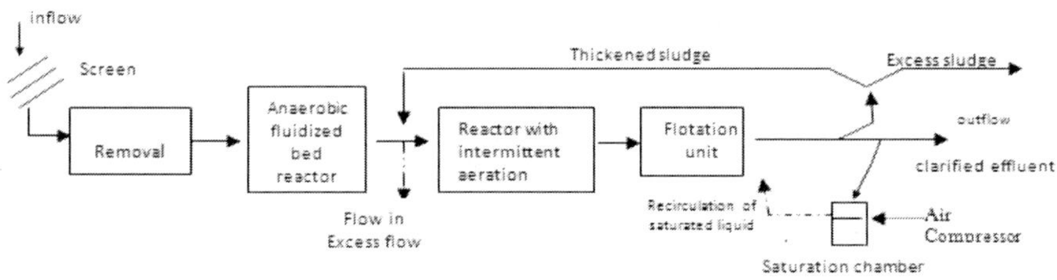

Figure 1: Schematic flowchart of the experimental system.

An effluent portion of the anaerobic reactor flowed by the unit of IAR in which DO concentration was controlled. The aerated reactor had a useful volume of 0.092 m^3, (width: 0.25 m, length: 0.82 m and height of 0.50 m). The effluent of the aerobic reactor was directed to a subsequent system of DAF, presenting width: 0.25 m, length: 0.10 m and height of 0.5 m with useful area about 0.025 m^2. The pressure was maintained in the strip of 456 ± 10 kPa in the saturation camera and the air amount was supplied with temperature and recirculation flow function. Part of the accumulated sludge in the flotation unit surface was returned to the beginning of the aeration system and the sludge excess was discarded by the system. The clarified liquid (effluent) was returned to feed the saturation camera, that, after having saturated with air, it was directed to the DAF. In the IAR, the total hydraulic detention time was maintained in 8 h, for 194 days, and of 6 h in the of 41 days, with an average flow of 15.3 and 11.5 L / h, respectively.

The affluent and effluent characteristics were monitored in each unit: pH, DO, temperature, Biochemical Oxygen demand (BOD), (COD), orthophosphate (PO_4^{-3}-P), total solids (TS) and total suspended solids (TSS), total Kjeldahl nitrogen (TKN), ammonia-nitrogen (NH_3-N), nitrite-nitrogen (NO_2^- -N), nitrate-nitrogen (NO_3^- -N) in the " liquor ". Besides those, were also monitored: MPN of denitrifying and nitrifying bacteria, Total organic carbon (TOC) and metals. All determinations were followed the rules of Standard Methods.

3. DISCUSSION

System constituted by anaerobic reactor, followed by IAR associated with DAF, is a good alternative to promote biological phosphorus removal and also nitrogen and COD. That configuration constitutes quite attractive, capable to promote high a degree of domestic wastewater nutrient removal, without needing external carbon source and alkalinity addition. The intermittent aeration is a resource that results in operational cost reduction for activated sludge reactors fed with anaerobic reactor effluent.

Operational changes in existing reactors can bring economy and in certain cases even efficiency increase in the N and P removal. Marchetto [7] observed that anaerobic reactor in series with IAR associated with DAF, presented a good performance in relation to the COD and phosphorus removal, mainly for a total treatment time (θ_h) of 8:00 and of 6:00 hours. The system can be an economical alternative for the biological phosphorus removal due to the short detention period needed.

The AEBR effluent presented, on average COD of samples without filtering: 183 mg/L, COD of the filtered samples: 76 mg/L; NTK: 26 mg/L; ammoniacal-N: 16 mg/L.

Working with θ_h on 6:00h was obtained PO_4^{-3}-P removal of 89% and 94% for the effluent without filtering and filtrates, respectively. In addition, for θ_h of 8:00h was obtained maximum removal of PO4 - 3-P, about 95%. Sasaki Kousei et al. [13], aiming at the simultaneous nitrogen and phosphorus removal, of a sanitary sewer, in system of activated sludge aerated intermittently (aerobic-anoxic), used θ_h on 20:00h with the residues temperature varying among 9 to 33°C. The TSS concentration of the mixed liqueur varied between 3100 to 4300 mg/L and the reason of sludge recirculation among 150 to 200%. During the aerobic period, the DO concentration was maintained in 2.5 mg/L. The pilot installation was shown stable and with high BOD removal efficiency: (98%); KTN: (92%) and total phosphorus: (85%). Sasaki Kousei et al. [13] almost used θ_h three times superior to the used in this research; even so, they obtained inferior phosphorus removal.

In relation to the PO_4^{-3}–P concentration the results showed that, even for the period without aeration, the phosphorus residual of samples without filtering presented medium value of 1.5 mg/L, even for the period without aeration, the PO_4^{-3}–P residual concentration was of 4 mg/L, for raw samples and, 2.5 mg/L for the filtered samples. At the end of the period without aeration, that situation corresponds to the smallest removal efficiency of the (PO_4^{-3} -P), (80%) of the samples without filtering and (85%) of the samples filtered. The cycles were varied from 2:00 h, with aeration and 4 h, without aeration. Ikemoto et al. [14] observed that denitrification was dominant as a result of the competition between bacteria phosphate accumulative and bacteria sulfate reductors, in the organic matter degradation. In anaerobic conditions, the release of phosphate and the sulfate reduction happened simultaneously. The accumulative bacteria of phosphate competed with the bacteria sulfate reductors for organic acids, as propionate. However, the phosphate accumulative bacteria used the acetate produced by the sulfate redactors' bacteria.

It is interesting to observe when the DO concentration increased, the phosphate concentration decreased. It was verified that the values of the concentration of PO_4^{-3} –P began rising, in the corresponding time of 240 min, after 2:00 without the DO supply, (redox potential: -190 mV). This fact shows clearly that, after a certain period without the DO supply, the organisms tend to liberate phosphorus, unlike when it happens in aerobic phase, when phosphorus decrease gradation is verified along the period. It can be inferred that, only when the interior of the flake becomes anaerobic, it should happen the phosphate liberation for the liquid middle.

Finger and Cybis [15] observed that using two distinct conditions (anaerobic-aerobic) was efficient in the removal of phosphorus (81%); however, effluent presented high nitrate concentrations. Sorm et al. [16] also compared the rates of phosphorus removal in anoxic and aerobic conditions. The phosphorus removal rate in anoxic condition was harmed when concentration relatively high, of nitrogen in the nitrate form around 25 mg/L.

For Maurer and Gujer [17] the release of phosphate and substratum accumulation quickly biodegradable only happened under strict anaerobic condition (when there was not nitrate, nitrite or presence of oxygen); however, in agreement with the results of the present researches obtained with intermittent aeration followed by DAF, it was verified that in the periods without aeration, even with the presence of nitrate in the reactor with intermittent aeration, phosphate release was observed.

The organic matter absorbed by the bacteria is used by both purposes, what becomes an advantage, when the organic matter concentration is low, as it happens when the wastewater goes by a pre-treatment system. The anaerobic pre-treatment can reduce the competition significantly for organic substratum between the bacteria poly-P and the denitrifying.

4. FINAL THOUGHTS

The system constituted by (AEBR), followed by (IAR) associated with (DAF) showed a better alternative to promote biological phosphorus removal and also nitrogen and COD. That configuration constitutes quite attractive, capable to promote a high degree of domestic wastewater nutrient removal, without needing external carbon source and alkalinity addition. The intermittent aeration is a resource that results in operational cost reduction for activated sludge reactors fed with effluent of anaerobic reactor. Operational changes in existing reactors can bring economy and in certain cases even efficiency increase in the N and P removal. This system can be an economical alternative for the biological phosphorus removal due to the short detention period needed. It was obtained efficiency of COD removal of 92% and of 95% for the samples filtered and without filtering, respectively.

The maximum removal of PO_4^{-3}–P in the system, of 95% and 96% of the raw samples and filtered samples, with average removal of 84%, for the raw samples and 88% of the filtered samples.

It can be verified that besides the alternate conditions of anaerobic and aerobic, the phosphate removal is directly related with the sludge recirculation, confirming the several authors' report.

Until 2005 the São Paulo state designed and constructed two wastewater treatment plants with an intermittent system for cities having population around 200,000 inhabitants. In this case, it was employed total hydraulic detention time (with and without aeration) in the range of 12 to 14 hr. Currently is necessary to do a survey to verify the number of existing plants.

Reuse of greywater has also been widely disseminated. Treatment technologies must be easily applied, low-cost implementation, operation and maintenance, aiming at environmental restoration of water bodies and improve the living conditions of the population. Thus, the use of recycled water and rainwater are alternative solutions to increase the supply of water, obeying specific standards and laws.

The reuse of gray water produced in homes, are alternatives to meet the needs in water supply, especially in activities that do not require noble quality standards as in domestic toilet discharges, washing floors and gardens.

ACKNOWLEDGEMENTS

This research was developed at the Department of Hydraulics & Sanitation, EESC/USP–Sao Carlos School of Engineering, Sao Paulo University. The author thanks the PRONEX (Program of Support for Nucleuses of Excellence CNPq e FINEP) and PROSAB (Program of Basic Sanitation FINEP e CEF) for the financial support. The author also would like to thank the professors Dr José R. Campos and Dr Marco A. P. Reali by guidance and collaboration.

REFERENCES

1. UNEP (2013) Water and Wastewater Reuse. An Environmentally Sound Approach for Sustainable Urban Water Management.
2. Carol et al. (1990) Health and Aesthetic Aspects of Water Quality. In Pontius F. Water quality and Treatment. A Handbook of Community Water Supplies/ AWWA (4th edn).

3. Kerrn-Jespersen JP, Henze, M, Strube R (1994) Biological Phosphorus Release and Uptake Under Alternating Anaerobic and Anoxic Conditions in a Fixed-film Reactor. Water Research 28: 1253-1255.
4. Osada T, Haga KE, HaradaY (1991) Removal of Nitrogen and Phosphorus from Swine Wastewater by the Activated Sludge units with the Intermittent Aeration Process. Water Research 25: 1377-1388.
5. Van Haandel, Marais O (1999) Behavior of Activated Sludge System-Theory and Application for Design and Operation ABES. Campina Grande Federal University of Paraíba. 448.
6. Campos JR (1989) Removal of COD and Nitrogen in a System of Three Fixed Film Bioreactors in series. Habilitation Thesis - School of Engineering of São Carlos-USP. 295.
7. Marchetto M (2001) Nutrient removal from effluents Reactor Anaerobic Reactors Using Microaerado and Intermittent Aeration followed by Dissolved Air Flotation Doctoral Thesis. São Carlos School of Eng -USP.
8. Choi E, Lee HS, Lee JW (1996) Another Carbon Source for BNR System. Water Science and Techenology 34: 363-369.
9. Rittmann BE, Langeland WE (1985) Simultaneous Denitrification with Nitrification in Single- Channel Oxidation Ditches. Journal WPCF 57: 300-308.
10. Kerrn-Jespernen JP, Henze M (1993) Biological Phosphorus Uptake under Anoxic and Aerobic Conditions. Water Research 27: 617-624.
11. Mendonça N (1999) Material Characterization and Study Support Match-Bed Reactor Anaerobic Expanded Used in Sewage Treatment. Dissertation SHS-EESC/USP.
12. Pereira J.A R. (2000) Design, Construction and Operation of Anaerobic Expanded Bed, Real Scale for Sewage Treatment. Doctoral Thesis - EESC-USP 339.
13. Sasaki Kousei et al. (1996) Development of 2-Reactor Intermittent-Aeration Activated Sludge Process for Simultaneous Removal of Nitrogen and Phosphorus. Water Science and Technology 34: 111-118.
14. Ikemoto YR, Matsui S, Kiomori T, Hamilton BEJ (1996) Symbiosis and Competition among Sulfate Reduction, Filamentous Sulfur, Denitrification, and Poly-P Accumulation Bacteria in the Anaerobic-oxic Activated Sludge of a Municipal Plant. Water Scence and Technology 34:119-128.
15. Finger JLE, Cybis FL (1999) Biological Phosphorus Removal in Sequential Batch Reactors In: 20 ° Brazilian Congress of Sanitary and Environmental Engineering ABES Rio de Janeiro.
16. Sorm et al. (1996) Phosphate Uptake under Anoxic Conditions and Fixed-Film Nitrification in Nutrient Removal Activated Sludge System. Water Science and Technology 30:1573-1596.
17. Maurer ME, Gujer W (1994) Prediction of Performance of Enhanced Biological Phosphorus Removal Plants. Water Science and Technology 30: 333-343.
18. Asano T, Levine A (1998) Wastewater Reclamation, Recycling and Reuse: Introduction. In: Asano T (ed), Wastewater Reclamation and Reuse, CRC Press, Boca Raton, Florida, USA.1-55.
19. Campos JR, Reali M, Dombroski S, Marchetto M, Lima F (1996) Physico-chemical treatment of Flotation of Effluents of anaerobic Ballasts, In: XXV Congresso Inter-Ingenieria Y Sanitary Environmental, Mexico. Proceedings electronics AIDIS.

CHAPTER 3

Removal of nitrate and sulphate from biologically treated municipal wastewater by electrocoagulation

Arun Kumar Sharm and A. K. Chopra

Department of Zoology and Environmental Science, Gurukula Kangri University, Haridwar, Uttarakhand 249404, India

ABSTRACT

The present investigation observed the effect of current density (*j*), electrocoagulation (EC) time, inter electrode distance, electrode area, initial pH and settling time on the removal of nitrate (NO_3^-) and sulphate (SO_4^{2-}) from biologically treated municipal wastewater (BTMW), and optimization of the operating conditions of the EC process. A glass chamber of two-liter volume was used for the experiments with DC power supply using two electrode plates of aluminum (Al–Al). The maximum removal of NO_3^- (63.21 %) and SO_4^{2-} (79.98 %) of BTMW was found with the optimum operating conditions: current density: 2.65 A/m^2, EC time: 40 min, inter electrode distance: 0.5 cm, electrode area: 160 cm^2, initial pH: 7.5 and settling time: 60 min. The EC brought down the concentration of NO_3^- within desirable limit of the Bureau of Indian Standard (BIS)/WHO for drinking water. Under optimal operating conditions, the operating cost was found to be

1.01\$/m^3 of water in terms of the electrode consumption (23.71 × 10^{-5} kg Al/m^3) and energy consumption (101.76 kWh/m^3).

Keywords: Biologically treated municipal wastewater, Electro-coagulation, Nitrate, Operating cost, Sulphate

1. INTRODUCTION

Due to increased water demand, water shortage is going to get worse in the forthcoming years. Water scarcity is already a fact of life in arid and semi-arid regions where agricultural, domestic and industrial water demands compete for this limited resource. Continuous population growth, rising living standards, rapid industrialization and urbanization have limited the freshwater sources available for agricultural practices (Katz and Dosoretz 2008). Conventionally biologically treated sewage water may not meet the requirements for the discharge and permissible limit of wastewater for its reuse. These processes are, therefore, being ruled out due to requirement of large space, long residence time and skilled technicians. However, even after treatment, secondary treated sewage (STS) effluent contains organic matter and nutrients which cause eutrophication and increase the oxygen demand of the surface water (Sharma 2013).

Sewage treatment deserves ample documentation due to the environmental impact caused by such wastewater if directly discharged into water bodies. In addition, due to an increase in the scarcity of clean water (Aiyuk et al. 2006) there is an urgent need for proper management of available water resources. Some of the goals of environmental protection and resource conservation concepts are the use of treated wastewater, residues emanating therefrom and other treatment by-products (Lettinga et al. 2001; Yi 2001).Conventional treatment methods often induce chemical reactions with the use of coagulants, flocculants and other additives that aid in the removal or sedimentation of inorganic and organic contaminants present in wastewater. It has been reported in the literature that nitrates can be removed from wastewater through their adsorption onto the surfaces of hydroxide precipitates, which are generated from metals and released by the electrodes (Hua et al. 2003; Huanga et al. 2009).

In the electrolytic process of the treatment of biologically treated municipal wastewater (BTMW), Al electrodes are dissolved during electrocoagulation (EC) to produce hydroxides and poly hydroxides of aluminum as coagulants which combine and subvert the organic and inorganic impurities in BTMW.

The electrochemical dissolution has been reported to have stronger affinity to capture the pollutants in the wastewater, causing more coagulation than that from conventional Al coagulants. Additionally, the gas bubbles that evolve due to the water electrolysis can cause flotation of the pollutants and the coagulated materials. Therefore, electro-flotation may also play an important part in an electrolytic cell. Although the $Al(OH)_3$ produced by the anodic Al dissolution is more effective in coagulating the pollutants in wastewater, the passivation of Al anodes and impermeable film formed on cathodes may and does interfere with the performance of electro-coagulation and electro-flotation (Holt et al. 2005). The present study has been focused to find out the treatability of BTMW in terms of $NO_3^-$and SO_4^{2-} removal by EC under various operating conditions using Al electrodes and also to assess the economics of electrode and energy consumption during the EC.

2. MATERIALS AND METHODS

2.1 Collection of BTMW samples

The samples of BTMW were collected from the outlet of activated sludge process (ASP) of the sewage treatment plant (STP), Jagjeetpur, Haridwar (Uttarakhand), India, and brought to the laboratory and then used for EC using Al–Al electrode combination. The pH of BTMW was adjusted before the electrochemical process and was maintained by adding the required amount of H_2SO_4 (1 M) or NaOH (1 M).The characteristics of BTMW are given in Table 1.

Table 1 Characteristics of BTMW

Parameter	**Mean ± SD**
pH	7.4 ± 0.52
Conductivity (μS)	735 ± 58.08
TDS (mg/l)	445.6 ± 67.3
Nitrate (mg/l)	55.6 ± 10.44
Sulphate (mg/l)	136.33 ± 12.43

2.2 Electrolytic experimental set up

A rectangular Reactor with external dimensions of height = 30 cm, width = 7 cm, length = 11 cm and wall thickness = 10 mm was constructed with glass. The experimental set up consisted of a glass chamber as a Reactor with a capacity of 2.0 L sample. Each time, the BTMW sample of 2.0 L was collected and placed in an electrolytic cell (Chopra and Sharma 2012). Al–Al electrode combination was connected to their respective anode and cathode leading to the D.C Power supply (LMC electronics, India 0–500 V and 0–2 A) and energized for a required duration of time at different voltages and currents. All the experiments were performed at room temperature (30 ± 2 °C) and at a constant stirring speed (100 rpm) to maintain the uniform mixing of BTMW sample during the EC. Before conducting the experiment, the electrodes were washed with water, dipped into diluted HCl (5 % v/v) for 5 min, thoroughly washed with water and then finally rinsed twice with distilled water. Electrodes were dipped into BTMW samples with different electrode areas (80, 120 and 160 cm^2) and different inter electrode distances (0.5, 1.0, 1.5, 2.0 and 2.5). The different voltages (5–40 V) were applied for different operating times (10–80 min). After applying the particular voltage for a particular time period, i.e., after each batch experiment, the treated samples were allowed to settle for different times (30, 60 and 90 min).

2.3 Analytical methods

$NO_3^-$and $SO_4^{2-}$of BTMW were analyzed before and after EC following the Standard methods (APHA 2005). The calculation of $NO_3^-$and SO_4^{2-} removal efficiency after EC was carried out using the following formula:

$$\text{CR}\ \% = \frac{C_0 - C}{C_0} \times 100 \tag{1}$$

where C_0 and C are concentrations of wastewater before and after electrolysis.

2.4 Calculation of operating cost

The cost of energy and electrode material was taken into account for the calculation of the operating cost (US $/m^3) of EC using the following formula described by Ghosh et al. (2008):

$$\text{Operating cost} = a\, C_{energy} + b\, C_{electrode} \tag{2}$$

where C_{energy} (kWh/m^3) and $C_{electrode}$ (kg Al/m^3) are the consumption quantities for the $NO_3^-$and SO_4^{2-} removal. "*a*" electrical energy price 0.01 US$/kWh, "*b*" electrode material price 3.5 US$/kg for Al electrode cost due to electrical energy (kWh/m^3) was calculated as:

$$C_{energy} = \frac{U \times I \times t_{EC}}{v} \quad (3)$$

Cost for electrode (kg Al/m^3) was calculated using the following equation:

$$C_{electrode} = \frac{I \times t \times M_w}{z \times F \times v} \quad (4)$$

where U is cell voltage (V), I is current (A), t_{EC} is time of electrolysis (s) and v the volume (m^3) of BTMW water, M_W the molecular mass of aluminum (26.98 g/mol), *z the* no of electrons transferred (z = 3 for Al) and F is Faraday's constant (96,487 C/mol).

3. RESULTS AND DISCUSSION

The results on $NO_3^-$and SO_4^{2-} removal efficiency of BTMW by EC at different operating conditions like voltage/current density, EC time, inter electrode distance, electrode area, pH and settling time are shown in Figs. 1, 2, 3, 4, 5 and 6.

3.1 Effect of current density

Current density was found to be one of the most important parameter of electrolytic process that could drastically affect the removal efficiency of nitrate and sulphate from BTMW. It not only determines the coagulant dose but also the bubble size generation and floc formation which can influence the treatment efficiency of electrolytic technology. As the rate of bubble-generation increases, the bubble size decreases with increase in current density and both of these trends are beneficial in terms of high pollutant removal efficiency by H_2 flotation (Can et al. 2003; Kobya et al. 2006; Tezcan Un et al. 2009; Song et al. 2008).With high current, the coagulant dosage rate increased resulting in a greater amount of precipitate in the stable stage. Likewise, the bubble density increases resulting in a greater upward momentum flux and thus there is faster removal of pollutant and coagulant by flotation (Holt et al. 2001). The higher production rate of hydrogen allowed by higher current favored the flotation of the flocculated matter (Khemis et al. 2006).

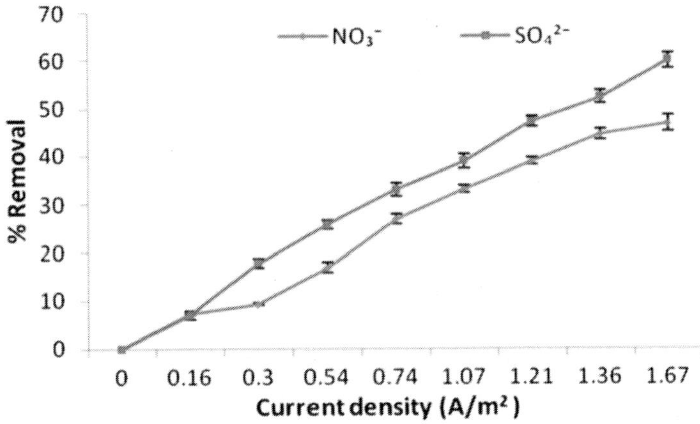

Figure. 1 % removal (with standard error) of NO_3^- and SO_4^{2-} using Al–Al electrode combination with different current densities at constant operating conditions (EC time: 30 min, inter electrode distance: 1.0 cm, electrode area: 80 cm^2, pH: 7.5, settling time: 30 min)

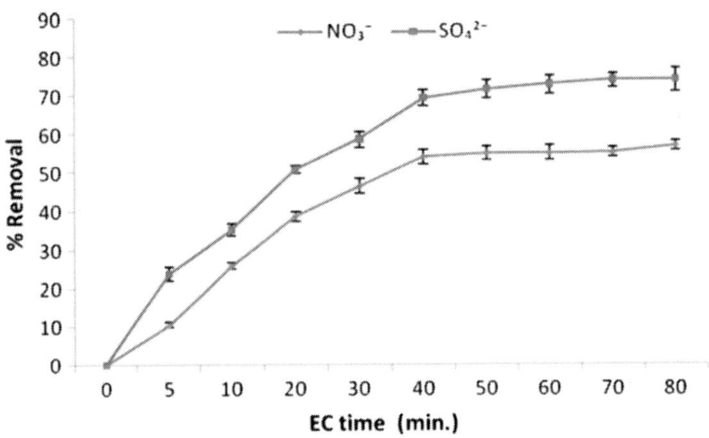

Figure. 2 % Removal (with standard error) of NO_3^- and SO_4^{2-} using Al–Al electrode combination with different EC at constant operating conditions (current density: 1.68 A/m^2, inter electrode distance: 1.0 cm, electrode area: 80 cm^2, pH: 7.5, settling time: 30 min)

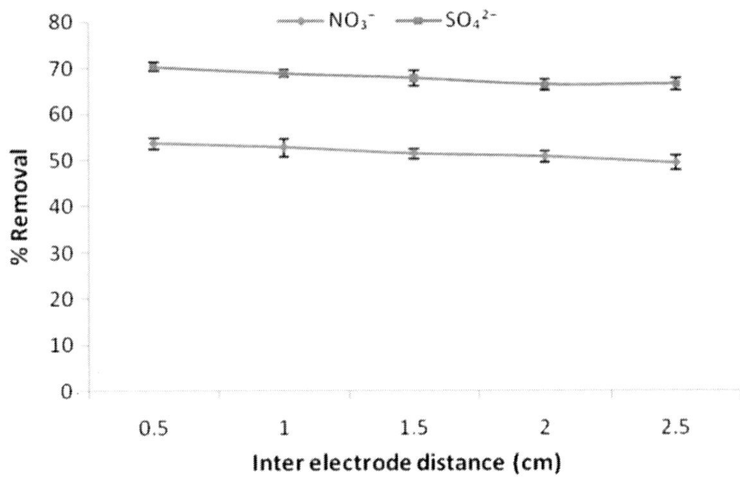

Figure. 3 % Removal (with standard error) of NO_3^- and SO_4^{2-} using Al–Al electrode combination with different inter electrode distance at constant operating conditions (current density: 1.68 A/m^2, EC time: 40 min, electrode area: 80 cm^2, pH: 7.5, settling time: 30 min)

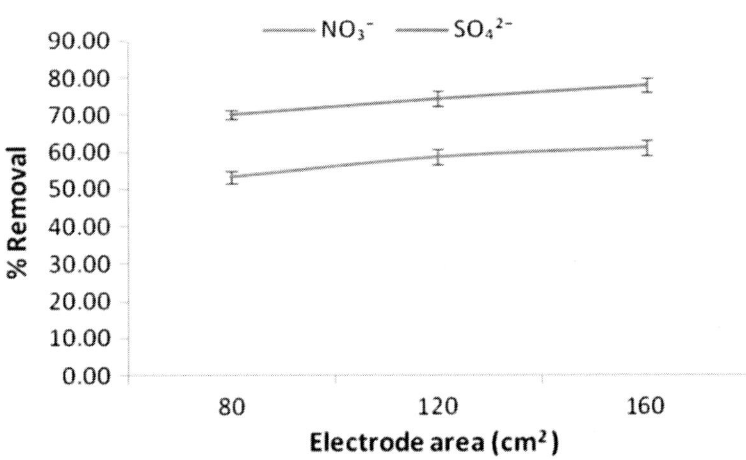

Figure. 4 % Removal (with standard error) of NO_3^- and SO_4^{2-} using Al–Al electrode combination with different electrode area at constant operating conditions (current density: 1.68 A/m^2, EC time: 40 min, inter electrode distance: 0.5 cm, pH: 7.5, settling time: 30 min)

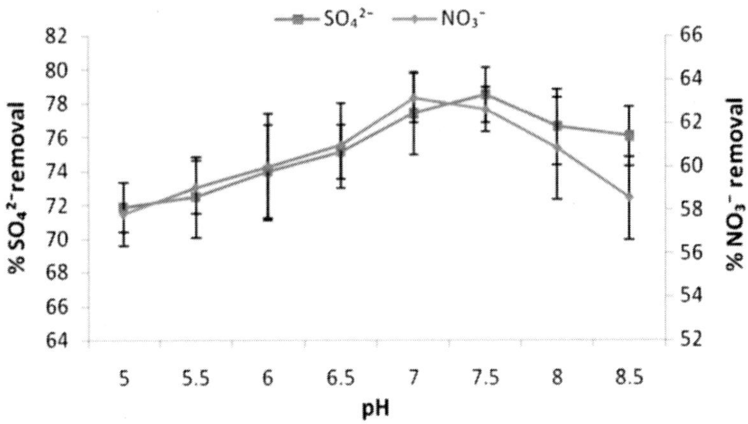

Figure. 5 % Removal (with standard error) of NO_3^- and SO_4^{2-} using Al–Al electrode combination with different pH at constant operating conditions (current density: 2.65 A/m^2, EC time: 40 min, inter electrode distance: 0.5 cm, electrode area: 160 cm^2, settling time: 30 min)

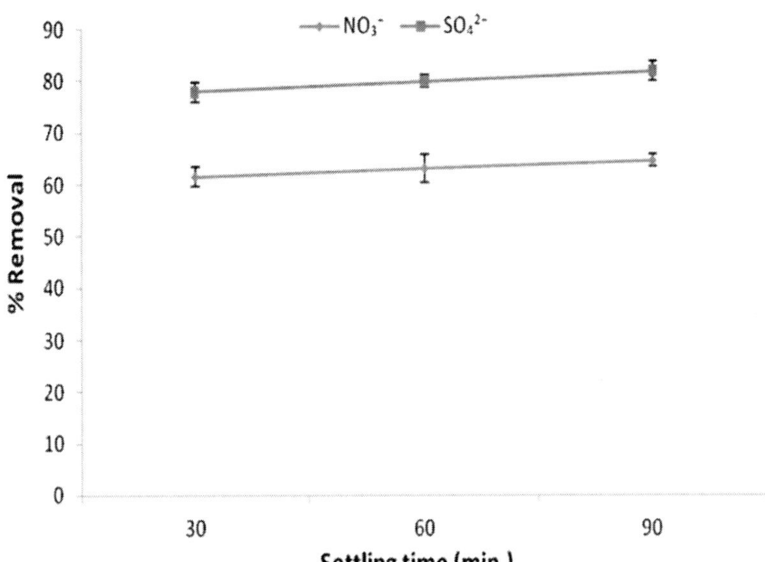

Figure. 6 % Removal (with standard error) of NO_3^- and SO_4^{2-} using Al–Al electrode combination with different settling times at constant operating conditions (current density: 2.65 A/m^2, EC time: 40 min, inter electrode distance: 0.5 cm, electrode area: 160 cm^2, pH: 7.5)

We observed that the removal percentage of $NO_3^-$and SO_4^{2-}increased progressively with an increase in the current density from 0.16 to 1.68 A/m^2 corresponding to its constant voltages (5–40 V), and it was indicated that the maximum removal of NO_3^-(46.7 %) and SO_4^{2-} (59.9 %) of BTMW effluent was with the operating conditions of EC time: 30 min, inter electrode distance: 1.0 cm, electrode area: 80 cm^2, pH: 7.5 and settling time: 30 min (Fig. 1). This was due to the fact that as the current increased, more $Al(OH)_3$ was produced, which contributed to greater removal of NO_3^- and SO_4^{2-} from BTMW by precipitation and flotation. Thus, the higher values of current electrolysed the higher mass of Al during EC, thereby indicating that applied current density to the EC was directly related to the electrocoagulation. (This was in accordance with Faraday's Law of Electrolysis). It was observed that applied voltage varying from 10 to 25 V, the maximum percent removal of NO_3 from drinking water was 84 % at 25 V (Kumar and Sudha 2010). The increase in NO_3 removal efficiency has been reported with an increase of the electrical potential from 10 to 40 V for aqueous solution (Malakootian et al. 2011). The cell voltage increases gradually with the increase in current densities as can be expected from the rate of metal oxidation resulting in a greater amount of precipitate being formed during the electrolytic treatment that increases the removal of pollutants from wastewater. There was a slight increase in the temperature with the increase in current densities because of poor conductivity of the solution as also reported by us (Chopra and Sharma 2015) and others (Asaithambi and Matheswaran 2011).

3.2 Effect of EC time

It has been observed that as the EC time increases, the concentration of metal ions and their hydroxide flocs increase (Daneshvar et al. 2006; Zaroual et al. 2006; Daneshvar et al. 2007). More bubbles are generated at higher currents that improve the degree of mixing of $Al(OH)_3$and phenol which enhance floatation ability of the cell with a consequent increase in the phenol removal efficiency (Golder et al. 2007).

The EC time influenced the treatment efficiency of EC. With an increase in EC time, the anodic electrode dissolution led to release of metal ions and the cathode released OH^-, which formed their hydroxides into BTMW. The removal of NO_3^- and SO_4^{2-} increased progressively with an increase in the EC time from 5 to 40 min with the operating conditions of current density: 1.68 A/m^2, inter electrode distance: 1.0 cm, electrode area: 80 cm^2, pH: 7.5 and settling time: 30 min. The maximum removal of NO_3^- (53.95 %) and SO_4^{2-} (69.15 %) was observed at optimum EC time of 40 min, beyond which there was no significant removal in NO_3^- and SO_4^{2-}(Fig. 2).

3.3 Effect of inter electrode distance

It has been reported that increasing the distance between electrodes increases electric resistance against the current flowing between anode and cathode (Andrzejm 1980; Dima et al. 2003*)*. As the electrode gap becomes less, there is a low mixing of the fluid between electrodes and that remains not sufficient to increase in concentration-polarization layer on electrode surface. With the result, these effects can increase electric potential or resistance of electrodes, thereby diminishing the NO_3^- removal efficiency subsequently (Tsai et al. 1997; Lee et al. 2006).

During the present study, inter electrode distance was an effective factor in the EC of BTMW. The percentage removal of NO_3^- and SO_4^{2-} increased progressively with decrease in inter electrode distance from 2.5 to 0.5 cm, whereby it exhibited the maximum removal of NO_3^-(53.50 %) and SO_4^{2-} (70.29 %) with a distance of 0.5 cm between the electrodes, each having an electrode area of 80 cm^2 with the operating conditions of current density: 1.68 A/m^2, EC time: 40 min, electrode area: 80 cm^2, pH: 7.5 and settling time: 30 min. After that the removal efficiency of NO_3^- and SO_4^{2-} was almost constant as the inter electrode distance increased from 0.5 to 1.5 cm (Fig. 3). Similar observations have also been reported by Li et al. (2008) for the removal of COD which decreases with the decrease in distance between electrodes of the same composition. This is because that the shorter distance speeds up the anion discharge on the anode and improves the oxidation. It also reduces resistance, electricity consumption and the cost of the wastewater treatment. Ghosh et al. (2008) have observed that with the increase of inter electrode distance, the percentage removal of dye products from wastewater decreased. At a lower inter electrode distance, the resistance encountered by the current flowing in the solution medium decreased, thereby, facilitated the electrolytic process resulting in enhanced dye removal. Nandi and Patel (2013) also observed that with weak inter electrode distance of 1 cm, there was maximum dye removal of 99.59, 89.98 and 76.14 % for three different current densities of 41.4, 27.8 and 13.9 A/m^2, respectively.

3.4 Effect of electrode area

An increase in the electrode area causes corresponding increase of coagulants. The entire effectiveness of the coagulation process depended on the appropriate amount of coagulant (Daneshava et al. 2005). Logistical relationship between electrode geometric area (AG) and copper removal efficiency indicated that an increase in copper removal was related to an increase in AG, reaching to an optimal value of 35 cm^2 with an asymptotic value of ≈80 % (Escobara et al. 2006). The removal efficiency of TD, BOD and COD has been attributed to a greater EA that produced larger amount of anions and cations from the anode and cathode. The greater the EA, the greater the rate of flock formation, which in turn influenced the removal efficiency of ET (Chopra and Sharma 2013). Similarly, in the present study it was observed that with an increase in electrode area from 80 to 160 cm^2, the current density increased from 1.68 to 2.65 A/m^2 with a corresponding constant voltage of 40 V which resulted in an increase in the removal % of NO_3^- and SO_4^{2-}. The maximum removal of NO_3^- (61.14 %) and SO_4^{2-} (77.94 %) was achieved with maximum electrode area of 160 cm^2 with the following operating conditions of current density: 1.68 A/m^2, EC time: 40 min, inter electrode distance: 0.5 cm, initial pH: 7.5 and settling time: 30 min (Fig. 4).

3.5 Effect of pH

pH is an important operating factor influencing the electro-coagulation process (Lin and Chen 1997; Chen et al. 2000; Gurses et al. 2002; Kobya et al. 2003). It is the removal efficiency of colloidal particles in the pH range of 4–7 that leads to the formation of amorphous hydroxide precipitates and other aluminum hydroxo complexes with hydroxide ions and polymeric species (Bayramoglu et al. 2004). It has been shown that pH is an important factor in EC and its variation is usually caused by the solubility of metal hydroxides. The pH of the effluent increases for acidic influent after electro-coagulation and decreases for alkaline influent. The hydroxy radicals which may be produced as a result of dissociation of water under certain electrochemical reactions are known to be one of the most reactive aqueous radical species and these radicals have the ability to oxidize almost all of the organic compounds because of their high affinity value of 136 kcal (Chen and Hung 2007). In fact, hydroxyl radical is an extremely potent oxidizing agent with a short half-life, and is able to oxidize organic compound by hydrogen abstraction reaction, by redox reaction and/or by electrophilic addition to pi systems resulting in a cascade of free radical reactions leading to oxidative degradation of pollutants (Oturan 2000). However, electrolysis using Al electrodes leads to the formation of ionic complexes, and formation of hydroxyl radical, to the best of our knowledge, has not been demonstrated. The Al_3^+ ions on hydrolysis generated the aqueous complex Al $(H_2O)_6^{3+}$, which was predominant at pH < 4. Between pH 5 and 6; the predominant hydrolysis products were $Al(OH)_2^+$ and $Al(OH)_2^+$ between pH 5.2 and 8.8, the solid $Al(OH)_3$ was more prevalent and above pH 9, the soluble species $Al(OH)_4$ were the predominant and were the only species present above the pH 10 (Gomes et al. 2007).

The removal of NO_3^- and SO_4^{2-} from BTMW effluent with its different initial concentrations of pH 5–8.5, operating conditions of current density: 2.65 A/m^2, EC time: 40 min, inter electrode distance: 0.5 cm, electrode area: 160 cm^2 and settling time: 30 min indicated that the removal of NO_3^- increased from 57.66 to 63.14 % with an increase in the pH from 5 to 7.0, while the removal of SO_4^{2-} increased from 71.78 to 78.49 % with an increase in pH from 5 to 7.5. However, the increase in pH of BTMW to more than 7.0 decreased the % removal of NO_3^-, whereas the removal % of SO_4^{2-} decreased with the increase in pH more than 7.5 (Fig. 5).

3.6 Effect of settling time

The effect of settling time on removal efficiency after EC has not been given due consideration so far. There appears to be little work with regard to the settling time on the BTMW treatment during electrolysis. During the present study, it was interesting to note that the removal efficiency with the constant operating conditions of current density: 2.65 A/m^2, EC time: 40 min, electrode area: 160 cm^2 and pH: 7.5 improved with an increase of settling time from 30 to 90 min of BTMW. The maximum removal of NO_3^- (63.21 %) and SO_4^{2-} (79.98 %) was found to be at the settling time of 60 min. Beyond which there was almost no more significant removal (Fig. 6). However, our earlier study has also reported the maximum removal of COD (92.35 %) at the settling time of 60 min and that of BOD (84.88 %) at the settling time of 30 min (Sharma and Chopra 2013). It was also interesting to note that the EC of BTMW brought down the concentration of NO_3^- from 123.45 to 36.47 mg/l to the desirable limit (45 mg/l) of BIS (1991) standard of drinking water. Though the concentration of SO_4^{2-} of BTMW was already within permissible limit of BIS, EC further brought down the removal of SO_4^{2-} from 142.34 to 28.92 mg/l.

3.7 Economic evaluation

Electrical energy and electrode consumption are the important parameters in the economic evaluation of EC process. In EC process, the operating cost mainly included the cost of electrodes and electrical energy as well as labor, maintenance, sludge dewatering and their disposal. The removal of NO_3^- and SO_4^{2-} during EC, energy consumption increased from 64.32 to 101.76 kWh/m^3 with an increase in electrode area (80–160 cm^2) and current density (1.68–2.65 A/m^2) that resulted in increase of the electrode consumption from 14.985×10^{-5} to 23.71×10^{-5} kg/m^3. The cost was found to be 1.01 \$ on the basis of energy and electrode consumption with the optimum operating conditions (Table 2).

Table 2 Operating cost using Al–Al electrode combination at different electrode area and current density (EC time: 40 min, inter electrode distance: 0.5 cm, pH: 7.5 and settling time: 60 min.)

Electrode area (cm^2)	**Al–Al electrode combination**				
	Current density (A/m^2)	**Final pH**	**Electrode consumption (kg/m^3)**	**Energy consumption (kWh/m^3)**	**Total OC ($\$/m^3$)**
80	1.68	9.21	14.99×10^{-5}	64.32	0.64
120	2.09	10.89	18.68×10^{-5}	80.16	0.80
160	2.65	11.23	23.71×10^{-5}	101.76	1.01

4. CONCLUSION

The removal of $NO_3^-$and SO_4^{2-} from BTMW using Al electrodes was found to depend on voltage/current density, EC time, inter electrode distance, electrode area and initial pH during the EC. The maximum removal of NO_3^- (63.21 %) and SO_4^{2-} (79.98 %) was with the optimal operating conditions of current density: 2.65 A/m^2, EC time: 40 min, electrode area: 160 $cm^2$and pH: 7.5.

The higher values of current electrolysed the higher mass of Al during EC, thereby indicating that applied current density to the EC was directly related to the electrolysis. The EC removed the concentration of NO_3^- from the BTMW below to the desirable limit of the BIS/WHO standards of drinking water. The operating cost was 1.01 US $\$/m^3$ in terms of energy and electrode consumption. There was no need of pH adjustment of the BTMW during EC as the optimal removal of NO_3^- and SO_4^{2-} was close to the neutral pH of 7.0

ACKNOWLEDGMENTS

The University Grants Commission, New Delhi, India, is acknowledged for providing the financial support in the form of UGC Research Fellowship (F.4-1/2006 (BSR) 7-70/2007 BSR) to Mr. Arun Kumar Sharma.

REFERENCES

1. Aiyuk S, Forrez I, Kempeneer DL, Haandel A, Verstraete W (2006) Anaerobic and complementary treatment of domestic sewage in regions with hot climates—A review. Bioresour Technol 97:2225–2241
2. Andrzejm B (1980) Electrocoagulation of biologically treated sewage. In: 35th Industrial waste conference proceeding, p 541
3. APHA (2005) Standard methods for the examination of water and wastewater, 21st edn. American Public Health Association, Washington
4. Asaithambi P, Matheswaran M (2011) Electrochemical treatment of simulated sugar industrial effluent: optimization and modeling using a response surface methodology. Arab J Chem. doi:10.1016/j.arabjc.2011.10.004
5. Bayramoglu M, Kobya M, Can OT, Sozbir M (2004) Operating costs analysis of electrocoagulation of textile dye wastewater. Sep Purif Technol 37:117–125
6. BIS (1991) Specifications for Drinking Water, IS: 10500. Bureau of Indian Standards, New Delhi
7. Can OT, Bayramoglu M, Kobya M (2003) Decolorization of reactive dye solutions byelectrocoagulation using aluminum electrodes. Ind Eng Chem Res 42:3391–3396
8. Chen G, Hung YT (2007) Electrochemical wastewater treatment processes, in handbook of environmental engineering, volume 5: advanced physicochemical treatment technologies, L.K. Wang, Y-T. Hung, and N.K. Shammas, The Humana Press Inc., Totowa, pp 57–105
9. Chen X, Chen G, Yue PL (2000) Separation of pollutants from restaurant waste water by electrocoagulation. Sep Purif Tech 19:65–76
10. Chopra AK, Sharma AK (2012) Efficiency of turbidity and BOD removal from secondarily treated sewage by electrochemical treatment. J Appl Nat Sci 4(2):304–309
11. Chopra AK, Sharma AK (2013) Removal of turbidity, COD and BOD from secondarily treated sewage water by electrolytic treatment. Appl Water Sci 3(1):125–132
12. Chopra AK, Sharma AK (2015) Effect of electrochemical treatment on the COD removal from biologically treated municipal wastewater. Desalin Water Treat 53(1):41–47
13. Daneshava N, Oladegaragoze A, Djafarzadeh N (2005) Decolorization of basic dye solutions by electrocoagulation: an investigation of the effect of operational parameters. J Hazard Mater 129:116–122
14. Daneshvar N, Oladegaragoze A, Djafarzadeh N (2006) Decolorization of basic dye solutions by electro coagulation: an investigation of the effect of operational parameters. J Hazard Mater B 129:116–122
15. Daneshvar N, Khataee AR, Amani Ghadim AR, Rasoulifard MH (2007) Decolorization of C.I. acid yellow 23 solution by electrocoagulation process: investigation of operational parameters and evaluation of specific electrical energy consumption (SEEC). J Hazard Mater 148:566–572
16. Dima GE, de Vooys ACA, Koper MTM (2003) Electro catalytic in acid solutions. J Electro anal Chem 15:554–555
17. Escobara C, Cesar SS, Toral M (2006) Optimization of the electrocoagulation process for the removal of copper, lead and cadmium in natural waters and simulated wastewater. J Environ Manag 81(4):384–391
18. Ghosh D, Medhi CR, Solanki H, Purkait MK (2008) Decolorization of crystal violet solution by electro coagulation. J environ prot sci 2:25–35
19. Golder AK, Samanta AN, Ray S (2007) Removal of trivalent chromium by electrocoagulation. Sep Purif Technol 53:33–41
20. Gomes JAG, Daida P, Kesmez M, Weir M, Moreno H, Parga JR, Irwin G, Mc Whinney H, GradyT Peterson E, Cocke DL (2007) Arsenic removal by electrocoagulation using combined Al-Fe electrode system and characterization of products. J Hazard Mater 139(2):220–231

21. Gurses A, Yalcin M, Dogar C (2002) Electro coagulation of some reactive dyes: a statistical investigation of some electrical variables. Waste Manag 22:491–499
22. Holt PK, Barton G W, Mitchell C A (2001) The role of current in determining pollutant removalin a batch electrocoagulation reactor. In: 6th World congress of chemical engineering, conference media CD, Melbourne, Australia
23. Holt PK, Barton GW, Mitchell CA (2005) The future for: electrocoagulation as a localized water treatment technology. Chemosphere 59:355–367
24. Hua CY, Loa SL, Kuan WH (2003) Effects of co-existing anions on fluoride removal in electrocoagulation (EC) process using aluminum electrodes. Water Res 37:4513–4523
25. Huanga CH, Chen L, Yang CL (2009) Effect of anions on electrochemical coagulation for cadmium removal. Sep Purif Technol 65:137–146
26. Katz I, Dosoretz CG (2008) Desalination of domestic wastewater effluents: phosphate removal as pretreatment. Desalination 222:230–242
27. Khemis M, Leclerc JP, Tanguy G, Valentin G, Lapicque F (2006) Treatment of industrial liquid wastes by electro-coagulation: experimental investigations and an overall interpretation model. Chem Eng Sci 61:3602–3609
28. Kobya M, Can OT, Bayramoglu M (2003) Treatment of textile waste waters by electrocoagulation using iron and aluminum electrodes. J hazard Mat B 100:163–178
29. Kobya M, Demirbas E, Can OT, Bayramoglu M (2006) Treatment of levafix orange textile dye solution by electrocoagulation. J Hazard Mat 132:183–188
30. Kumar NS, Sudha Goel (2010) Factors influencing arsenic and nitrate removal from drinking water in a continuous flow electro coagulation (EC) process. J Hazard Mat 173:528–533
31. Lee SM, Maken S, Jang JH, Park KN, Park JW (2006) Development of physico chemical nitrogen removal process for high strength industrial waste water. Water Res 40:975–980
32. Lettinga G, Lier Van, Van JB, Buuren JCL, Zeeman G (2001) Sustainable development inpollution control and the role of anaerobic treatment. Water Sci Technol 44:181–188
33. Li Xu, Wei Wang, Mingyu Wang, Yongyi Cai (2008) Electrochemical degradation of tridecanedicarboxylic acid wastewater with tantalum-based diamond film electrode. Desalination 222:388–393
34. Lin SH, Chen ML (1997) Treatment of textile wastewater by electrochemical methods for reuse. Water Res 31:868–876
35. Malakootian M, Yousefi N, Fatehizadeh A (2011) Survey efficiency of electrocoagulation on nitrate removal from aqueous solution. Int J Environ Sci Tech 8:107–114
36. Nandi BK, Patel S (2013) Effects of operational parameters on the removal of brilliant green dye from aqueous solutions by electrocoagulation. Arab J Chem. doi:10.1016/j.arabjc.2013.11.032
37. Oturan MA (2000) An ecologically e€ective water treatment technique using electrochemically generated hydroxyl radicals for in situ destruction of organic pollutants: application to herbicide 2,4-D. J Appl Electrochem 30:475–482
38. Sharma AK (2013) A study on electrolytic technology for purification of secondarily treated sewage-waste effluents. Ph.D. Thesis, Gurukula Kangri University, Haridwar
39. Sharma AK, Chopra AK (2013) Removal of COD and BOD from biologically treated municipal wastewater by electrochemical treatment. J Appl Nat Sci 5(2):475–481
40. Song S, Yao J, He Z, Qiu J, Chen J (2008) Effect of operational parameters on thedecolorization of C.I. Reactive blue 19 in aqueous solution by ozone-enhanced electrocoagulation. J Hazard Mater 152:204–210
41. Tezcan Un U, Koparal AS, Bakir Ogutveren U (2009) Electrocoagulation of vegetable oil refinery wastewater using aluminum electrodes. J Environ Manage 90:428–433
42. Tsai CT, Lin ST, Shue YC, Su PL (1997) Electrolysis of soluble organic matter in leachate from landfills. Water Res 31(12):3073–3081
43. Yi Q (2001) A sustainable technology for developing country anaerobic digestion. In: Proc. 9th World Congress on anaerobic digestion––anaerobic conversion for sustainability (Part 1), Antwerp, Belgium, pp 23–30

44. Zaroual Z, Azzi M, Saib N (2006) Contribution to the study of electrocoagulation mechanism inbasic textile effluent. J Hazard Mat 131(1–3):73–78

CHAPTER 4

Land Application-Based Olive Mill Wastewater Management

Iosif Kapellakis [1,3,*], Vasileios A. Tzanakakis [2] and Andreas N. Angelakis [3]

[1]*Kapellakis Technical and Real Estate Firm, Venerato Palianis 70011, Greece*

[2]*Region of Crete, Directorate of Agricultural Economy, Iraklion 71201, Greece*

[3]*National Agricultural Research Foundation (N.AG.RE.F.), Institute of Iraklion, Iraklion 71307, Greece*

ABSTRACT

Land application of olive mill wastewater (OMW) is considered a promising low-cost practice for olive-oil producing countries. The objectives of this work were to investigate: (i) OMW treatment potential of a land treatment system (LTS), planted with a *E. camaldulensis species*, regarding N, P, C, and phenols; (ii) the effects of OMW on chemical properties of soil and soil solution characteristics; and (iii) the performance of *E. camaldulensis* in terms of biomass production and N and P recovery. *E. camaldulensis* received OMW for two growing seasons at rates based on maximum organic loading. These rates were almost equivalent to the reference evapotranspiration of the area. Soil solution and soil samples were collected from three different depths (15, 30 and 60 cm) at specified time intervals. - Also, samples of plant tissues were collected at the end of application periods. OMW land application resulted in significant reduction in inorganic and organic constituents of OMW. At 15 cm of soil profile, the average removal of COD, TKN, NH_4^+-N, TP, In-P, and total phenols approached 93%, 86%, 70%, 86%, 82%, and 85%, respectively, while an increase in soil depth (30 and 60 cm) did not improve significantly treatment efficiency.

Furthermore, OMW increased soil organic matter (SOM), total kjeldahl nitrogen (TKN), and available P, particularly in the upper soil layer. In contrast, low inorganic N content was observed in the soil throughout the study period caused probably by increased competition among soil microorganisms induced by the organic substrate supply and high C/N ratio. Also, electrical conductivity (EC) and SAR increased by OMW addition, but at levels that may do not pose severe risk for soil texture. Enhancement of soil fertility due to OMW application sustained eucalyptus trees and provided remarkable biomass yield. In conclusion, land application of OMW has a great potential for organic matter and phenol assimilation and can be effectively used for OMW detoxification.

Keywords: olive mill wastewater; *E. camaldulensis*; irrigation; land treatment; land application; biomass production; nutrient uptake; soil impacts; organic matter; phenols; salinity

1. INTRODUCTION

Olive mill wastewater (OMW) is the liquid by-product generated mainly in three-phase (olive oil-OMW-pomace) olive mills during olive oil production. Around 6×10^6 m^3 of OMW is produced worldwide at a short period, between November and March, of which 98% is produced in the Mediterranean basin [1]. In areas that have not yet adopted the newer two-phase extraction technique (olive oil-liquid pomace) the most common adopted practices to manage OMW remain the storage in evaporation ponds [2] and the direct application onto land [3]. Spreading of OMW onto land is considered a promising low-cost practice since most of the Mediterranean olive-oil producing countries suffer from water deficiency and deterioration of terrestrial and aquatic environment [3,4]. Italy, Portugal, and Catalonia have already regulated land application of OMW allowing, however, low application rates in order to prevent toxicity phenomena in soil and crops [3,5].

Earlier study demonstrate the high potential of soil in terms of the assimilation and removal of both inorganic and organic OMW constituents at application rates up to 1000 mm/year [6]. This is efficiency due to presence of easily decomposable organic compounds in OMW that stimulate soil microorganisms and their activities [7,8] inducing decomposition and mineralization processes. In addition to biodegradation part of the applied organic substrate and nutrients may be retained into the soil matrix [9,10], leading to enhancement of soil fertility [6,7,11,12]. Also, OMW application may cause alterations in soil biological properties

[7,13,14,15,16] associated with the changing nutrient availabilities and/or toxicity phenomena. With regard to physical properties of soil, there is evidence of changes due to OMW application, but the results so far are contradictory [6,17,18]. Several factors are considered important to the response of soil to OMW application including organic substrate, soil-climatic conditions, management practices, and vegetation. With regard to vegetation, recent studies have evidenced that plant species, via litter decomposition and rhizodeposition, affect soil physicochemical properties, soil microbial communities, and C and N turnover rates [19,20,21,22]. Given the high organic loading in OMW and the presence of vegetation the optimization of OMW application rates and treatment potential—still remains a challenge.

To date available data on OMW land application, particularly for eucalyptus species is limited. These species are widely used in land application schemes due to several desirable attributes, such as adaptability to soil salinity, treatment potential, and biomass production [23], however, their performance in cases of OMW application remains unknown. Moreover, despite the plethora of work dealing with OMW effect on soil properties, little information is available with regard the distinct effect of OMW on the quality and characteristics of soil solution. The composition of this fraction is critical as it can be easily moved through soil profile and threaten underlying and surface waters. Thus, the objectives of this work were to investigate: (i) treatment potential of a land application scheme planted with *E. camaldulensis* species especially with regard to N, P, C, and phenols; (ii) the effects on soil chemical properties and soil solution characteristics; and (iii) the performance of *E. camaldulensis* in terms of biomass production and nutrient recovery. The knowledge provided is expected to contribute towards a better understanding of the capacity of land treatment systems (LTS) to assimilate OMW constituents, optimization of their operation, and elimination of the environmental risk.

2. MATERIALS AND METHODS

The study was carried out at the experimental field of National Agricultural Research Foundation at Skalani village, located 10 km south of Iraklio City, Greece. The map reference co-ordinates of the experimental field are latitude 35°16'51" N and longitude 25°10'51" E, and the elevation is 233 m a.s.l. The experimental field, the soil sampling and analysis, and the biomass production are discussed as follows.

A pilot LTS was established in April 2003, including 40 *E. camaldulensis* seedlings planted at two double rows of 20 seedlings each. The seedlings were planted at distances of 1 m within and between rows in each of the double row. The distance between double rows was 1 m. The total covered area was 44 m^2. The soil was conventionally tilled before planting the *E. camaldulensis* trees. The LTS operated from June to October 2003, and from April to October 2004. OMW from a nearby olive mill was stored in a retention pond and applied through furrow irrigation to the LTS. Olive mill wastewater was applied at such rates so that the organic load did not exceed 500 kg BOD_5/ha·day, the highest organic loading rate recommended for LTS [24]. Based on the COD/BOD ratio of the applied wastewater, which is approximately 2.5, the average loading of LTS was approximately 1250 kg COD/ha·d. Effluent was applied approximately on fortnightly cycles to provide adequate time for organic matter degradation and to avoid clogging of soil pores. The hydraulic load applied in the field, was derived from the following Equation:

$$\frac{V \cdot C}{T \cdot A} \leq 1250 \frac{kgCOD}{ha \cdot d} \tag{1}$$

where, V: Loading volume (m^3); C: Influent COD concentration (mg/L); T: Time intervals (typically 15 d); A: Area of the field (0.0044 ha).

The reference evapotranspiration (ETo) was calculated using Cropwat 8.0 software [25], based on FAO Penman-Monteith method. The soil of the study area was clay loam with high calcium content (55% $CaCO_3$); pH: 7.58; electrical conductivity (ECsp): 0.55 dS/m; organic matter content: 0.79%; total kjeldahl nitrogen (TKN): 0.049%; and Olsen-P: 14 ppm. OMW samples before the fortnightly application were being collected and analyzed for pH, EC, COD, TKN, NH_4^+-N, NO_3^-N, TP, inorganic P (In-P), and total phenols. The average composition of OMW applied to LTS is presented in Table 1. During the study soil solution samples were collected at 15 cm, 30 cm, and 60 cm soil depth with soil water samplers (Soil Moisture Equipment Corp. model 1920F1l24-B0,5 M2; Santa Barbara, CA, USA) 24 h after OMW application, to allow the applied effluent to equilibrate with soil and were analyzed for the parameters previously reported. The soil solution samplers were installed at the center of a double row of the LTS, and were stabilized with a thin layer of betonite. Tensiometers were installed at depths of 30 and 60 cm to provide indications of soil moisture during the OMW application period. Preparation and analyses of the samples was undertaken according to the Standard Methods for the Examination of Water and Wastewater [26]. Total phenols were assessed according to the Folin-Ciocalteu method [27]. Soil samples were collected in April 2003 (beginning of the study), in October 2003 (end of first year application), and in October 2004 (end of the experiment) from 0 to 15, 15 to 30, 30 to 45, and 45 to 60 cm soil depth. Samples were also collected from 60 to 75 cm soil layer in October 2003 and 2004. Samples were analyzed for soil organic matter (SOM), TKN, available P (Olsen-P), pH, ECsp, Ca^{2+}, Mg^{2+}, and SAR according to Methods of Soil Analysis [28].

Table 1. Composition of olive mill wastewater (OMW) (average values) in the retention pond (in mg/L; pH is unitless; EC in dS/m). In parenthesis is shown the standard error.

Parameters	Applied OMW
pH	6.22 (0.28)
EC	7.15 (0.25)
COD	21,862 (2,92)
NH_4^+-N	17.61 (2.48)
TKN	167.8 (29.16)
Org-N	150.2 (22.16)
NO_3^-	2.11 (0.04)
In-P	21.5 (3.65)
TP	63.87 (16.89)
Total phenols	1,386 (151.67)

Biomass accumulation was assessed in October 2003 and 2004. In October 2003 three representative trees were harvested and separated into individual organs. Dry weight of total biomass was assessed after drying representative sub-samples at 65 °C to constant weight. Then dry samples were grounded to pass through a 1-mm screen and stored for elemental analysis. Similarly, at the end of the second experimental period, seven representative trees were harvested and the dry weight and nutrient tissue concentrations were assessed similar to the previous growing season. In the present study, water use efficiency (WUE), nutrient used efficiency for nitrogen (NUE-N) and phosphorus (NUE-P) were determined by taking into account biomass yield, OMW loading, and the amounts of N and P recovered, respectively. Specifically, WUE of plant species was estimated by dividing the dry weight of the produced hypergeous biomass with the applied amount of effluent. NUE-N and NUE-P were determined by hypergeous biomass divided by the amount of N and P recovered, respectively.

Nutrient removal refers to the percentage of the applied, via OMW application, nutrient that assimilated in plant biomass. Analysis of variance (ANOVA) was performed to assess the effect of soil depth on the parameters examined using SPSS software [29].

3. RESULTS

3.1. Storage of Olive Mill Wastewater

Storage in retention pond affected the physico-chemical properties of OMW. The pH of the raw OMW increased from 5.04 in April to 7.47 in the middle of October 2004. The EC also showed an increasing trend from the initial value of 5.75 dS/m to 8.00 dS/m in October, due to the high ET values prevailing in the study area and the concentration of the effluent volume (Figure 1). The degradation of organic compounds and the release of electrolytes may have also contributed to this increase. In contrast, COD concentration decreased from the initial values of 30,000 mg/L to 5200 mg/L in September, due to the sedimentation of organic matter present in particulate form and the biodegradation of organic compounds. Decreasing trends were also observed for total phenols concentration from 2600 mg/L to 1200 mg/L, in accordance with previous study [4], attributable to microbiota activity present in OMW [30,31]. Significant fluctuations were observed in organic-N concentration, throughout the experimental period without a clear trend. TP concentration in raw OMW followed a decreasing trend over time. In April 2004, TP concentration was 129.01 mg/L and by the end of the experiment, it was 18.67 mg/L. Composition of OMW applied in the field is shown in Table 1.

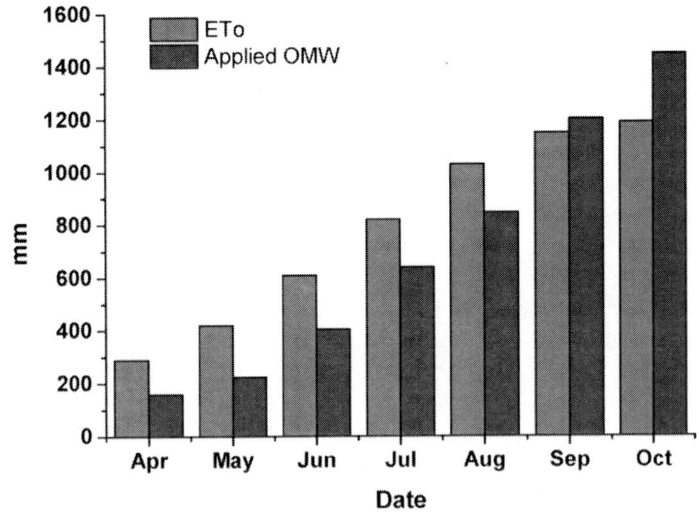

Figure 1. Reference evapotranspiration (ETo) and applied OMW during 2004 application period.

3.2. Applied Hydraulic and Nutrient Loads

The cumulative OMW and N and P loads applied at LTS are shown in Table 2. Variations in cumulative hydraulic load and reference evapotranspiration (ET_0) of the area throughout the experimental period are also shown in Figure 1. The OMW applied at a relatively stable rate throughout the second application period in contrast to ET_0 rate, which showed a gradual decline particularly from August to the end of the season due to lower temperatures. As a result the hydraulic load of OMW exceeded the ET_0 at the beginning of October (Figure 1).

Table 2. Cumulative hydraulic, nitrogen, and phosphorus loads at the end of 2003 and 2004, applied at *E. camaldulensis* trees.

Year	Hydraulic Load (mm)	Nitrogen (kg/ha)	Phosphorus (kg/ha)
2003	318.18	531.36	176.59
2004	1767.97	2952.50	981.22

3.3. Soil Solution Samples

The data of the soil solution samples regards the period April to October 2004. pH values in soil solution samples increased gradually until late September and then almost leveled off by the end of the experiment (Figure 2a). Lower pH values were observed with increasing soil depth and at 20 September soil pH was 7.65 at 15 cm, 7.45 at 30 cm, and 7.25 at 60 cm soil depth.

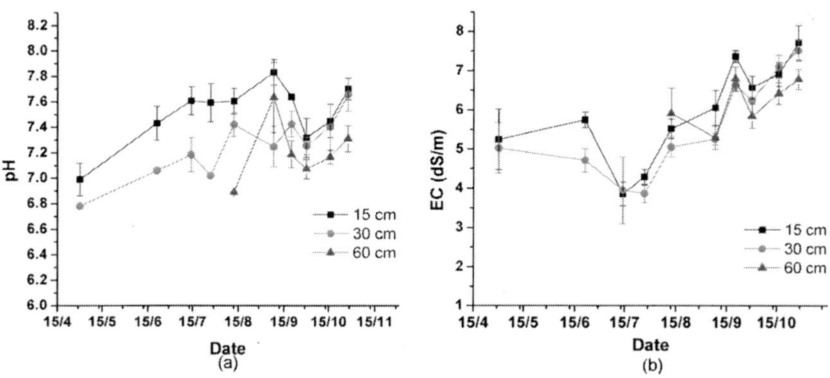

Figure 2. Values in soil solution of (**a**) Ph and (**b**) EC during the second application period.

EC in soil solution samples had an increasing trend throughout the OMW application period reaching at 7.1 dS/m at the end of October, 2004 (Figure 2b).

Deeper soil layers tended to have lower EC values, but the differences were proved not statistically significant. Variations in COD concentration of the soil solution samples were observed during the second application period (2004), showing an decreasing trend from 5000 mg/L at the beginning of the period to 200 mg/L toward the end of September (Figure 3a).

The removal of organic matter (represented here as COD) occurred at the upper 15 cm of the soil profile (93%) and no further reduction was observed with increasing soil depth. Total phenols concentration in the soil solution samples followed a similar trend with COD, with average reduction at 86% at 15 cm, while an increase was observed after mid-September. An increase in soil depth did not contribute further to reduction of phenols in the soil solution (Figure 3b).

With regard to TP in soil solution samples, variations were observed throughout the study period between 20 and 1 mg/L, at all soil profiles (Figure 3f).

Similarly to COD, TKN, and NH_4^+-N concentrations significantly decreased at the 0–15 cm layer (86% and 70%, respectively) and no further reduction was observed with the increase of soil depth (Figure 3c,d). Their values in soil solution varied during OMW application without showing a characteristic pattern.

With regard to NO_3^--N concentration, low values were observed in soil solution samples during the examined period, below the level of 2.5 mg/L, showing a decreasing trend with the season's progress (Figure 3e).

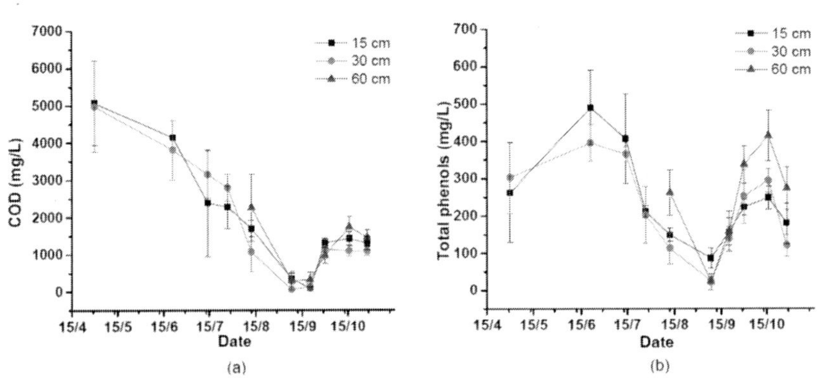

(a) (b)

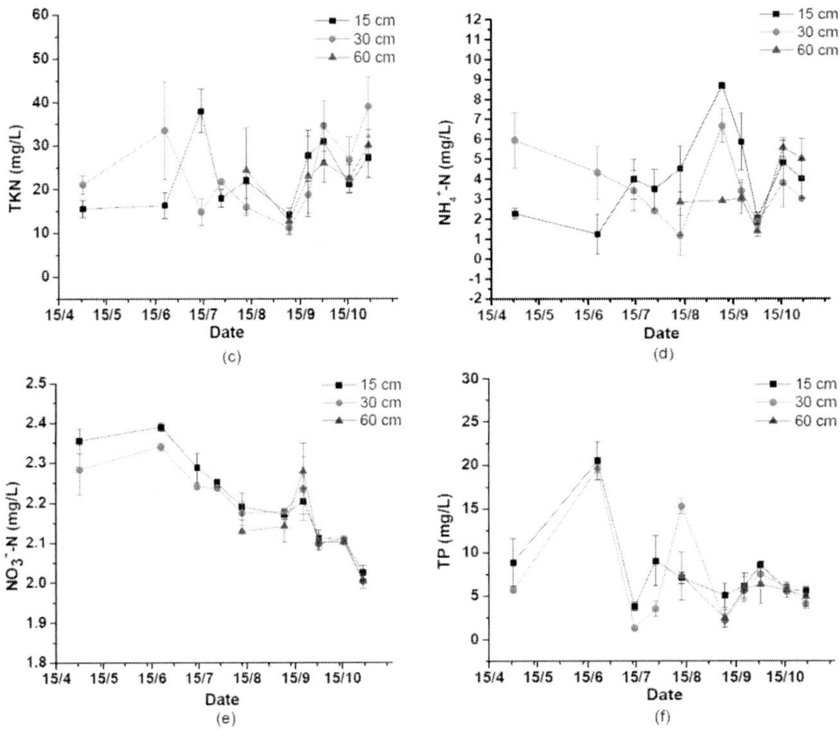

Figure 3. Concentration in soil solution of (**a**) COD; (**b**) total phenols; (**c**) total kjeldahl nitrogen (TKN); (**d**) NH_4^+-N; (**e**) NO_3^--N; and (**f**) total phosphorus (TP).

3.4. Soil Samples

OMW application gradually increased SOM mainly at the upper soil layers. A slight increase in SOM occurred only at the first 30 cm of soil depth in 2003, but the following year SOM increased up to 10-fold at the 0–15 cm soil depth (Figure 4a). A similar to SOM trend was observed in TKN; approximately 2.5-fold higher than the initial value at the end of the study period in the 0–15 cm and 15–30 cm soil layers (Figure 4b). OMW application increased substantially the concentration of available P in the soil. This effect was stronger at the surface soil layer and showed a decreasing trend toward the deeper soil layers (Figure 4c).

The OMW application resulted in gradual increase of EC throughout the two application periods, as shown in Figure 4d, associated with the accumulation of salts in the soil. SAR values followed a similar pattern with that of EC and increased during OMW application periods, from 1.7 at the beginning of the first period (April 2003) to 6.7 at the end of the experiment (October 2004) in 0–15 cm of the soil depth resulting from gradual domination of Na^{+} over the Ca^{2+} and Mg^{2+}(Figure 4e).

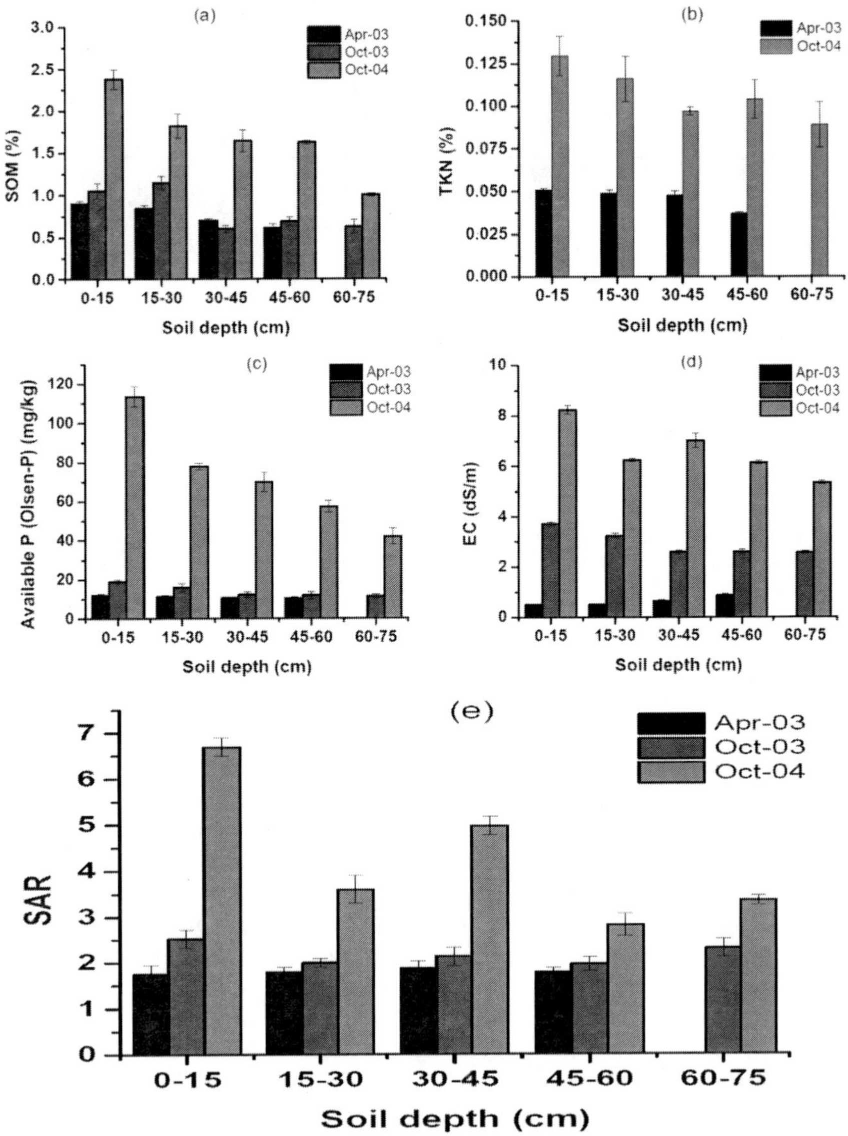

Figure 4. Concentrations in soil of (**a**) soil organic matter (SOM); (**b**) TKN (%); (**c**) TP; (**d**) electrical conductivity (EC); and (**e**) SAR.

3.5. Biomass Production and Nutrient Recovery

The total hypergeous biomass yield of eucalyptus plantation and allocation to different plant organs is shown in Table 3. The N content in trunk, shoots, and leaves in eucalyptus trees was 1.17%, 1.37%, and 4.08%, respectively, in 2003 and 0.55%, 0.58%, and 2.04% in the following year. P content in plant organs was 0.09% for trunk 0.09% for shoots, and 0.25% for leaves in 2003 and 0.07%, 0.09%, and 0.15%, respectively, in 2004. The higher plant nutrient tissue contents in the first growing season are due to the higher portion of juvenile growth which is characterized by a greater content of nutrients.

Table 3. Cumulative biomass production, accumulated N and P (and removal percentage), nutrient used efficiency for nitrogen (NUE-N), nutrient used efficiency for phosphorus (NUE-P), and water use efficiency (WUE) in different plant organs of *E. camaldulensis* (Kg/ha) trees received OMW at the end of the two growing seasons (2003 and 2004). In parenthesis is shown the standard error.

Parameter	Year	Trunk	Shoots	Leaves	Total	Removal (%)
Biomass yield (kg/ha)	2003	1,676 (201)	993 (138)	2,500 (358)	5,170 (694)	-
	2004	10,389 (1,078)	5,389 (475)	6,993 (930)	22,771 (2,316)	-
N (kg/ha)	2003	19.64 (2.7)	13.77 (2.5)	101.20 (12.7)	134.61 (17.8)	25.33
	2004	51.89 (7.3)	29.53 (5.1)	134.74 (27.6)	216.16 (38.2)	7.32
P (kg/ha)	2003	1.51 (0.2)	0.92 (0.2)	6.02 (0.6)	8.45 (0.9)	4.79
	2004	6.81 (1.4)	4.80 (0.9)	10.64 (2.4)	22.26 (4.5)	2.27
NUE-N (kg/kg)	2003	85.80 (3.9)	73.28 (3.3)	24.58 (0.9)	38.38 (1.1)	-
	2004	185.07 (7.8)	194.83 (30.7)	51.39 (4.1)	102.97 (5.6)	-
NUE-P (kg/kg)	2003	1,109.03 (12.2)	1,131.37 (139.1)	411.43 (26.1)	608.00 (17.8)	-
	2004	1,980.96 (409.7)	1,168.09 (119.6)	680.89 (48.4)	1,132.40 (119.7)	-
WUE (g/kg)	2003	-	-	-	1.62	-
	2004	-	-	-	1.29	-

Eucalyptus trees accumulated in 134.61 kgN/ha and 8.63 kgP/ha in hypergeous biomass in 2003 and 216 kgN/ha and 22.6 kgP/ha after the following growing season (2004) (Table 3). Plant organs had different contribution to total nutrients accumulation in hypergeous biomass due to their differences in nutrients content and the amount of biomass produced. The distribution of N and P to plant organs is shown in Table 3. Taking into account the applied N and P loads, eucalyptus trees accumulated the 25% of the applied N and 5% of the applied P at the end of two growing seasons. NUE-N and NUE-P increased significantly during OMW application, as shown in Table 3. In contrast, WUE had lower value at the end of second growing season period compared to that of 2003.

4. DISCUSSION

Soil planted with eucalyptus species showed great assimilation potential for the organic C, particularly at the first 15 cm of soil profile, which is consistent with the presence of labile organic compounds in the OMW. Such an organic fraction, having also high C/N ratio, stimulates the indigenous heterotrophic microorganisms inducing organic matter mineralization [7,13,15]. COD concentration in soil solution decreased gradually until late the second application period (September) probably following the pattern of COD content in the applied OMW. It is also possible a gradual adaptation to OMW application for soil microbial communities. Land application of OMW enhanced SOM, in agreement with the findings of previous studies [3,6], an effect that was observed throughout the examined soil profile (up to 75 cm soil depth). This effect is attributed to the presence of recalcitrant compounds in OMW and/or the effect of certain retention mechanisms, such as the adsorption onto minerals and/or occlusion in clay microstructures that are inaccessible to decomposers and enzymes [9,10]. Our finding indicates the great potential of OMW in terms of the organic matter assimilation and increase of SOM content in the soil, an effect with great importance for the olive-oil producing countries in most of which soils suffer from low organic matter content and gradual degradation as a result of the favorable environmental conditions and thee applied agronomic practices.

Phenols in soil solution followed almost a similar pattern with that of SOM. Their reduction occurred mainly at the first 15 cm of soil, following in parallel a decreasing trend until late August. Several factors are responsible for the depletion of soluble phenols in the soil, including microbial degradation, polymerization to recalcitrant high-molecular mass polyphenols, incorporation into the SOM, and adsorption into soil particles. In this study, the high removal rates observed from the beginning of OMW application provide evidence of the effective adsorption of phenols onto clay minerals, induced by the clay-rich experimental soil [32]. Moreover, these rates continued until late August probably following the decreasing trend of phenol concentration in the storage pond and the stimulated phenols biodegradation processes. Previous studies report changes in the composition of soil microorganisms [7,15,33] and enzymes activities [13,34] involved in phenol degradation, indicating that OMW may favor soil microorganisms and activities specialized to degradation of phenolic compounds. Similarly to our study, under field conditions, it has been reported significant reduction of phenols, described by a first order kinetics with the constant value (K) to range from −0.014 to −0.018 per day [3]. However, these values correspond to half-life time of about 15 to 32 days which are considerably lower to the removal rates observed in the present study in which 15-day interval resulted in approximately 80% removal. Differences in environmental conditions prevailing between study areas and the presence of vegetation may explain this variation. Surprisingly, phenols increased toward the end of the study, similarly with COD pattern, which is probably associated with less favorable environmental conditions that possibly retarded mineralization processes.

Land application of OMW resulted in significant reduction in TKN, as indicated by low TKN content in soil solution, effect that is largely attributed to organic N mineralization induced by stimulated heterotrophic microorganisms. The increase in TKN content in soil samples observed at the end of the experimental period suggest that a part of the organic N remained in the soil, consistent with SOM findings and the findings of previous study [6]. Despite the effective organic N mineralization NH_4^+-N content in solution did not increase throughout the OMW application period. This is probably due to intense microbial immobilization caused by increased competition for N among soil microorganisms [34]. The low levels of nitrates observed in this study support further this hypothesis. Similar result regarding nitrates is reported by a recent microcosm study attributing it to potential inhibition of nitrifying microorganisms by low mineral N availability, caused by increased inorganic N microbial immobilization, rather than toxicity [34]. With regard to P, it had also a remarkable reduction through soil infiltration due to adsorption and precipitation phenomena induced by the clay-rich and calcareous soil of the present study.

Application of OMW increased substantially soil salinity at levels which may have affected crop performance and microbial activity. However, in terms of the former, this might be unexpected considering great adaptability and tolerance of eucalyptus species to soil salinity [23]. In addition, rainfall during winter period leached effectively salts accumulated during irrigation with effluent of approximately 2 dS/m electrical conductivity- [35]. Increase of SAR throughout the experimental period does not also imply adverse impact on structure of soil, given the increased EC values in the soil. Likewise, in a three year OMW treated soil no sodification hazard was observed despite the increase in SAR values [6]. In terms of the effect of salinity on soil microorganisms, adverse effects are suggested by previous studies, indicated by decreases in soil respiration, microbial biomass C, and activities of the exocellular enzymes, such as β-glucosidase, alkaline phosphatase and arylsulphatase [36,37]. Moreover, there are reports for shifts in microbial community composition associated with changes in status of salinity in the soil [38,39]. However, in this study only chemical parameters were examined and the response of soil microorganisms to OMW additions and/or to eucalyptus activity remains questionable.

Short rotation coppice is a practice highly compatible with the concept of land treatment because it stimulates nutrient recycling, minimizes environmental risks, and results to economic benefits. In previous study dealing with meatwork irrigation on different Eucalyptus spp, *E. globules* produced 35 t/ha at the age of two years [40], which is approximately 50% higher than the yield of the present study. Even higher dry biomass is reported for 24 months old *E. camaldulensis* seedlings, grown under municipal effluent application at rates equivalent to evapotranspiration needs, at 7000 g per seedling [41]. This yield corresponds to about 60 t/ha, when the spacing of the present study is considered. In contrast, *E. camaldulensis* seedlings irrigated for 18 months with industrial effluent achieved extremely low hypergeous biomass at yield 300 per seedling [42]. Finally, in an adjacent field of the present study, biomass produced by *E. camaldulensis* irrigated with municipal effluent at ET rates reached at 28.4 t/ha at the end of the second application period [43]. The lower yield obtained in the present study compared to the previous mentioned study is probably due to water deficit developed early in the growing season (Figure 1) and to low inorganic N availability rather than to a toxic effect of OMW. Phytotoxicity is negatively related to plant age and mainly occur at stage of seed break [4,16]. Exposure to phenol concentration of up to 250 mg/L did not induce toxicity development of *S. viminalis* seedlings grown hydroponically [44].

Plant uptake is an important pathway for nutrients removal in LTS. In the present study only a minor portion of the applied N assimilated in plant biomass, lower than 10% of applied loads, but this is primarily due to high nutrient loadings (Table 2) and less to nutrient assimilation potential of eucalyptus plantation. Similar findings to our study in terms of nutrient uptake potential of eucalyptus plantations have been reported by a previous study [40]. On the contrary, lower N and P uptake potential was achieved by *E. camaldulensis* in the present study compared to that of the adjacent field study [43], at 30% and 42%, respectively, attributed to lower biomass yield and N and P contents in plant tissues probably caused by low inorganic N availability in the soil.

5. CONCLUSIONS

Based on the findings of the present study we can conclude:

- Land application has a great potential for organic matter and phenol assimilation and they can be effectively used for OMW detoxification;
- OMW can efficiently contribute to the long-term increase in soil organic matter and nutrients (particularly P) which is of great importance for

the Mediterranean olive-oil producing countries where soils have relatively low organic matter content and prone to progressive degradation;

- Enhancement of soil fertility resulting from OMW application can sustain eucalyptus trees and provide remarkable biomass yield. This yield, however, may be negatively affected by low inorganic N availability in the soil and/or possible toxicity phenomena in the early growth stages of vegetation;
- The organic matter and phenols concentrations in the soil appear to be affected by their initial contents in the applied OMW. This suggests the application of appropriate strategies before the OMW application in the field to reduce organic matter and phenols loadings;
- Further work is needed to elucidate: (i) the potential effects on physical properties of soil; and (ii) the potential effect of OMW and vegetation on microbial community and activity involved in critical processes in the soil, particularly those included in C and N cycling and phenol degradation.

ACKNOWLEDGMENTS

This study was sponsored by the Greek General Secretariat for Research and Technology. The authors would like to acknowledge the critical contribution of Nikolaos V. Paranychianakis throughout the preparation of this study.

REFERENCES

1. Nassar, N.N.; Arar, L.A.; Marei, N.N.; Abu Ghanim, M.M.; Dwekat, M.S.; Sawalha, S.H. Treatment of olive mill based wastewater by means of magnetic nanoparticles: Decolourization, dephenolization and COD removal. *Environ. Nanotechnol. Monit. Manag.* **2014**, *1–2*, 14–23.
2. Kavvadias, V.; Doula, M.K.; Komnitsas, K.; Liakopoulou, N. Disposal of olive oil mill wastes in evaporation ponds: Effects on soil properties. *J. Hazard. Mater.* **2010**, *182*, 144–155.
3. Sierra, J.; Martí, E.; Garau, M.A.; Cruañas, R. Effects of the agronomic use of olive oil mill wastewater: Field experiment. *Sci. Total Environ.* **2007**, *378*, 90–94.
4. Saadi, I.; Laor, Y.; Raviv, M.; Medina, S. Land spreading of olive mill wastewater: Effects on soil microbial activity and potential phytotoxicity. *Chemosphere* **2007**, *66*, 75–83.
5. Casa, R.; D'Annibale, A.; Pieruccetti, F.; Stazi, S.R.; Giovannozzi Sermanni, G.; Lo Cascio, B. Reduction of the phenolic components in olive-mill wastewater by an enzymatic treatment and its impact on durum wheat (Triticum durum Desf.) germinability. *Chemosphere* **2003**, *50*, 959–966.
6. Cabrera, F.; López, R.; Martinez-Bordiú, A.; Dupuy de Lome, E.; Murillo, J.M. Land treatment of olive oil mill wastewater. *Int. Biodeterior. Biodegrad.* **1996**, *38*, 215–225.

7. Mekki, A.; Dhouib, A.; Sayadi, S. Changes in microbial and soil properties following amendment with treated and untreated olive mill wastewater. *Microbiol. Res.* **2006**, *161*, 93–101.
8. Piotrowska, A.; Iamarino, G.; Rao, M.A.; Gianfreda, L. Short-term effects of olive mill waste water (OMW) on chemical and biochemical properties of a semiarid Mediterranean soil. *Soil Biol. Biochem.* **2006**, *38*, 600–610.
9. Dungait, J.A.J.; Hopkins, D.W.; Gregory, A.S.; Whitmore, A.P. Soil organic matter turnover is governed by accessibility not recalcitrance. *Glob. Change Biol.* **2012**, *18*, 1781–1796.
10. Von Lützow, M.; Kögel-Knabner, I.; Ludwig, B.; Matzner, E.; Flessa, H.; Ekschmitt, K.; Guggenberger, G.; Marschner, B.; Kalbitz, K. Stabilization mechanisms of organic matter in four temperate soils: Development and application of a conceptual model. *J. Plant Nutr. Soil Sci.* **2008**, *171*, 111–124.
11. López-Piñeiro, A.; Albarrán, A.; Rato Nunes, J.M.; Peña, D.; Cabrera, D. Long-term impacts of de-oiled two-phase olive mill waste on soil chemical properties, enzyme activities and productivity in an olive grove. *Soil Tillage Res.***2011**, *114*, 175–182.
12. Chartzoulakis, K.; Psarras, G.; Moutsopoulou, M.; Stefanoudaki, E. Application of olive mill wastewater to a Cretan olive orchard: Effects on soil properties, plant performance and the environment. *Agric. Ecosyst. Environ.* **2010**, *138*, 293–298.
13. Bodini, S.F.; Cicalini, A.R.; Santori, F. Rhizosphere dynamics during phytoremediation of olive mill wastewater. *Bioresour. Technol.* **2011**, *102*, 4383–4389.
14. Gamba, C.; Piovanelli, C.; Papini, R.; Pezzarossa, B.; Ceccarini, L.; Bonari, E. Soil microbial characteristics and mineral nitrogen availability as affected by olive oil waste water applied to cultivated soil. *Commun. Soil Sci. Plant Anal.* **2005**, *36*, 937–950.
15. Karpouzas, D.G.; Rousidou, C.; Papadopoulou, K.K.; Bekris, F.; Zervakis, G.I.; Singh, B.K.; Ehaliotis, C. Effect of continuous olive mill wastewater applications, in the presence and absence of nitrogen fertilization, on the structure of rhizosphere–soil fungal communities. *FEMS Microbiol. Ecol.* **2009**, *70*, 388–401.
16. Barbera, A.C.; Maucieri, C.; Cavallaro, V.; Ioppolo, A.; Spagna, G. Effects of spreading olive mill wastewater on soil properties and crops, a review. *Agric. Water Manag.* **2013**, *119*, 43–53.
17. Mahmoud, M.; Janssen, M.; Haboub, N.; Nassour, A.; Lennartz, B. The impact of olive mill wastewater application on flow and transport properties in soils. *Soil Tillage Res.* **2010**, *107*, 36–41.
18. Mahmoud, M.; Janssen, M.; Peth, S.; Horn, R.; Lennartz, B. Long-term impact of irrigation with olive mill wastewater on aggregate properties in the top soil. *Soil Tillage Res.* **2012**, *124*, 24–31.
19. Osanai, Y.; Flittner, A.; Janes, J.; Theobald, P.; Pendall, E.; Newton, P.D.; Hovenden, M. Decomposition and nitrogen transformation rates in a temperate grassland vary among co-occurring plant species. *Plant Soil* **2012**, *350*, 365–378.
20. Vesterdal, L.; Elberling, B.; Christiansen, J.R.; Callesen, I.; Schmidt, I.K. Soil respiration and rates of soil carbon turnover differ among six common European tree species. *Forest Ecol. Manag.* **2012**, *264*, 185–196.
21. Haichar, F.E.Z.; Santaella, C.; Heulin, T.; Achouak, W. Root exudates mediated interactions belowground. *Soil Biol. Biochem.* **2014**, *77*, 69–80.
22. Shi, S.; Richardson, A.E.; O'Callaghan, M.; DeAngelis, K.M.; Jones, E.E.; Stewart, A.; Firestone, M.K.; Condron, L.M. Effects of selected root exudate components on soil bacterial communities. *FEMS Microbiol. Ecol.* **2011**, *77*, 600–610.
23. Paranychianakis, N.; Angelakis, A.; Leverenz, H.; Tchobanoglous, G. Treatment of wastewater with slow rate systems: A review of treatment processes and plant functions. *Crit. Reviews Environ. Sci. Technol.* **2006**, *36*, 187–259.
24. Reed, S.C.; Crites, R.W.; Middlebrooks, E.J. *Natural Systems for Waste Management and Treatment*; McGraw-Hill, Inc.: New York, NY, USA, 1995.
25. Swennenhuis, J. *CROPWAT 8.0.*; Water Resources Development and Management Service of FAO: Rome, Italy, 2009.
26. Apha, A. *Standard Methods for the Examination of Water and Wastewater*, 19th ed.; American Public Health Association: Washington, DC, USA, 1995.

27. García García, I.; Jiménez Peña, P.R.; Bonilla Venceslada, J.L.; Martín Martín, A.; Martín Santos, M.A.; Ramos Gómez, E. Removal of phenol compounds from olive mill wastewater using Phanerochaete chrysosporium, Aspergillus niger, Aspergillus terreus and Geotrichum candidum. *Process Biochem.* **2000**, *35*, 751–758.
28. Page, A.L. *Methods of Soil Analysis. Part 2. Chemical and Microbiological Properties*; American Society of Agronomy, Soil Science Society of America: Madison, WI, USA, 1982.
29. Statistics, I.S., Ed.; *SPSS Version 21.0 for Microsoft Windows Platform*; SPSS Inc.: Chicago, IL, USA, 2012.
30. Millán, B.; Lucas, R.; Robles, A.; García, T.; Alvarez de Cienfuegos, G.; Gálvez, A. A study on the microbiota from olive-mill wastewater (OMW) disposal lagoons, with emphasis on filamentous fungi and their biodegradative potential. *Microbiol. Res.* **2000**, *155*, 143–147.
31. Robles, A.; Lucas, R.; de Cienfuegos, G.A.; Gálvez, A. Biomass production and detoxification of wastewaters from the olive oil industry by strains of Penicillium isolated from wastewater disposal ponds. *Bioresour. Technol.* **2000**, *74*, 217–221.
32. Cecchi, A.; Koskinen, W.; Cheng, H.H.; Haider, K. Sorption-desorption of phenolic acids as affected by soil properties. *Biol. Fertil. Soils* **2004**, *39*, 235–242.
33. Karpouzas, D.G.; Ntougias, S.; Iskidou, E.; Rousidou, C.; Papadopoulou, K.K.; Zervakis, G.I.; Ehaliotis, C. Olive mill wastewater affects the structure of soil bacterial communities. *Appl. Soil Ecol.* **2010**, *45*, 101–111.
34. Tsiknia, M.; Tzanakakis, V.; Oikonomidis, D.; Paranychianakis, N.; Nikolaidis, N. Effects of olive mill wastewater on soil carbon and nitrogen cycling. *Appl. Microbiol. Biotechnol.* **2014**, *98*, 2739–2749.
35. Tzanakakis, V.A.; Paranychianakis, N.V.; Londra, P.A.; Angelakis, A.N. Effluent application to the land: Changes in soil properties and treatment potential. *Ecol. Eng.* **2011**, *37*, 1757–1764.
36. Rietz, D.N.; Haynes, R.J. Effects of irrigation-induced salinity and sodicity on soil microbial activity. *Soil Biol. Biochem.* **2003**, *35*, 845–854.
37. Setia, R.; Marschner, P.; Baldock, J.; Chittleborough, D.; Smith, P.; Smith, J. Salinity effects on carbon mineralization in soils of varying texture. *Soil Biol. Biochem.* **2011**, *43*, 1908–1916.
38. Wichern, J.; Wichern, F.; Joergensen, R.G. Impact of salinity on soil microbial communities and the decomposition of maize in acidic soils. *Geoderma* **2006**, *137*, 100–108.
39. Mohamed, D.J.; Martiny, J.B.H. Patterns of fungal diversity and composition along a salinity gradient. *ISME J.* **2011**, *5*, 379–388.
40. Guo, L.B.; Sims, R.E. H.; Horne, D.J. Biomass production and nutrient cycling in Eucalyptus short rotation energy forests in New Zealand. I: Biomass and nutrient accumulation. *Bioresour. Technol.* **2002**, *85*, 273–283.
41. Singh, G.; Bhati, M. Growth, biomass production, and nutrient composition of eucalyptus seedlings irrigated with municipal effluent in loamy sand soil of Indian desert. *J. Plant Nutr.* **2003**, *26*, 2469–2488.
42. Shah, F.U.R.; Ahmad, N.; Masood, K.R.; Peralta-Videa, J.R.; Zahid, D.M.; Zubair, M. Growth, biomass production, and nutrient composition of eucalyptus seedlings irrigated with municipal effluent in loamy sand soil of Indian desert. *Int. J. Phytoremed.* **2010**, *12*, 343–357.
43. Tzanakakis, V.A.; Paranychianakis, N.V.; Angelakis, A.N. Nutrient removal and biomass production in land treatment systems receiving domestic effluent. *Ecol. Eng.* **2009**, *35*, 1485–1492.
44. Ucisik, A.S.; Trapp, S. Uptake, removal, accumulation, and phytotoxicity of phenol in willow trees (Salix viminalis). *Environ. Toxicol. Chem.* **2006**, *25*, 2455–2460.

CHAPTER 5

Treatment of Organic Recalcitrant Contaminants in Wastewater

Asmita Gupta* and Indu Shekhar Thakur

School of Environmental Sciences, Jawaharlal Nehru University, New Delhi, India

ABSTRACT

Research has shown that a myriad of contaminants enter the environment through industrial and domestic sources on a daily basis. The biodegradable compounds often get degraded or mineralized by various physical, chemical or biological processes, whereas the recalcitrant organic contaminants either are transformed or get dispersed and persist in the receiving environments, and to an extent much greater than was earlier estimated. Many chemical compounds that were not previously included as pollutants can now be detected at much higher concentrations globally. The effect of most of these emerging contaminants on human and environment health is still unknown. Therefore, there is an urgent need to study the fate of these persistent compounds so as to better understand and manage their ecological and health effects.

Keywords: Wastewater, organic contaminants, recalcitrant, biodegradation, sorption

1. INTRODUCTION

Water adversely affected in quality by anthropogenic activities is, typically called wastewater. Wastewater is generally collected and treated by various processes at centralized facilities, referred to as wastewater treatment plants (WWTPs). There can be several sources contributing towards wastewater generation, including domestic, industrial and agricultural. As there are various sources of wastewater generation, so are the compounds present in them. Wastewater, thus, is a cocktail of chemicals—the class, structure, biodegradability, toxicity and human and environmental impact of most of which are still unknown.

Some of the wastewater contaminants, including aromatics, pharmaceuticals, pesticides,33 chlorinated congeners and plasticizers, pose deleterious effects on human and environmental health, even at trace levels [1]. Some of their harmful effects include impairment and/or abnormality in physiological processes, including reproductive impairment, increased risk of cancer in aquatic and terrestrial species, development of antibiotic-resistant bacterial strains and increase in effluent toxicity post-treatment plausibly owing to the synergistic or antagonistic toxic effects of such recalcitrant chemical mixtures. Still unknown are the environmental effects of many emerging contaminants.

While most of the easily degradable wastewater contaminants are removed by conventional treatment methods, compounds that remain even in the treated effluent are recalcitrant and hence persist in the receiving environments, causing environmental and health problems. Low concentrations of such recalcitrants in large volumes of wastewater make their efficient treatment and removal very difficult by the conventional treatment processes including activated carbon, chemical precipitation, ionic exchange resins and membrane filtration [2]. Such processes have other disadvantages such as high plant operation and maintenance cost, accumulation and disposal issues of concentrated sludge, use of excessive chemicals, low sensitivity towards target compounds and accumulation of concentrated sludge and their disposal problems [3]. Removal of some of the organic recalcitrants is not effected even by the traditional biological processes, including activated sludge and trickling filters, employing microorganisms as these biorecalcitrants may result in death of the microbial population, thus reducing the efficiency of or halting the treatment process. Advanced treatment methods such as a pre-separation step or post-treatment of recalcitrants using potent and specialized microbial strains need to be employed for the efficient removal of such persistent organic pollutants from effluent [2].

Hence, there is a need for better understanding of the occurrence, behaviour and fate of organic contaminants during sewage treatment processes. The present paper reviews literature about the fate of some of the recalcitrant organic contaminants during the various treatment processes.

2. STATUS OF WASTEWATER GENERATION

Better management of wastewater at regional and global level requires up-to-date information on the status of sewage generation and treatment. Globally, a complete sewage generation and treatment data are available for only 55 countries, 37% of it being recent (2008–2012) [4]. There is a generation of about 15, 644 millions litre per day (MLD) of sewage from 35 metropolitan cities in India, out of which only 8040 MLD (51.4%) is the existing treatment capacity. While 3800 MLD is the municipal sewage generation in the national capital region of Delhi, the city has a treatment capacity of only about 2300 MLD. Rest 31% sewage is discharged into the environment untreated [5].

3. WASTEWATER TREATMENT PROCESSES

Various processes are employed for the removal of wastewater contaminants depending on their type and level in the influent. Municipal wastewater is mostly treated in sewage treatment plants (STPs) which use various treatment processes including physical, chemical and biological. Wastewater treatment and discharge are done according to regional and national regulations and standards. Wastewater treatment is done with the purpose of producing a pollutant- and toxicity-free effluent which can safely be discharged into the environment [6]. Three main stages are involved in wastewater treatment, viz., primary or physic-chemical, secondary or biological and tertiary or advanced treatment.

a) *Primary treatment* involves physical separation of heavy solid particles gravimetrically and oil and other lighter floating materials mechanically in settling basins called primary-settling tanks or primary clarifiers. The remaining liquid wastewater is pumped to the next treatment tank for secondary treatment.

b) *Secondary treatment* involves the removal of dissolved and suspended biological component by means of an indigenous microbial population, which is removed prior to release of the treated water into the environment or tertiary treatment stage. It is carried out in secondary treatment chambers such as aeration tanks or bioreactors. In the presence of sufficient oxygen supplied through aeration pumps, the indigenous microflora degrades the soluble organic fractions while segregating the less soluble components into flocs. Secondary treatment may include either *fixed-film* or **attached growth** systems such as trickling filters, rotating biological contactors and bio-towers, where the sewage passes over the surface of attached biomass, or *suspended-growth* systems including activated-sludge process, where sewage is mixed with microbial biomass. While the latter type of secondary treatment system has a lower space requirement for wastewater treatment, requires less space for treatment, the fixed-film systems are better able to acclimatize to sudden microbial changes and have a higher removal rate of organic matter and suspended solids [7–9].

c) *Tertiary treatment* includes any advanced wastewater treatment methods beyond the primary and secondary treatment, before discharge of wastewater in the receiving environment.

The most important aerobic treatment system is the activated-sludge process, based on the maintenance and recirculation of a complex biomass composed by micro-organisms able to absorb and adsorb the organic matter carried in the wastewater. Other biological treatment processes such as expanded granular sludge bed (EGSB) reactor and upflow anaerobic sludge blanket (UASB) are also employed for wastewater treatment. Synthetic membranes and micro-filtration are now commonly being used as tertiary treatment technologies.

4. FATE OF ORGANIC RECALCITRANT CONTAMINANTS IN WASTEWATER TREATMENT

4.1. Pathways of contaminant removal

There has been a radical increase in the occurrence and concentration of organic contaminants in wastewater and sludge as a result of an increase in the demand and industrial production of synthetic organic chemicals. Point discharge sources including discharges from industrial users or manufacturers and diffuse discharge sources such as commercial and domestic premises or run-off after aerial deposition are some of the major contributors to the loading of organic contaminant in sewage. The following are some of the pathways (**Figure 1**) through which organic contaminants may be transformed or degraded during sewage treatment:

- Air stripping
- Biodegradation
- Chemical degradation
- Sorption
- Volatilization

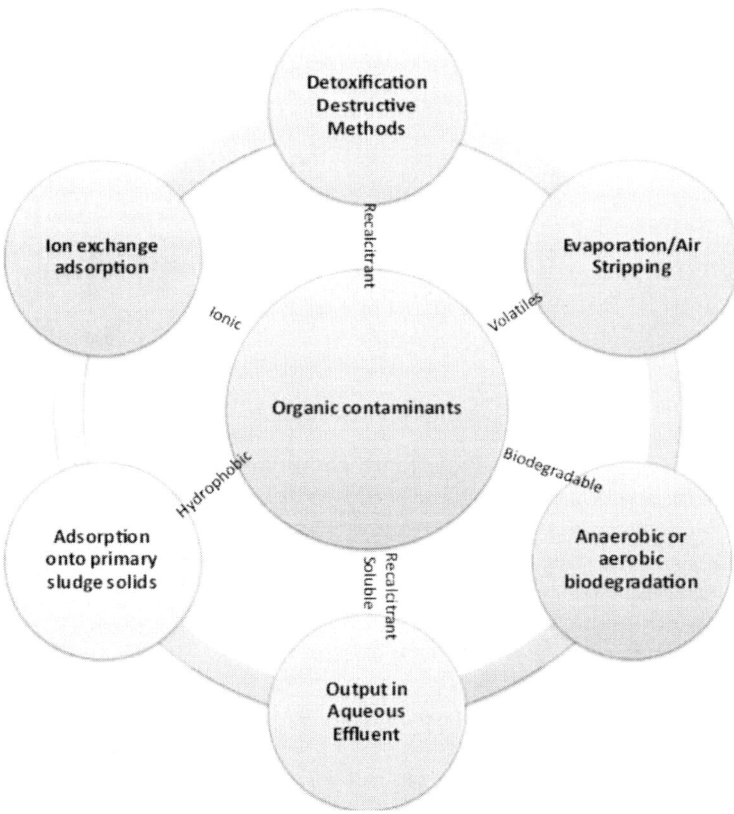

Figure 1. Some of the pathways involved in transformation of organic contaminant in wastewater treatment.

While some compounds may get completely degraded or mineralized in the process of treatment, some others are partially degraded and form breakdown products and a few other recalcitrant compounds may remain unaffected and persist in the effluent even after treatment. The occurrence of these synthetic organic contaminants in wastewater may be either in solution or sorbed onto solids. The hydrophobic or lipophilic nature of many organic contaminants result into their getting adsorbed on solid particles during wastewater treatment, eventually resulting in their accumulation in the sludge solids, sometimes at concentrations much higher than in the untreated wastewater [10, 11].

Structural composition of the organic residues may also provide information about their biodegradation pathways. For instance, biodegradation of unbranched and long-chained hydrocarbons is easier as compared to the short-chained or highly branched molecules. Biodegradation of unsaturated aliphatic compounds is generally more favoured than their saturated analogues. Molecules having highly polar groups and linkages tend to react by nucleophilic displacement (such as hydrolysis) [12]. Petrasek et al. [13] reported the association of recalcitrant and toxic chloro-organic pentachlorophenol (PCP) with the sludge solids, and considerable degradation of phenolic compounds having polar groups.

4.2. Processes involved in contaminant removal

Several researches have been made to study the removal efficiency of various contaminants by different wastewater treatment processes. Partitioning of hydrophobic contaminants of influent onto settled primary sludge solids may take place during the primary sedimentation process in the primary clarifiers. Bulk organic components of wastewater such as cellulose, proteins and carbohydrates get biodegraded during the secondary treatment involving aerobic processes such as trickling filters, activated-sludge process, oxidation ponds or anaerobic processes resulting in sludge digestion. Transformation or loss of some of the synthetic recalcitrant organic contaminants may also take place during the secondary treatment processes. Polysaccharides, proteins and fats occur in two phases during the anaerobic digestion process. First phase (acid phase) involves hydrolysis of polysaccharides to form mono- and disaccharides, of proteins to form amino acids, and of fats resulting in the formation of long-chain fatty acids, and volatile acids such as formic, acetic and butyric acid. Second phase (methanogenic phase) results in the reduction of the volatile acids to methane and carbon dioxide [12, 14]. In one study involving a generalized model for the presentation of fate of organic compounds in an activated-sludge process, it was demonstrated that the phase distribution of xenobiotic chemicals depended quantitatively upon their physico-chemical properties and the operating conditions of wastewater treatment. The study also showed the removal of hydrophobic chemicals of wastewater, mostly by the process of sorption onto sludge particles followed by their transfer to the sludge-processing units. Meanwhile, advective transport into the final effluent and biodegradation was shown to be the common mechanism for the removal of hydrophilic compounds of wastewater. The model also predicted an increase in the effluent concentration of complex organics such as substituted phthalates, high molecular weight (HMW) polycyclic aromatic hydrocarbons (PAHs) and dioxins with increasing solids retention time (SRT) during the operation of wastewater treatment plant [15].

4.3. Common classes of contaminants found in wastewaters

Although wastewaters contain a multitude of contaminants, yet they can be broadly grouped under different classes on the basis of their chemical structure. A total of 129 specific pollutants including heavy metals and specific organic chemicals have been defined by the US Clean Water Act as "Priority Pollutants". Municipal Environmental Research Laboratory (MERL), US EPA, conducted a comprehensive research programmes on the occurrence and fate of priority pollutants present in wastewater and sludge. The study assessed the fate and behaviour of 22 harmful organics including phenols, pesticides, poly aromatic hydrocarbons and phthalates in the conventional water treatment systems and demonstrated up to 95–98% removal of organic compounds from the liquid phase. Many such organic compounds were found to have been partitioned onto the solid phases of primary and return activated sludges. Similar results were reported in other studies as well [16, 17]. In one study, the highest degree of enrichment of PAHs was observed in the primary sludge and phthalates such as bis-(ethylhexyl) and di-n-octyl phthalate were found to be among the most recalcitrant compounds present in wastewaters [13]. Wild and Jones [18] reported the occurrence of volatile chemicals, such as benzene, in sewage sludge, possibly as a result of their sorption over organic substances present in the sludge. Based on the reported literature, the following description discusses the fate of some common classes of organic compounds occurring in wastewaters (**Figure 2**).

4.3.1. Phthalic acid esters

Phthalates have a high environmental significance owing to their high production volumes as well as their eco-toxicological effects especially on aquatic fauna including molluscs, crustaceans and amphibians. They have been reported to cause biological effects even at very low levels of exposure, varying in the range of ng L^{-1} to μg L^{-1} [19, 20]. Microbial degradation of phthalates under aerobic and anaerobic conditions has been previously reported [21]. The difference in the biodegradability of various phthalates could possibly be due to the steric effect of their side ester chains that hinders the binding of hydrolytic enzymes to the phthalates thus inhibiting their hydrolysis [22]. In a previous study on the occurrence of phthalates in raw and treated wastewater of WWTPs, it was found that most of the studied phthalates were present in post-treated water samples, bis(2-ethylbenzyl) phthalate (DEHP) being the most abundant. Also, biotransformation and adsorption onto sludge solids (that directly depend on the molecular weight and lipophilic nature of the compound) were shown to be the possible pathways of phthalate removal from liquid phase during wastewater treatment [23]. Roslev et al. [24] studied the degradation of four different phthalic acid esters in an activated-sludge process, and showed an almost 96% association of DEHP (showing the least biodegradation among the four phthalates) with the wastewater suspended solids. The study also revealed a 7–9% recovery of the influent phthalate esters in the effluent. Also, aerobic and anoxic-denitrifying conditions were found to be less favourable for biodegradation of phthalate esters as compared to the alternating aerobic-anoxic conditions.

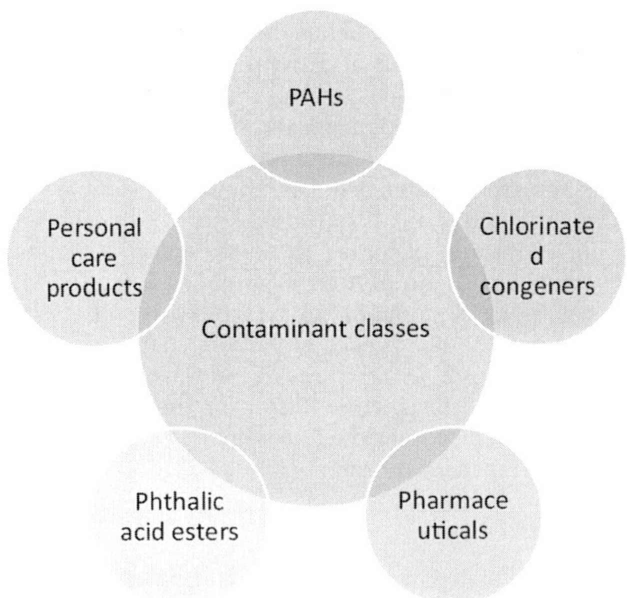

Figure 2. Classes of organic contaminants commonly found in wastewater.

4.3.2. Polycyclic aromatic hydrocarbons

PAHs are among the most mutagenic, carcinogenic and toxic class of organic contaminants some of which have also been included in the US-EPA and EU list of priority pollutants [25]. The presence of PAHs in the environment is commonly attributed to various anthropogenic activities such as petroleum refining, power and heat generation from coal production, and chemical manufacturing [26]. A study on the fate of PAHs and other volatile organic compounds (VOCs) during wastewater treatment by the conventional activated-sludge process (CASP) and the membrane bioreactors (MBRs) concluded that aromatic VOCs were removed mainly by volatilization and with comparable removal efficiencies for both treatment processes, that is, CASP and MBRs. On the other hand, removal efficiency for PAHs was found to be enhanced in case of MBRs [27]. In another study conducted by Zhang et al. [28], the occurrence, behaviour and fate of 18 PAHs in a coking wastewater treatment plant was investigated and it was found that mostly high molecular weight PAHs were present in the raw coking wastewater, while 3–6 ring PAHs were the predominant PAHs detected in the effluent. There was detection of PAHs such as pyrene, phenanthrene and fluoranthene in the gas samples and pyrene, fluoranthene, chrysene and benzo[k]fluoranthene in sludge. While there was almost 97% removal for all the PAHs during treatment, the percent removal of PAHs from the liquid phase varied in a range of 47–92% in the biological stage. It was also observed that low molecular weight (LMW) PAHs were mostly removed in the aerobic tanks and following the mechanism of transformation, whereas their HMW counterparts were mainly removed in anaerobic tank. While transformation was observed to be the most common mechanism of removal of LMW PAHs from wastewaters, adsorption onto sludge solids was mainly responsible for the removal of HMW PAHs from the liquid phase.

4.3.3. Chlorinated congeners

Chlorinated congeners including polychlorinated biphenyls and polychlorinated pesticides are very toxic to human and environment health and are mostly added into the environment by industrial and domestic sources. Their presence has commonly been reported in wastewater, surface water bodies as well as in sediments. Biologically mediated reductive dehalogenation process is one of the common pathways of degradation of these chlorinated contaminants during wastewater treatment. The less investigated reductive dechlorination process has also been identified as one of the possible pathways for the transformation of specific contaminants during anaerobic digestion of sludge. Previous studies have reported the formation of intermediates such as 1, 2 ,4-trichlorobenzene and pentachlorobenzene, 1, 2, 4, 5-tetrachlorobenzene and final products such as dichlorobenzene isomers and 1, 3 ,5-trichlorobenzene during the reductive dechlorination of hexachlorobenzene. The formation of 2, 4-dichlorophenol and 4-chlorophenol as intermediates and phenol as the end product during reductive dechlorination of 2, 4-dichlorophenoxy acetate has similarly been reported [29].

While some of the chlorinated congeners such as polychlorinated biphenyls, have been known for long [30], some others have recently been documented as toxic contaminants including pharmaceuticals such as diclofenac and pesticides 4-hydroxychlorothalonil and clomazone [31, 32]. The detection of such chlorinated contaminants, some of which are also endocrine-disrupting and toxic to biota, in effluent and receiving water bodies is a matter of concern [33]. The concentrations of chlorinated congeners in effluent have been reported to be much lower than in the influent, indicating their efficient removal by various physical, chemical or biological processes operational during the treatment of wastewater [34]. Nevertheless, there have been reports indicating the presence of chlorinated contaminants such as triclosan and triclocarban in effluent of STPs, and eventually in the downstream water bodies and sediments [35, 36], thus pointing towards a need for upgradation of treatment mechanisms for their efficient removal. In a study conducted on the efficiency of aerobic and anaerobic processes in organic contaminant removal during treatment processes, it was concluded that a sequential system using a combination of both oxidative and reductive processes was probably the most efficient for the removal of recalcitrant organics. Highly chlorinated and volatile organohalogen compounds were found to degrade appreciably only under anaerobic conditions, while being resistant to oxidative degradation under aerobic conditions [37].

4.3.4. Pharmaceutical compounds

Pharmaceutical compounds are another class of emerging contaminants that have gained growing concerns in the past two decades mostly because of their less known health and environmental effects and ever-increasing usage and unchecked release into the environment. Metabolic excretion post consumption and improper disposal techniques are the main sources of these compounds in the environment. In a study conducted to investigate the presence of some common pharmaceutical compounds and fluoroquinolones (one of the "priority pollutants" having potential hazardous effects on the aquatic life) in two wastewater treatment plants in Spain, frequent detection of pharmaceuticals such as analgesics, anti-inflammatories and lipid regulators in effluent and incomplete elimination of most of the fluoroquinolones posttreatment was observed. The results also demonstrated higher efficiency of membrane bioreactor technique in removing pharmaceutical compounds as compared to the activated-sludge process [38]. Similar findings have been reported by other workers as well [39, 40].

4.3.5. Personal care products

There has been a recent concern over the toxic and ecological impact of personal care products (PCPs). Although there have been several reports on the assessment of concentrations of these chemicals in the environment [41–43], less work has been done to know their fate in the environment. In one assessment of the efficiency of various treatment processes for the removal of pharmaceuticals and personal care products, it was concluded that membrane bioreactor and activated-sludge process with nitrogen treatment were the most efficient processes for the treatment of such compounds [44].

5. CONCLUSION

Wastewater treatment facilities such as wastewater treatment plants, or domestic septic systems, which have been operating on the conventional technologies, are often inefficient in treating such a cocktail of compounds ranging from simple to complex and recalcitrant organic compounds. Thus, these centralized facilities, discharging treated effluent, which may still be contaminated with household chemicals, pharmaceuticals and biogenic hormones, into the environment end up being a source of pollutants for the receiving water bodies. Also, the sewage sludge generated at the STPs, often having a high accumulation of recalcitrant and hydrophobic contaminants, acts as a sink of such contaminants in the treatment facilities but a major source of organic recalcitrants when directly used as manure.

Such unchecked disposal and use of sewage and sludge into the environment or their direct application for domestic or agriculture purposes could lead to exposure of toxic contaminants to biological systems, possibly resulting in adverse metabolic responses. Advanced treatment technologies such as membrane bioreactors and sequential system using a combination of both oxidative and reductive processes were found to be more effective in the removal of various organic recalcitrant compounds. Therefore, implementation of such treatment technologies and addition of tertiary treatment techniques to the conventional methods, for the removal of such persistent contaminants,have become quintessential.

Thus, the occurrence of persistence organic contaminants in the effluent and sludge posttreatment and ambiguity about their fate pose a serious environmental challenge. Therefore, much research is still needed to identify the source, behaviour and sink as well as their ecological and health effects.

ACKNOWLEDGEMENTS

This paper was supported by the research grants of the Jawaharlal Nehru University, New Delhi, India. The author (Gupta A) also thanks the University Grants Commission (UGC), New Delhi, Government of India, for providing Research Fellowship.

REFERENCES

1. Lai HT, Hou JH, Su CI, Chen CL. Effects of chloramphenicol, florfenicol, and thiamphenicol on growth of algae *Chlorella pyrenoidosa, Isochrysis galbana*, and *Tetraselmis chui.* Ecotoxicol. Environ. Safe. 2009; 72:329–334.
2. Padmavathy S, Sandhya S, Swaminathan K, Subrahmanyam YV, Chakrabarti T, Kaul SN. Microaerophilic-aerobic sequential batch reactor for treatment of azo dyes containing simulated wastewater. Process Biochem. 2005; 40(2):885–890.
3. Balaji T, Matsunaga H. Adsorption characteristics of As(III) and As(V) with titanium dioxide loaded Amberlite XAD-7 resin. Anal. Sci. 2002; 18:1345–1349.
4. Sato T, Qadir M, Yamamoto S, Endo T, Zahoor A. Global, regional, and country level need for data on wastewater generation, treatment, and use. Agricul. Water Manage. 2013; 130:1–13. ISSN 0378-3774
5. CPCB. Inventorization of sewage treatment plants, Control of urban pollution Series: CUPS//2015. Central Pollution Control Board, India. 2015. Available from: http:// cpcb.nic.in/upload/NewItems/NewItem_210_Inventorization_of_Sewage-Treatment_ Plant.pdf
6. Khopkar SM. Environmental Pollution Monitoring and Control. New Delhi: New Age International; 2004. p. 299. ISBN81-224-1507-5.
7. EPA. Package Plants. Wastewater Technology Fact Sheet. Washington, DC. Document no. EPA. 2000; 832-F-00-016.
8. EPA. Primer for Municipal Waste Water Treatment Systems. Washington, DC. Document no. EPA. 2004; 832-R-04-001. 12 Biological Wastewater Treatment and Resource Recovery
9. Randall CW, Sen D. Full-scale evaluation of an integrated fixed-film activated sludge (IFAS) process for enhanced nitrogen removal. Water Sci. Technol. Elsevier; 1996; 33(12): 155–162. doi: 10.1016/0273-1223(96)00469-6
10. Katsoyiannis A, Zouboulis A, Samara C. Persistent organic pollutants (POPs) in the conventional activated sludge treatment process: model predictions against experimental values. Chemosphere. 2006; 65(9):1634–1641.
11. Rogers HR. Sources, behaviour and fate of organic contaminants during sewage treatment and in sewage sludges. Sci. Total Environ. 1996; 185:3–26.
12. Scow KM. Rate of biodegradation. In: Lyman WJ, Reehl WF and Rosenblatt DH (eds). Handbook of Chemical Property Estimation Methods. Environmental Behaviour of Organic Compounds. New York, NY: McGraw Hill; 1982.
13. Petrasek AC, Kugelman IJ, Austern BM, Pressley TA, Winslow LA, Wise RH. Fate of toxic organic compounds in wastewater treatment plants. J. Water Poll. Contr. Fed. 1983; 55(10):1286–1296.
14. Mason CF. Biology of Freshwater Pollution, Harlow: Longman; 1981.
15. Byrns G. The fate of xenobiotic organic compounds in wastewater treatment plants. Water Res. 2001; 35(10):2523–2533.
16. Gupta A, Thakur IS. Biodegradation of wastewater organic contaminants using Serratia sp. ISTVKR1 isolated from sewage sludge. Biochem. Eng. J. 2015; 102:115–124.

17. Harrison EZ, Oakes SR, Hysell M, Hay A. Organic chemicals in sewage sludges. Sci. Total Environ. 2006; 367(2–3):481–497.
18. Wild SR, Jones KC. Organic chemicals entering agricultural soils in sewage sludges: screening for their potential to transfer to crop plants and livestock. Sci. Total Environ. 1992; 119:85–119.
19. Call DJ, Cox DA, Geiger DL, et al. An assessment of the toxicity of phthalate esters to freshwater benthos: 1. Aqueous exposures. Environ. Toxicol. Chem. 2001; 20:1798–1804.
20. Oehlmann J, Schulte-Oehlmann U, Kloas W, et al. A critical analysis of the biological impacts of plasticizers on wildlife. Philos. Trans. R. Soc. B. 2009; 364:2047–2062.
21. Staples CA, Peterson DR, Parkerton TF, Adams WJ. The environmental fate of phthalate esters: a literature review. Chemosphere. 1997; 35:667–749.
22. Liang D-W, Zhang T, Fang HHP, He J. Phthalates biodegradation in the environment. Appl. Microbiol. Biotechnol. 2008; 80:183–198.
23. Clara M, Windhofer G, Hartl W et al. Occurrence of phthalates in surface runoff, untreated and treated wastewater and fate during wastewater treatment. Chemosphere. 2010; 78(9):1078–1084. Treatment of Organic Recalcitrant Contaminants in Wastewater http://dx.doi.org/10.5772/66346 13
24. Roslev P, Vorkamp K, Aarup J, et al. Degradation of phthalate esters in an activated sludge wastewater treatment plant. Water Res. 2007; 41(5):969–976.
25. Arfsten DP, Schaeffer DJ, Mulveny DC. The effects of near ultraviolet radiation on the toxic effects of polycyclic aromatic hydrocarbons in animals and plants: A review. Ecotox. Environ. Safe. 1996; 33:1–24.
26. Suess MJ. The environmental load and cycle of polycyclic aromatic hydrocarbons. Sci. Total Environ. 1976; 6:239–250.
27. Fatone F, Di Fabio S, Bolzonella D, Cecchi F. Fate of aromatic hydrocarbons in Italian municipal wastewater systems: an overview of wastewater treatment using conventional activated-sludge processes (CASP) and membrane bioreactors (MBRs). Water Res. 2011; 45(1):93–104.
28. Zhang W, Wei C, Chai X et al. The behaviours and fate of polycyclic aromatic hydrocarbons (PAHs) in a coking wastewater treatment plant. Chemosphere. 2012; 88:174– 182.
29. Schwartzenbach RP. Groundwater contamination by volatile halogenated alkanes: abiotic formation of volatile sulphur compounds under anaerobic conditions. Environ. Sci. Technol. 1985; 19:322–327.
30. Gutierrez AG, McIntyre AE, Perry R, Lester JN. Behaviour of persistent organochlorine micropollutants during primary sedimentation of waste water. Sci. Total Environ. 1984; 39:27–47.
31. Hug C, Ulrich N, Schulze T, Brack W, Krauss M. Identification of novel micropollutants in wastewater by a combination of suspect and non-target screening. Environ. Pollut. 2014; 184:25–32.
32. Nelson ED, Do H, Lewis RS, Carr SA. Diurnal variability of pharmaceutical, personal care product, estrogen and alkylphenol concentrations in effluent from a tertiary wastewater treatment facility. Environ. Sci. Technol. 2011; 45:1228–1234.
33. Krzmarzick MJ, Novak PJ. Removal of chlorinated organic compounds during wastewater treatment: achievements and limits. Appl. Microbiol. Biotechnol. 2014; 98(14): 6233–6242.
34. Langford KH, Lester JN. Fate and behavior of endocrine disrupters in wastewater treatment processes. In: Lester JN, Birkett JW (eds) Endocrine Disrupters in Wastewater and Sludge Treatment Processes. London: CRC Press; 2003. 103–143.
35. Zhang Q-Q, Zhao J-L, Liu Y-S, Li B-G, Ying G-G. Multimedia modeling of the fate of triclosan and triclocarban in the Donjian River Basin, South China and comparison with field data. Environ Sci: Process Impacts. 2013; 15(11):2142–2152. 14 Biological Wastewater Treatment and Resource Recovery
36. Gautam P, Carsella JS, Kinney CA. Presence and transport of the antimicrobials ticlocarban and triclosan in a wastewater-dominated stream and freshwater environment. Water Res. 2014; 48(1):247–256.

37. Zitomer DH, Speece RE. Sequential environments for enhanced biotransformation of aqueous contaminants. Environ. Sci. Technol. 1993; 27:226–243.
38. Guedes-Alonso R, Afonso-Olivares C, Montesdeoca-Esponda S, Sosa-Ferrera Z and Santana-Rodríguez JJ. An Assessment of the Concentrations of Pharmaceutical Compounds in Wastewater Treatment Plants on the Island of Gran Canaria (Spain). Springer Plus. 2013; 2:24. http://www.springerplus.com/content/2/1/24
39. Ö llers S, Singer HP, Fässler P, Müller SR. Simultaneous quantification of neutral and acidic pharmaceuticals and pesticides at the low-ng/l level in surface and waste water. J. Chromatogr. A. 2001; 911:225–234.
40. Ternes TA. Occurrence of drugs in German sewage treatment plants and rivers. Water Res. 1998;32(11):3245–3260.
41. Fent K, Weston AA, Caminada D. Ecotoxicology of human pharmaceuticals. Aquat. Toxicol. 2006; 7:122–159.
42. Garric J, Ferrari B. Pharmaceuticals in aquatic ecosystems. Levels of exposure and biological effects: A review. Revue des Sciences de l'Eau/Journal of Water Science. 2005; 18(3): 307–330.
43. Kummerer K. Drugs in the environment: emission of drugs, diagnostic aids and disinfectants into wastewater by hospitals in relation to other sources – a review. Chemosphere. 2001; 45:957–969.
44. Miège C, Choubert JM, Ribeiro L, Eusèbe M, Coquery M. Fate of pharmaceuticals and personal care products in wastewater treatment plants – Conception of a database and first results. Environ. Pollut. 2009; 157:1721–1726. Treatment of Organic Recalcitrant Contaminants in Wastewater http://dx.doi.org/10.5772/66346

CHAPTER 6

Seasonal Patterns of Gastrointestinal Illness and Streamflow along the Ohio River

Jyotsna S. Jagai [1,*], Jeffrey K. Griffiths [2,4,5], Paul K. Kirshen [3], Patrick Webb [4] and Elena N. Naumova [2,5]

[1] *U.S. Environmental Protection Agency, Research Triangle Park, NC 27711, USA*

[2] *Tufts University School of Medicine, Boston, MA 02111, USA*

[3] *Institute for the Study of Earth, Oceans, and Space, University of New Hampshire, Durham, NH 03824, USA*

[4] *Friedman School of Nutrition Science and Policy, Tufts University, Boston, MA 02111, USA*

[5] *Tufts University School of Engineering, Medford, MA 02155, USA*

ABSTRACT

Waterborne gastrointestinal (GI) illnesses demonstrate seasonal increases associated with water quality and meteorological characteristics. However, few studies have been conducted on the association of hydrological parameters, such as streamflow, and seasonality of GI illnesses. Streamflow is correlated with biological contamination and can be used as proxy for drinking water contamination. We compare seasonal patterns of GI illnesses in the elderly (65 years and older) along the Ohio River for a 14-year period

(1991–2004) to seasonal patterns of streamflow. Focusing on six counties in close proximity to the river, we compiled weekly time series of hospitalizations for GI illnesses and streamflow data. Seasonal patterns were explored using Poisson annual harmonic regression with and without adjustment for streamflow. GI illnesses demonstrated significant seasonal patterns with peak timing preceding peak timing of streamflow for all six counties. Seasonal patterns of illness remain consistent after adjusting for streamflow. This study found that the time of peak GI illness precedes the peak of streamflow, suggesting either an indirect relationship or a more direct path whereby pathogens enter water supplies prior to the peak in streamflow. Such findings call for interdisciplinary research to better understand associations among streamflow, pathogen loading, and rates of gastrointestinal illnesses.

Keywords: drinking water quality; gastrointestinal infections; hydrology; pathogens; seasonality; streamflow

1. INTRODUCTION

Waterborne pathogens are a significant health concern worldwide, with an estimated 88% of diarrheal diseases attributable to unsafe water supplies [1]. These food and waterborne pathogens are a concern even in the United States, with approximately 76 million cases of illness and 5,000 deaths caused by gastrointestinal infections annually [2]. A recent assessment of waterborne disease outbreaks in the U.S. demonstrated that 780 of the 833 outbreaks that occurred between 1971 and 2006 were associated with drinking water and of these 87.2% occurred in public water supplies [3].

Disease causing waterborne protozoa, such as *Giardia* and *Cryptosporidium* spp., are ubiquitous in water supplies and have been detected in both surface water sources and finished, post-treatment, drinking water supplies [4,5,6]. These protozoa were detected in 97% of samples of raw surface waters tested in Eastern and Midwestern states in the United States [5] and *Cryptosporidium oocysts* were detected in 34 of 35 samples of river waters in Washington State [7]. These pathogens are not only common in the environment, the concentration of the pathogens demonstrate seasonal patterns in surface waters [7,8,9,10].

The waterborne diseases caused by these protozoa in humans also demonstrate seasonal patterns in incidence. Increases in cryptosporidiosis incidence are seen during the warm, rainy season in tropical climates [11,12,13]. In temperate climates, increases in incidence of cryptosporidiosis are seen in the spring and fall in [14,15,16,17]. Fewer studies have explored the seasonal patterns in incidence of giardiasis; however similar patterns have been seen. In the tropical climate of southern India, giardiasis demonstrated higher rates in months with higher rainfall [18] while the arid climate of Lebanon, giardiasis demonstrated consistent rates throughout the year with little seasonal variation [19].

Increase in diarrheal disease incidence has been shown to be associated with water quality parameters, such as turbidity [20,21,22,23]. However, limited studies have been conducted on the association between rates of gastrointestinal illness and streamflow, or river discharge, which is highly correlated with turbidity ($|r| > 0.4$; $p < 0.01$) [24]. A study conducted in England and Wales demonstrated that the incidence rate of cryptosporidiosis was positively related to the maximum average streamflow in that month for months between April and July. Between August and November, the cryptosporidiosis incidence rate was also positively associated with maximum river flows but only after adjusting for the previous month's temperature, rainfall, and cryptosporidiosis rate [25].

Environmental conditions can affect the distribution of these protozoa, and other pathogens, in surface drinking water supplies. For example, *Cryptosporidium* oocysts are transmitted via the feces of infected animals, primarily cattle. The manure may be spread on land where the oocysts may survive for over a year [26] and may be transferred into rivers waters under the correct conditions, such as being within the 100 year flood plain and including a large enough farm area [27]. Sewage effluent has also been recognized as a source of pathogen contamination of surface waters [27,28]. Once the pathogens are washed into rivers, specific climatic and soil conditions allow them to survive for extended periods of time and be re-suspended at a later time.

Each river has unique hydrological characteristics, such as streamflow and water temperature, which may affect the rates and seasonal patterns of waterborne diseases in communities that rely on drinking water from that particular river. While most pathogens are removed from drinking water supplies in the treatment process through coagulation, sedimentation, and filtration [29] some pathogens, such as *Cryptosporidium*, are resistant to the disinfectants typically used in drinking water treatment [30]. In general, water treatment only reduces pathogen loads rather than completely removing pathogens and higher pathogen concentrations are expected in the spring time with increased runoff and snowmelt. A study of gastrointestinal outbreaks between 1948 and 1994 in the U.S. demonstrated that 51% of outbreaks were preceded by heavy rainfall events [31].

This demonstrates that treatment facilities can be overburdened by upstream sewage discharges or increases in streamflow due to rainfall. Streamflow, a commonly monitored measure, is highly correlated with biological contamination [24], and therefore can be used as a proxy for drinking water contamination for which monitoring data is not systematically maintained.

In this study, we assessed seasonal patterns for gastrointestinal illness in the elderly for counties along the Ohio River and compared them to the seasonal patterns of streamflow. The elderly are a vulnerable subpopulation for gastrointestinal infections. They often have cardiac, renal, or other illnesses which affect their ability to compensate for the fluid shifts seen with gastroenteritis and rates of hospitalization for gastroenteritis rise with age [32]. The Ohio River was selected as a case study because it serves as a source of drinking water for approximately five million people in 29 public drinking water utilities [33].

The most recent assessment of water quality along the Ohio River demonstrated that the entire river fully supports public water supply [33]. However based on the Clean Water Act Standards, several areas were listed on the 303(d) impairment list due to violations of criteria for iron, temperature, and dissolved oxygen [33]. In fact, two-thirds of the river was listed as impaired for contact recreation due to the presence of bacteria [33]. Additionally, the Ohio River is fairly homogenous in meteorological characteristics, such as temperature and precipitation (Figure 1).

2. METHODS

2.1. Location Selection

Cities in close proximity to the Ohio River were selected for this analysis using ArcGIS 9.1 (ESRI, Redlands, CA, USA). A 10-mile buffer was created along the river and overlaid with a layer of cities with populations greater than 50,000 people.

Those cities which fell within the river buffer were selected for analysis and we analyzed data for the counties in which each of these cities fell. Figure 1 shows the 7 cities from 6 counties in the Ohio River watershed that were used for the analysis along with annual average temperature, annual cumulative precipitation, and the proportion of public water supplied by surface water among these communities.

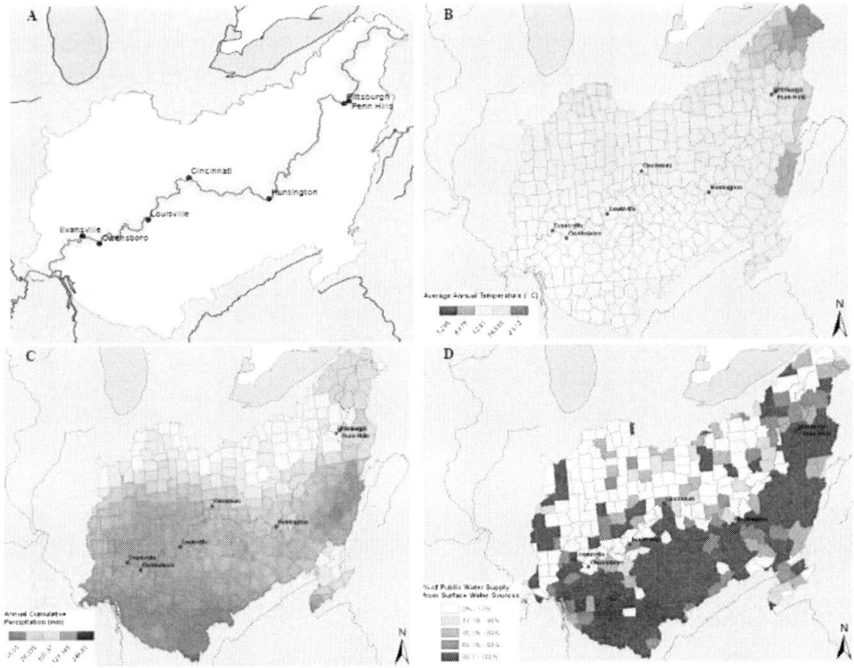

Figure 1. Selected cities and characteristics of the Ohio River watershed. Cities (populations >50,000 persons) which fall within a 10-mile buffer of the Ohio River (Panel A). Average annual temperature (°C) by county (Panel B) and annual cumulative precipitation (mm) by county in the Ohio River watershed (PRISM Data - http://www.prism. oregonstate. edu/) (Panel C). Percent of county public water supply from surface water sources in the Ohio River watershed (Panel D) (USGS Data, http:// water.usgs.gov/watuse/data/2000/index.html).

2.2. Outcome Data

Hospitalization records for persons aged ≥65 years were abstracted for each of the six selected counties for a 14-year period (1 January 1991–31 December 2004) from the Centers for Medicare and Medicaid Services (CMS). About 96% of all adults aged ≥65 years are CMS beneficiaries, therefore their hospitalization charges are reflected in this dataset [34,35]. Each hospitalization record contains individual patient information including state of residence, sex, age at admission, dates of admission and discharge, and ten ICD-9-CM system diagnosis codes. For this analysis, we considered records with the following diagnoses in any of these ten diagnosis codes: cryptosporidiosis (ICD 007.2, 007.4) [36], giardiasis (ICD

007.1), other protozoa (ICD 007.8, 007.9), all protozoa (ICD 007.1, 007.2, 007.4, 007.8, 007.9), viral GI (ICD 008.6), ill-defined GI infections (ICD 008.5, 008.8, 009), GI symptoms (ICD 558.9, 787) and all GI infections without Clostridium difficile (ICD 001-009 excluding 008.45).

We removed *Clostridium difficle* from this outcome as it is primarily a nosocomial infection. Hospitalization records were aggregated according to each patient's diagnosis code, location of residence, and date of admission. Annual rates for each outcome were calculated for each county using the linearly interpolated elderly population for 1997 (midpoint of data timeframe) from 1990 and 2000 U.S. Census Data as the denominator. We also created a weekly time series of rates for each county for each outcome of interest for seasonality assessment.

2.3. Exposure Data

We abstracted daily streamflow data for each of the six counties selected along the Ohio River, as described above, from publicly available U.S. Geological Survey (USGS) 37] databases for the study period (1991–2004). In cases when more than one monitoring station provided data for the entire study period the station which was closest to the main river stem (Ohio River), based on latitude and longitude coordinates of the station, was chosen. The daily streamflow data was aggregated on a weekly level. We did not have any missing data and therefore, no interpolation was needed.

2.4. Exploratory Assessment of Correlation

Spearman cross-correlations were calculated for weekly streamflow between the six selected counties to assess similarities in seasonal pattern of streamflow along the river. Spearman cross-correlations were also calculated between weekly outcome rates and streamflow for each selected county along the Ohio River for outcomes with sufficient disease outcome counts; ill-defined GI infections, GI symptoms, and all GI infections.

We considered time-lagged correlations with weekly GI disease outcome rates lagged by a week after peak streamflow for up to twenty four weeks, 6 months, to assess associations with the nadir or seasonal minimum in outcome.

We also considered the correlation between weekly outcome rate and streamflow for the previous week, two weeks previous, three weeks previous, *etc.* up to twenty five weeks previous.

2.5. Seasonality Assessment

We assessed seasonal patterns for each outcome and each county selected along the Ohio River using Poisson harmonic regression. Seasonality is characterized as systematic, periodic fluctuations within the course of a year. It is assessed by several parameters: (1) the time when the seasonal curve reaches its maximum; (2) annual maximum value (peak); and (3) annual minimum value (nadir) [16]. These seasonal parameters are calculated based on values predicted by the harmonic regression (Equation 1):

$$\ln(E[y_t]) = \beta_0 + \beta_1 \sin(2\pi\omega t) + \beta_2 \cos(2\pi\omega t) + \varepsilon_t \qquad (1)$$

where y_t is a time-series of rates for a specific outcome, t is time in weeks, ω is frequency (ω = 1/52.25), β_0 is intercept, β_1, and β_2 are regression parameters, and ε_t is the error term. The relative intensity, a measure of the shape of the seasonal pattern from peak to nadir, is calculated by dividing the estimated seasonal maximum value by the estimated seasonal minimum value.

This model was used to assess the seasonal peak timing of streamflow, and the seasonal peak timing of each outcome, GI symptoms, ill-defined GI infections and all GI infections. The other outcomes of interest which we had defined had a low number of reported counts by county, with many weeks of zero counts, for which seasonality cannot be assessed using Poisson harmonic regression. Based on exploratory data analysis, we found that each time series, streamflow and all outcomes, demonstrated only one seasonal periodic cycle therefore a single harmonic term was used for seasonality assessment.

Seasonality was also assessed adjusting for weekly average streamflow by county using the Poisson harmonic regression equation (Equation 2):

$$\ln(E[y_t]) = \beta_0 + \beta_1 \sin(2\pi\omega t) + \beta_2 \cos(2\pi\omega t) + \beta_3 x_t + \varepsilon_t \qquad (2)$$

In this case–x_t is the weekly time series for average streamflow for a given location. Adjusting for streamflow allows us to take into account any seasonal variation which may be due to streamflow and only assess the seasonality of the outcome. The results of the models are presented in terms of peak timing with confidence intervals, estimated from regression parameters and relative intensity, or amplitude. Details of the derivation of these seasonal curve properties predicted by the Poisson harmonic regression are given in the supplemental material (Supplemental Material: Figure A).

3. RESULTS

We selected six counties with population centers with more than 50,000 residents within a 10-mile buffer of the Ohio River (Figure 1) and compared the seasonal patterns for rates of hospitalization for gastrointestinal infections with the seasonal patterns in streamflow in the selected counties. The streamflow data for Jefferson Co, KY and Daviess Co, KY were collected from Ohio River monitoring stations, which reported higher values for streamflow since it is a larger water body, whereas data for the other counties were gathered from tributaries which flowed into the Ohio River. Regardless of where the streamflow data were collected, the variability is similar as suggested by the coefficent of variation which was close to one for all sites (Table 1).

Table 1. Descriptive statistics for streamflow (ft^3/sec). Weekly mean, standard deviation, minimum, maximum and interquartile range over the 14-year period 1991–2004. Counties are listed in order based on the flow of the river, east to west.

	Streamflow (ft^3/sec)					
	Mean	Std. Dev	Min	Max	Interquartile Range	Coefficent of Variation
Allegheny Co, PA	19,588.48	15,728.91	2,431.43	80,185.71	6,349.29–28,803.57	0.803
Cabell Co, WV	50.93	82.47	0.22	780.86	4.72–65.21	1.619
Hamilton Co, OH	1,419.22	1,684.33	81.00	12,822.86	343.64–1,785.93	1.187
Jefferson Co, KY	124,019.44	108,094.86	8,595.71	620,571.43	37,382.14–183,607.14	0.872
Daviess Co, KY	135,325.20	112,379.97	9,528.57	627,142.86	45,875.00–193,714.29	0.830
Vanderburgh Co, IN	31,707.29	30,406.60	2,998.57	205,857.14	10,685.71–42,600.00	0.959

Weekly streamflow between the six selected counties is strongly correlated suggesting that the time series for streamflow demonstrate a similar pattern in each county (Supplemental Material: Table A). This analysis allowed us to confirm that regardless of where the data were collected (on the Ohio River or a tributary) the seasonal patterns in streamflow remain consistent.

Annual rates were calculated for each outcome of interest (Table 2). As documented elsewhere, rates for specific disease outcomes, such as cryptosporidiosis and giardiasis, were low because hospitalization and testing for specific pathogens is limited and under-reporting of these infections occurs [36]. Disease counts and annual rates were highest for GI symptoms, ill-defined GI infections, and all GI infections; thus, these three outcomes were used for correlation and seasonality assessment.

Table 2. Total cases and annual rate (per 10,000) for each outcome and for each selected county in the Ohio River Watershed over the 14-year period 1991–2004. Counties are listed in order based on the flow of the river, east to west. Those outcomes in bold provided sufficient counts for seasonality assessment.

County Name	Elderly Population	Crypto-sporidiosis *		Giardiasis *		Other Protozoa *		Viruses *		GI Symptoms *		Ill Defined GI Infections *		All GI Infections *	
	(65 + Y.O.)	Cases	Rate	Cases	Rate	Cases	Rate	Cases	Rate	Cases	Rate	Cases	Rate	Cases	Rate
Allegheny Co. PA	229,290	1	0.00	34	0.11	2	0.01	57	0.18	41,024	127.80	3,156	9.83	4,873	15.18
Cabell Co. WV	15,586	1	0.05	1	0.05	1	0.05	3	0.01	4,879	223.60	291	13.34	392	17.96
Hamilton Co. OH	114,280	1	0.01	22	0.14	1	0.01	15	0.05	26,264	164.16	1,242	7.76	1,850	11.56
Jefferson Co. KY	92,850	0	0.00	9	0.07	1	0.01	7	0.02	24,082	185.26	1,503	11.56	2,016	15.51
Daviess Co. KY	12,283	0	0.00	0	0.00	0	0.00	1	0.00	3,818	222.03	185	10.76	236	13.72
Vanderburgh Co. IN	26,233	0	0.00	2	0.05	2	0.05	10	0.03	7,002	190.65	364	9.91	527	14.35
Total (all counties)	490,522	3	0.00	68	0.10	7	0.01	93	0.14	107,069	155.91	6741	9.82	9,894	14.41

* ICD 9-CM code used for each outcome: cryptosporidiosis (ICD 007.4, 007.2); giardiasis (ICD 007.1); other protozoa (ICD 007.8, 007.9); viruses (ICD 008.6); GI symptoms (ICD 558.9, 787); ill-defined GI infections (ICD 008.5, 008.8, 009); all GI infections (ICD 001-009 W/O 008.45).

Cross-correlations between streamflow and all three outcomes, GI symptoms, ill-defined GI infections, and all GI infections, for each county revealed that in general, high correlations between these GI outcomes and streamflow were synchronous. With increasing lag periods the correlation weakened (Figure 2).

All counties, except Cabell Co, WV and Vanderburgh Co, IN, demonstrated significant synchronous (Lag0) correlation values for GI symptoms and ill-defined GI infections. These correlations remained significant when considering lagged relationships up to twelve weeks. All GI infections exhibited weaker associations and the correlations were only significant in Allegheny Co, PA, Jefferson Co, KY and Daviess Co, KY. For all three outcomes of interest the highest correlations were seen between 6 and 12 week lags. At a lag of 6 months, 24–25 weeks, the correlation becomes negative suggesting that the seasonal pattern for gastrointestinal infections and streamflow are similar with peaks and nadirs occurring at roughly the same time.

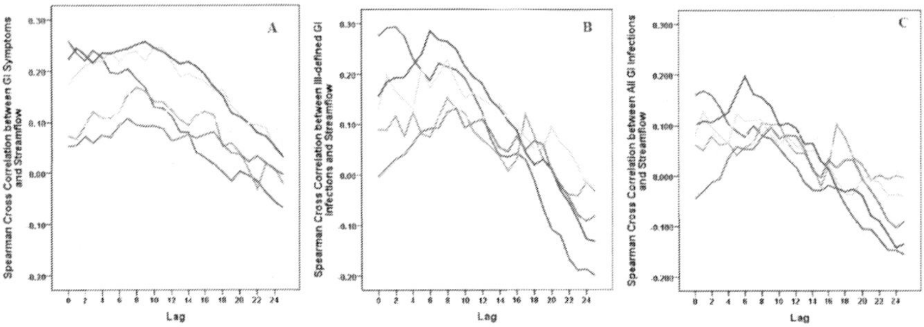

Figure 2. Cross correlations between streamflow and outcomes rates for GI symptoms (Panel A), Ill-defined GI infections (Panel B), and All GI infections (Panel C) with lags up to 25 weeks for each of the six selected counties. (Allegheny Co, PA, Cabell Co, WV, Hamilton Co, OH, Jefferson Co, KY, Daviess Co, KY, Vanderburgh Co, IN).

Seasonality assessment estimated the peak timing for streamflow between the 9th (~4th week of February) and 13th week (~4th week in March) for all selected counties and the relative intensity of the seasonal curve ranges from 3.26 to 10.18 (Table 3). The highest relative intensity of 10.18 was estimated for Cabell County, WV indicating that streamflow at this site has a greater difference between minimum and maximum values. GI symptoms demonstrated a significant seasonal pattern in counties in the Ohio River watershed except Vanderburgh Co, IN (Table 4).

Ill-defined GI infections (Table 5) and all GI infections (Table 6) demonstrated significant seasonal patterns in all counties except Hamilton Co, OH. The peak timing for all three disease outcomes was seen in the early part of the year, ranging from mid January to early March. For GI symptoms, the latest peak timing, ~1st week of March, was estimated for Allegheny Co, PA which is the upper-most site on the river. For most counties, the peak timing of disease rate preceded the peak timing in streamflow. Along the Ohio River, for all three outcomes of interest, adjustment for streamflow did not affect the point estimates of peak timing (Table 4–Table 6). With adjustment for streamflow, the point estimates for peak timing of all three disease outcomes remains within a week of the original estimate.

Table 3. Seasonal pattern for streamflow over 14 year-period for counties along the Ohio River with estimated peak timing, 95% confidence interval, relative intensity, and source of discharge data (main river stem or tributary). Counties are listed in order based on the flow of the river, east to west. Seasonal parameters which are significant ($p < 0.05$) are denoted with an asterisk (*).

County and Seasonal Pattern	Peak Estimate (95% CI)	Relative Intensity	Data Source
Allegheny Co, PA 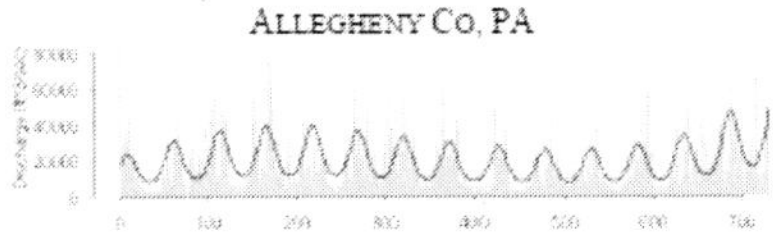	8.795 (8.793, 8.797) * ~4th week of February	3.56	Allegheny River (Tributary)
Cabell Co, WV 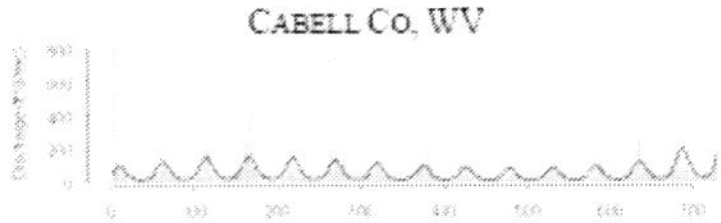	9.758 (9.724, 9.792) * ~1st week of March	10.18	East Fork Twelvepole Creek (Tributary)
Hamilton Co, OH 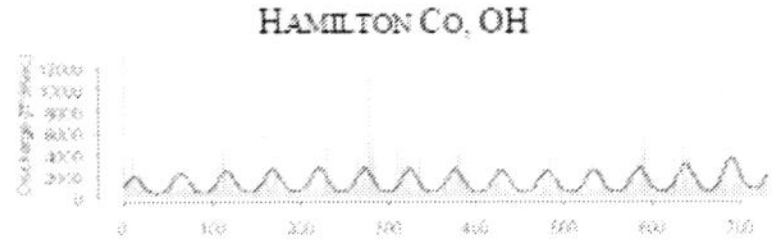	12.404 (12.402, 12.405) * ~3rd week of March	4.88	Little Miami River (Tributary)
Jefferson Co, KY 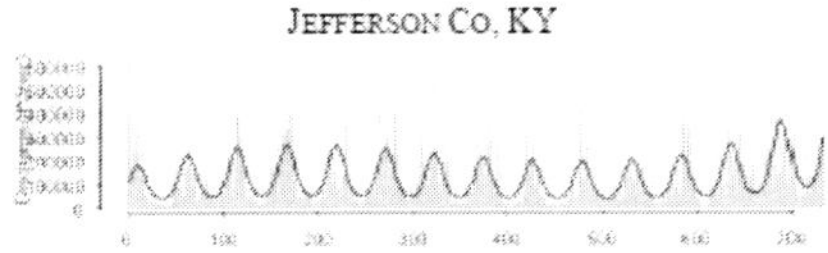	10.257 (10.256, 10.257) * ~1st week of March	5.27	Ohio River (Main Stem)
Daviess Co, KY 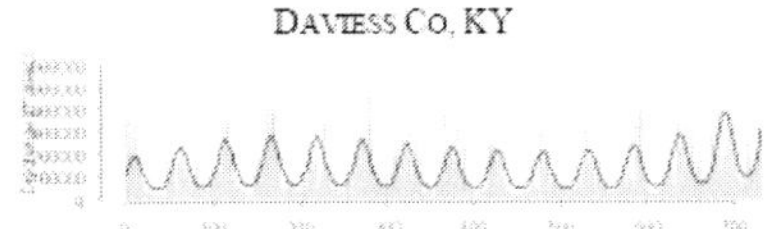	10.480 (10.480, 10.481) * ~1st week of March	4.83	Ohio River (Main Stem)
Vanderburgh Co, IN 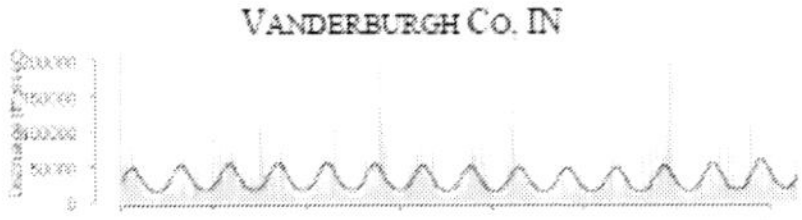	13.088 (13.088, 13.089) * ~4th week of March	3.26	Wabash River (Tributary)

Table 4. Seasonal pattern for GI symptoms over 14 year-period for counties along the Ohio River with estimated peak timing and relative intensity both with and without adjustment for streamflow. Counties are listed in order based on the flow of the river, east to west. Significant ($p < 0.05$) seasonal peak estimates are denoted with an asterisk (*).

County and Seasonal Pattern	Without Adjustment		Adjusting for Streamflow [†]	
	Peak Estimate (95% CI)	Relative Intensity	Peak Estimate (95% CI)	Relative Intensity
Allegheny Co, PA 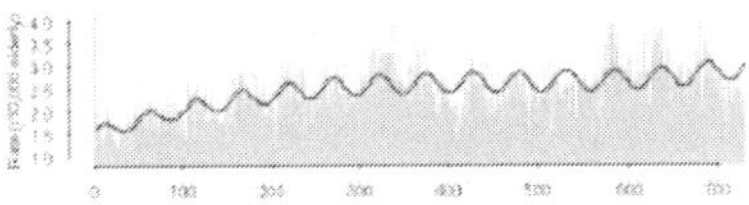	10.13 (9.95, 10.31) * ~1st week of March	1.19	10.28 (8.01, 12.54) * ~1st week of March	1.17
Cabell Co, WV 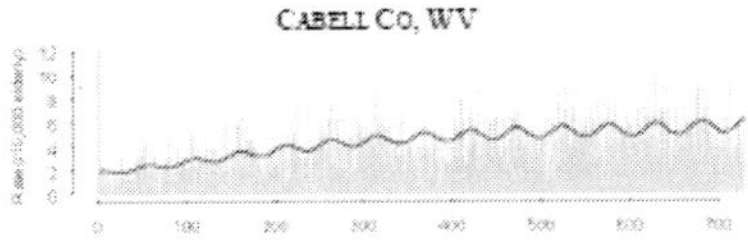	2.90 (2.69, 3.11) * ~3rd week of January	1.21	2.66 (1.75, 3.57) * ~3rd week of January	1.20
Hamilton Co, OH 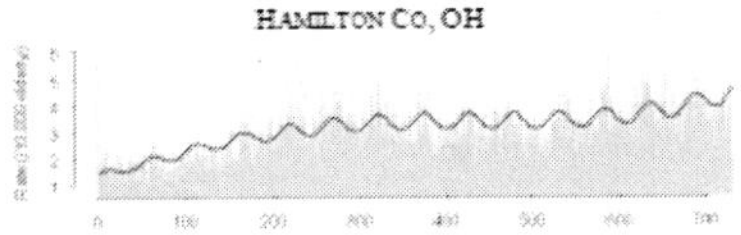	8.93 (8.61, 9.24) * ~4th week of February	1.20	8.85 (8.28, 9.43) * ~4th week of February	1.19
Jefferson Co, KY 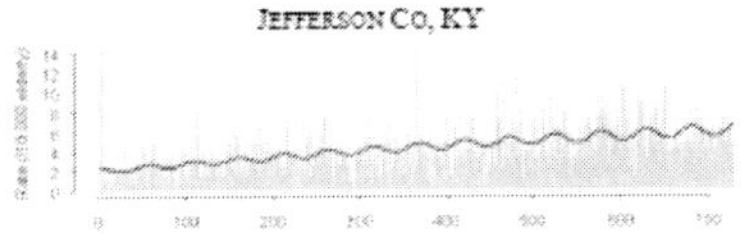	6.36 (6.02, 6.70) * ~2nd week of February	1.22	6.02 (4.03, 8.01) * ~2nd week of February	1.20
Daviess Co, KY 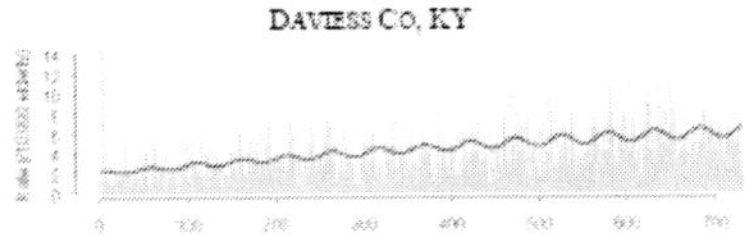	3.39 (3.15, 3.63) * ~3rd week of January	1.20	3.19 (1.68, 4.69) * ~3rd week of January	1.20
Vanderburgh Co, IN 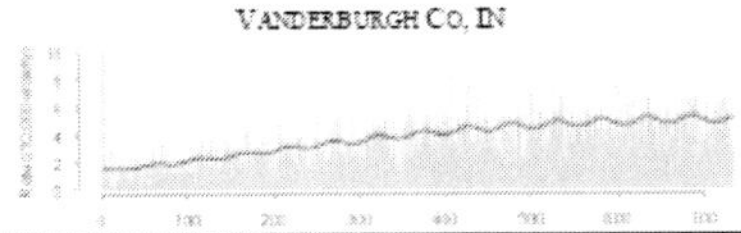	7.90 (7.38, 8.41) ~3rd week of February	1.11	7.55 (6.57, 8.53) ~3rd week of February	1.11

[†] Adjustments were based on weekly average streamflow.

Table 5. Seasonal pattern for Ill-defined GI infections over 14 year-period for counties along the Ohio River with estimated peak timing and relative intensity both with and without adjustment for streamflow. Counties are listed in order based on the flow of the river, east to west. Significant ($p < 0.05$) seasonal peak estimates are denoted with an asterisk (*).

County and Seasonal Pattern	Without Adjustment		Adjusting for Streamflow†	
	Peak Estimate (95% CI)	Relative Intensity	Peak Estimate (95% CI)	Relative Intensity
Allegheny Co, PA 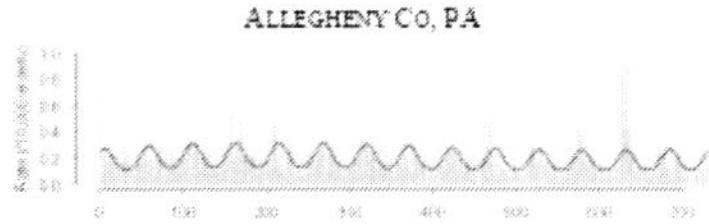	6.40 (5.97, 6.83) * ~2nd week of February	2.44	6.52 (4.93, 8.11) * ~2nd week of February	2.54
Cabell Co, WV 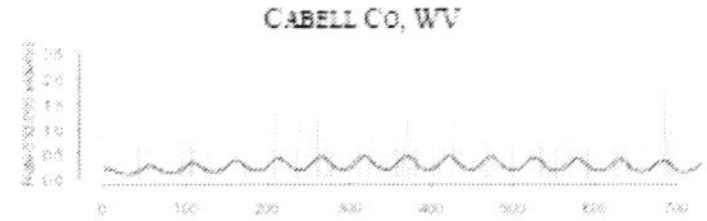	5.08 (4.64, 5.51) * ~1st week of February	2.62	4.93 (3.97, 5.89) * ~1st week of February	2.56
Hamilton Co, OH 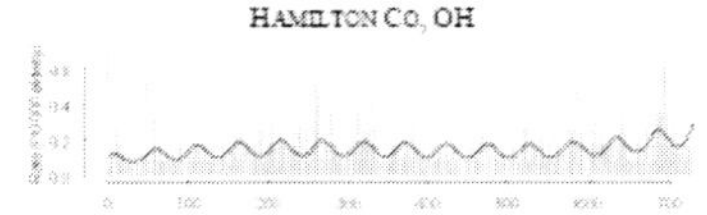	6.88 (6.32, 7.44) ~2nd week of February	1.76	6.69 (6.29, 7.09) ~2nd week of February	1.74
Jefferson Co, KY 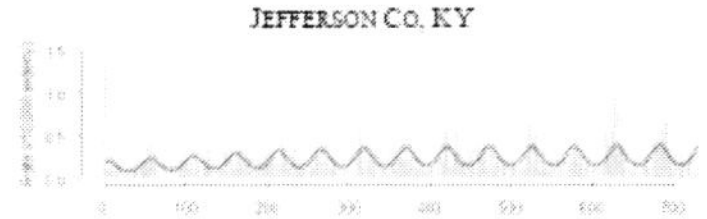	4.96 (4.52, 5.40) * ~1st week of February	2.57	5.52 (4.16, 6.87) * ~1st week of February	2.80
Daviess Co, KY 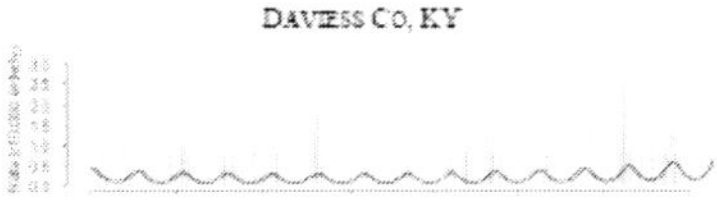	3.69 (3.27, 4.10) * ~3rd week of January	3.88	3.79 (2.84, 4.74) * ~3rd week of January	3.94
Vanderburgh Co, IN 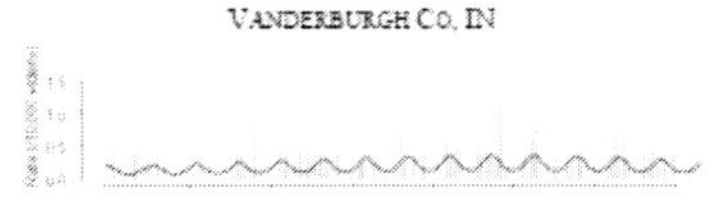	3.79 (3.51, 4.08) * ~3rd week of January	2.78	4.28 (3.90, 4.67) * ~3rd week of January	2.86

† Adjustments were based on weekly average streamflow.

Table 6. Seasonal pattern for All GI infections over 14 year-period for counties along the Ohio River with estimated peak timing and relative intensity both with and without adjustment for streamflow. Counties are listed in order based on the flow of the river, east to west. Significant ($p < 0.05$) seasonal peak estimates are denoted with an asterisk (*).

County and Seasonal Pattern	Without Adjustment		Adjusting for Streamflow[†]	
	Peak Estimate (95% CI)	Relative Intensity	Peak Estimate (95% CI)	Relative Intensity
Allegheny Co, PA 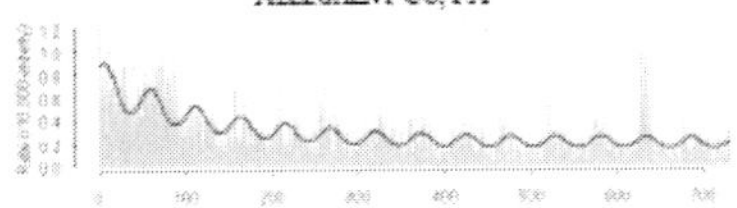	8.04 (7.36, 8.71) * ~3rd week of February	1.63	8.10 (5.87, 10.33) * ~3rd week of February	1.75
Cabell Co, WV 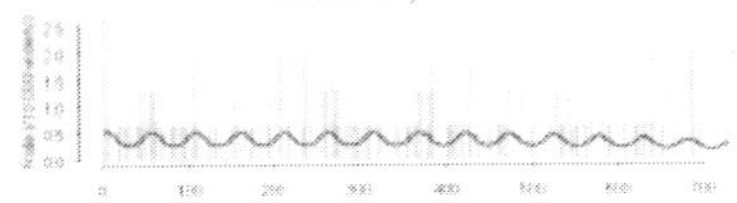	4.79 (4.70, 4.89) * ~1st week of February	1.90	4.69 (3.45, 5.92) * ~1st week of February	1.87
Hamilton Co, OH 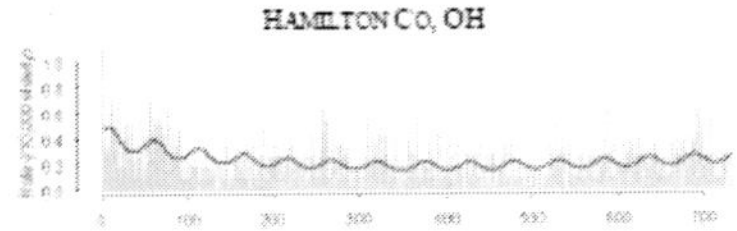	9.38 (8.24, 10.51) ~4th week of February	1.44	9.39 (8.61, 10.17) ~4th week of February	1.44
Jefferson Co, KY 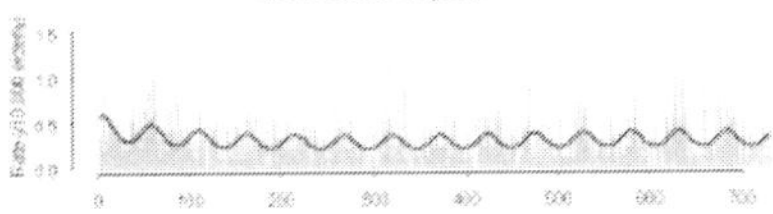	5.67 (5.49, 5.85) * ~1st week of February	1.82	6.47 (4.60, 8.33) * ~2nd week of February	2.04
Daviess Co, KY 	3.13 (2.99, 3.27) * ~3rd week of January	2.61	3.16 (2.03, 4.29) * ~3rd week of January	2.61
Vanderburgh Co, IN 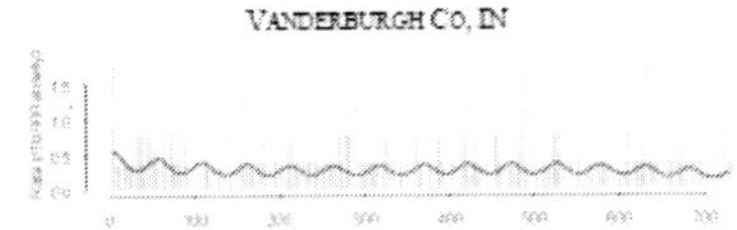	4.02 (3.55, 4.49) * ~4th week of January	1.76	5.00 (4.49, 5.51) * ~1st week of February	1.84

[†] Adjustments were based on weekly average streamflow.

4. DISCUSSION

Rivers have distinct characteristics which can affect the rate and seasonal pattern of waterborne diseases. In this paper, we compare the seasonal patterns of streamflow and various outcomes of gastrointestinal illness along the Ohio River. We demonstrated that both GI illnesses and streamflow exhibit strong seasonal patterns in the selected counties. For all outcomes the peak timing of disease preceded the peak timing seen in streamflow. We also demonstrate that the peak timing of GI illnesses did not change after adjustment for streamflow. This is the first study of its kind to address the seasonal patterns of disease outcomes along a river accounting for streamflow. Previous studies have primarily focused on one study location and measured the hydrological parameters of interest as part of the study itself [23] or had data from the water utility [21,22]. In this study we consider the entire river and utilize publicly available data to assess seasonal relationships.

The calculated rates of gastrointestinal illness in the elderly were similar to those seen in other studies [38,39,40]. It has been demonstrated that rural private supplies have a higher risk of contamination compared to public water supplies [41]. Our findings also demonstrated the highest rates for all gastrointestinal outcomes in the one of the least populated counties selected, Cabell Co, WV.

Our finding of peak timing of GI illness preceding the peak timing in streamflow may be because the level of pathogens in the water is diluted due to increased streamflow. Pathogens are thought to be flushed by rainfall and resulting runoff into surface waters [27,42]; however, research has demonstrated that the pathogen concentrations are lower during periods of higher streamflow due to dilution [43]. We did not observe this dilution effect when we considered the lagged cross-correlations up to six months. Alternatively, it is possible that pathogen concentration is higher preceding peak streamflow. Data on pathogen concentrations have demonstrated peak concentrations in the spring prior to peak streamflow [5]. The first flush phenomenon suggests this is possible as the majority, up to 90 percent, of pollutants are carried in the first 25 mm of runoff from a storm [44]. The data on gastrointestinal outbreaks is consistent with the concept of a first flush of pathogens and an increased likelihood of pathogens passing through treatment facilities as the majority of outbreaks are seen after severe precipitation events [31].

Additionally, pathogen concentrations are dependent on human and, to a degree, animal disease which demonstrate seasonal patterns as well [45]. For example, cryptosporidiosis concentrations are higher on the land during the spring months due to increased fecal contamination from new born calves which have higher concentrations of the pathogen [46]. Studies have also demonstrated a surge of pathogens with early snowmelt and runoff at the beginning of the rainy season with decreasing concentrations through the rainy season [10,47]. These variations in land use and topography may help explain our finding of a late peak for GI symptoms in the upper-most river site of Allegheny Co, PA. Future analyses will need to account for additional parameters which can affect runoff into surface waters, such as land use and land cover.

For this analysis, we used broad outcome categories which do not specify a particular waterborne pathogen therefore we may not have been able to detect a relationship between peak timing of streamflow and GI illness. Since testing for specific pathogens is not commonly practiced [36,48], reporting for these diseases is limited. Marjowicz *et al.*, for example, estimated that for each case in Ontario, Canada of a reportable gastrointestinal disease, there were between 105 and 1,389 cases (median 285) which went unreported, primarily because of a lack of testing [49]. With low counts of hospitalizations, we were not able to use pathogen specific outcomes and used broader outcome categories for seasonality assessment. These broad outcome categories include pathogens which have different incubation periods and seasonal patterns. The incubation period for viruses, such as norovirus and rotavirus, is only 24–28 hours whereas, the incubation period for protozoa, such as *Cryptosporidium* and *Giardia*, is typically much longer, averaging 7 days. Therefore, using these broad outcome categories that include multiple pathogens can mask the relationship with streamflow. Additionally, these different pathogens demonstrate differing seasonal patterns. Viral infections typically demonstrate higher incidence in colder, drier times of the year [50,51,52]. Our recent findings also demonstrate that the broad outcome categories for GI infections may be dominated by untested viral infections and therefore demonstrate peaks at colder times of the year [48]. A limitation of this study is that we are not able to tease apart GI illness due to non-waterborne transmission, such as food borne transmission. Future research should consider alternative modeling methods, such as a zero-inflated Poisson distribution, to assess seasonal patterns of the more specific disease outcomes with low counts.

Another limitation of the study was the lack of availability of hydrological data. Although the USGS has monitoring stations located throughout the United States the data collection has been irregular and sporadic over time. We considered examining several water quality parameters, such as turbidity, for this analysis however, only streamflow data were available for all counties of interest for the entire time period. While streamflow is highly correlated with water quality parameters [24] including turbidity, it is not a direct measure of microbiological water quality. Available streamflow data were not gathered from main Ohio River; but for four of the six counties only data from tributaries were available. Under the Clean Water Act, states are required to ensure that their water use for public water supplies protects the fish, wildlife and recreational uses of the water bodies [53]. Under this broad mandate, states regulate and monitor streamflow levels [53] which is why this parameter was the only one available for all counties for the length of this study. The local water utilities are required to monitor finished water supplies for water quality however; they are only required to report violations to the public and the Environmental Protection Agency (EPA). Therefore, in order to consider other hydrological parameters, such as turbidity, in this type of a long term time-series study would require the cooperation of local water utilities for all locations of interest. To compound the problem, government funding for monitoring of health and environmental measures has been reduced consistently over recent years and therefore, data collection has been reduced in almost all government agencies [54].

We selected counties with population centers (*i.e.*, large cities) which are in close proximity to the Ohio River as it was assumed they are more likely to use surface waters for public water supplies, as rural towns typically use ground water. However, upon further investigation, we found that the selected counties do not rely completely on surface water sources. We gathered water source information for each county and, according to information provided by the municipalities, most of the counties use surface or mixed (both ground and surface water) water supplies (5 of 6 counties along the Ohio River). Previous studies have shown varying seasonal patterns and rates for GI illnesses by water source [14] suggesting that the pathogen transmission and/or concentrations differ by water source. However, when assessing 548 reported gastrointestinal outbreaks researchers found a similar relationship between gastrointestinal illness and rainfall regardless of water source [31]. This finding is expected as disease transmission from within one population center to adjoining ones does not stop based on political boundary. Additionally, it has been established that one community can act as the initial site of infection with secondary transmission to adjacent communities and that community level characteristics, such as socio-demographic factors, are associated with rates of gastrointestinal illness [40]. Future research should strengthen this model by adjusting for community level characteristics, such type of water supply, percent of elderly population, and socio-demographic variables. The strength of our study is that utilizing publically available data, we are able to see a consistent pattern across all six counties when it would be expected that our results would be biased towards the null. Further interdisciplinary research is required to understand the transit time and residence time for various pathogens within ground water and surface water systems and how this can impact the seasonal patterns of the pathogens and the resulting GI illness. Rigorous modeling of streamflow, which capitalizes on advancements in hydrology modeling, will greatly add to the understanding of the associations between streamflow and gastrointestinal illness.

5. CONCLUSIONS

In this paper, we assess seasonal patterns of gastrointestinal illnesses in the elderly along the Ohio River and compare peak timing of illness to the peak timing of streamflow. We demonstrate that after adjusting for streamflow, the seasonal patterns of GI illness remain consistent. The peak timing of GI illness systematically precedes the peak timing of streamflow. Few studies have been conducted on the association between seasonality of hydrological parameters and GI illnesses. In the United States, hydrological data are incomplete for long time periods which limited these analyses. Given the limited availability of hydrological data, a study utilizing modeled hydrological parameters would allow for better understanding of the seasonal association in streamflow and gastrointestinal illness. Further research should investigate the timing of the first flush of runoff and associations with peak timing in GI illness since the first flush is expected to carry the majority of pathogens. Our findings suggest that pathogen loading from the land or other sources into watersheds precedes the time of peak flow resulting in the earlier peak in GI illness. However, it is necessary to conduct more interdisciplinary research to fully understand the hydrological associations with seasonal patterns of waterborne disease.

ACKNOWLEDGEMENTS

The authors thank the support of funding agencies: the National Institute of Environmental Health Sciences (Gastroenteritis and Extreme Weather Events in Elderly (GEWEL) Project (NIEHS ES 013171)) and the Ruth L. Kirschstein NIH National Research Trainee Fellowship (T90 DK070117-05) administered by the Tufts University Water: Systems, Science, and Society Program. The authors would also like to thank J. Michael Wright for comments and editorial suggestions on this manuscript.

REFERENCES

1. World Health Organization (WHO). *Burden of Disease and Cost-Effectivness Estimates*; WHO: Geneva, Switzerland, 2011. Available online: http://www.who.int/ water_sanitation_health/diseases/burden/en/index.html (accessed on 23 March 2012).
2. Centers for Disease Control and Prevention (CDC). *Enteric Diseases Epidemiology and Lab Branches*; CDC: Northeast Atlanta, GA, USA, 2006. Available online: http://www.cdc.gov/enterics/ (accessed on 10 November 2007).

3. Craun, G.F.; Brunkard, J.M.; Yoder, J.S.; Roberts, V.A.; Carpenter, J.; Wade, T.; Calderon, R.L.; Roberts, J.M.; Beach, M.J.; Roy, S.L. Causes of outbreaks associated with drinking water in the United States from 1971 to 2006. *Clin. Microbiol. Rev.* **2010**, *23*, 507–528.
4. LeChevallier, M.W.; Norton, W.D.; Lee, R.G. *Giardia* and *Cryptosporidium* spp. in filtered drinking water supplies. *Appl. Environ. Microbiol.* **1991**, *57*, 2617–2621.
5. LeChevallier, M.W.; Norton, W.D.; Lee, R.G. Occurrence of *Giardia* and *Cryptosporidium* spp. in surface water supplies. *Appl. Environ. Microbiol.* **1991**, *57*, 2610–2616.
6. Plutzer, J.; Tako, M.H.; Marialigeti, K.; Torokne, A.; Karanis, P. First investigations into the prevalence of *Cryptosporidium* and *Giardia* spp. in Hungarian drinking water. *J. Water Health* **2007**, *5*, 573–584.
7. Hansen, J.S.; Ongerth, J.E. Effects of time and watershed characteristics on the concentration of *Cryptosporidium oocysts* in river water. *Appl. Environ. Microbiol.* **1991**, *57*, 2790–2795.
8. Robertson, L.J.; Gjerde, B. Occurrence of *Cryptosporidium oocysts* and *Giardia cysts* in raw waters in norway. *Scand. J. Public Health* **2001**, *29*, 200–207.
9. Tsushima, Y.; Karanis, P.; Kamada, T.; Makala, L.; Xuan, X.; Tohya, Y.; Akashi, H.; Nagasawa, H. Seasonal change in the number of *Cryptosporidium parvum oocysts* in water samples from the rivers in Hokkaido, Japan, detected by the ferric sulfate flocculation method. *J. Vet. Med. Sci.* **2003**, *65*, 121–123.
10. Muchiri, J.M.; Ascolillo, L.; Mugambi, M.; Mutwiri, T.; Ward, H.D.; Naumova, E.N.; Egorov, A.I.; Cohen, S.; Else, J.G.; Griffiths, J.K. Seasonality of *Cryptosporidium oocyst* detection in surface waters of Meru, Kenya as determined by two isolation methods followed by pcr. *J. Water Health* **2009**, *7*, 67–75.
11. Adegbola, R.A.; Demba, E.; De Veer, G.; Todd, J. *Cryptosporidium* infection in Gambian children less than 5 years of age. *J. Trop. Med. Hyg.* **1994**, *97*, 103–107.
12. Newman, R.D.; Sears, C.L.; Moore, S.R.; Nataro, J.P.; Wuhib, T.; Agnew, D.A.; Guerrant, R.L.; Lima, A.A. Longitudinal study of *Cryptosporidium* infection in children in northeastern Brazil. *J. Infect. Dis.* **1999**, *180*, 167–175.
13. Perch, M.; Sodemann, M.; Jakobsen, M.S.; Valentiner-Branth, P.; Steinsland, H.; Fischer, T.K.; Lopes, D.D.; Aaby, P.; Molbak, K. Seven years' experience with *Cryptosporidium parvum* in Guinea-Bissau, West Africa. *Ann. Trop. Paediatr.***2001**, *21*, 313–318.
14. Naumova, E.N.; Chen, J.T.; Griffiths, J.K.; Matyas, B.T.; Estes-Smargiassi, S.A.; Morris, R.D. Use of passive surveillance data to study temporal and spatial variation in the incidence of *Giardiasis* and *Cryptosporidiosis*. *Public Health Rep.* **2000**, *115*, 436–447.
15. Naumova, E.N.; Christodouleas, J.; Hunter, P.R.; Syed, Q. Effect of precipitation on seasonal variability in cryptosporidiosis recorded by the North West England surveillance system in 1990—1999. *J. Water Health* **2005**, *3*, 185–196.
16. Naumova, E.N.; Jagai, J.S.; Matyas, B.; DeMaria, A., Jr.; MacNeill, I.B.; Griffiths, J.K. Seasonality in six enterically transmitted diseases and ambient temperature. *Epidemiol. Infect.* **2007**, *135*, 281–292.
17. Shepherd, R.C.; Sinha, G.P.; Reed, C.L.; Russell, F.E. Cryptosporidiosis in the west of Scotland. *Scott. Med. J.* **1988**, *33*, 365–368.
18. Shenoy, S.; Urs, S.; Prabhu, G.; Mathew, B.; Antony, G.; Bharati, B. Giardiasis in the adult population of Dakshina Kannada district of South India. *Trop. Dr.* **1998**, *28*, 40–42.
19. Araj, G.F.; Abdul-Baki, N.Y.; Hamze, M.M.; Alami, S.Y.; Nassif, R.E.; Naboulsi, M.S. Prevalence and etiology of intestinal parasites in Lebanon. *J. Med. Liban.-Leban. Med. J.* **1996**, *44*, 129–133.
20. Morris, R.D.; Naumova, E.N.; Levin, R.; Munasinghe, R.L. Temporal variation in drinking water turbidity and diagnosed gastroenteritis in Milwaukee. *Am. J. Public Health* **1996**, *86*, 237–239.
21. Schwartz, J.; Levin, R.; Hodge, K. Drinking water turbidity and pediatric hospital use for gastrointestinal illness in Philadelphia. *Epidemiology* **1997**, *8*, 615–620.
22. Schwartz, J.; Levin, R.; Goldstein, R. Drinking water turbidity and gastrointestinal illness in the elderly of Philadelphia. *J. Epidemiol. Commun. Health* **2000**, *54*, 45–51.

23. Egorov, A.; Naumova, E.N.; Tereschenko, A.; Kislitsin, V.; Ford, T. Daily variations in effluent water turbidity and diarrheal illness in a Russian city. *Int. J. Environ. Health Res.* **2003**, *13*, 81–94.
24. Atherholt, T.B.; LeChevallier, M.W.; Norton, W.D.; Rosen, J.S. Effect of rainfall on giardia and crypto. *J. AWWA* **1998**, *90*, 66–80.
25. Lake, I.R.; Bentham, G.; Kovats, R.S.; Nichols, G.L. Effects of weather and river flow on cryptosporidiosis. *J. Water Health* **2005**, *3*, 469–474.
26. Robertson, L.J.; Campbell, A.T.; Smith, H.V. Survival of *Cryptosporidium parvum oocysts* under various environmental pressures. *Appl. Environ. Microb.* **1992**, *58*, 3494–3500.
27. Graczyk, T.K.; Evans, B.M.; Shiff, C.J.; Karreman, H.J.; Patz, J.A. Environmental and geographical factors contributing to watershed contamination with *Cryptosporidium parvum oocysts*. *Environ. Res.* **2000**, *82*, 263–271.
28. Arnone, R.D.; Walling, J.P. Evaluating *Cryptosporidium* and *Giardia* concentrations in combined sewer overflow. *J. Water Health* **2006**, *4*, 157–165.
29. Rose, J.B.; Huffman, D.E.; Gennaccaro, A. Risk and control of waterborne cryptosporidiosis. *FEMS Microbiol. Rev.***2002**, *26*, 113–123.
30. Lisle, J.T.; Rose, J.B. Cryptosporidium contamination of water in the USA and UK: A mini-review. *J. Water Supply: Res. Technol. Aqua* **1995**, *44*, 103–117.
31. Curriero, F.C.; Patz, J.A.; Rose, J.B.; Lele, S. The association between extreme precipitation and waterborne disease outbreaks in the United States, 1948–1994. *Am. J. Public Health* **2001**, *91*, 1194–1199.
32. Cairns, S.; Reilly, J.; Stewart, S.; Tolson, D.; Godwin, J.; Knight, P. The prevalence of health care-associated infection in older people in acute care hospitals. *Infect. Control Hosp. Epidemiol.* **2011**, *32*, 763–767.
33. Ohio River Valley Water Sanitation Commision (ORSANCO), *Biennial Assessment of Ohio River Water Quality Conditions for 2005, 2006, 2007, 2008, and 2009*; The Ohio River Valley Water Sanitation Commission: Cincinnati, OH, USA, 2010.
34. Fisher, E.S.; Baron, J.A.; Malenka, D.J.; Barrett, J.; Bubolz, T.A. Overcoming potential pitfalls in the use of medicare data for epidemiologic research. *Am. J. Public Health* **1990**, *80*, 1487–1490.
35. Cohen, S.A.; Naumova, E.N. Population Dynamics in the Elderly: The Need for Age-Adjustment in National Biosurveillance Systems. In *Intelligence and Security Informatics: Biosurveillance: Second NSF Workshop, Biosurveillance 2007*; Zeng, D., Gotham, I., Komatsu, K., Lynch, C., Thurmond, M., Madigan, D., Lober, B., Kvach, J., Chen, H., Eds.; Springer: New York, NY, USA, 2007; Volume 1, pp. 47–58, New Brunswick, NJ, USA, 22 May 2007; 4506 ed.
36. Mor, S.M.; DeMaria, A., Jr.; Griffiths, J.K.; Naumova, E.N. Cryptosporidiosis in the elderly population of the United States. *Clin. Infect. Dis.* **2009**, *48*, 698–705.
37. United States Geological Survey (USGS). *National Water Information System*; USGS: Rolla, MO, USA, 2007. Available online: http://water.usgs.gov/ (accessed on 10 November 2008).
38. Hlavsa, M.C.; Watson, J.C.; Beach, M.J. Giardiasis surveillance—United States, 1998–2002. *MMWR Surveill Summ***2005**, *54*, 9–16.
39. Hlavsa, M.C.; Watson, J.C.; Beach, M.J. Cryptosporidiosis surveillance—United States 1999–2002. *MMWR Surveill. Summ.* **2005**, *54*, 1–8.
40. Cohen, S.A.; Egorov, A.I.; Jagai, J.S.; Matyas, B.T.; DeMaria, A., Jr.; Chui, K.K.; Griffiths, J.K.; Naumova, E.N. The seeds of two gastrointestinal diseases: Socioeconomic, environmental, and demographic factors related to *Cryptosporidiosis* and *Giardiasis* in Massachusetts. *Environ. Res.* **2008**, *108*, 185–191.
41. Strauss, B.; King, W.; Ley, A.; Hoey, J.R. A prospective study of rural drinking water quality and acute gastrointestinal illness. *BMC Public Health* **2001**, *1*, 8.
42. Gaffield, S.J.; Goo, R.L.; Richards, L.A.; Jackson, R.J. Public health effects of inadequately managed stormwater runoff. *Am. J. Public Health* **2003**, *93*, 1527–1533.
43. Cizek, A.R.; Characklis, G.W.; Krometis, L.A.; Hayes, J.A.; Simmons, O.D., 3rd; Di Lonardo, S.; Alderisio, K.A.; Sobsey, M.D. Comparing the partitioning behavior of *Giardia* and

cryptosporidium with that of indicator organisms in stormwater runoff. *Water Res.* **2008**, *42*, 4421–4438.
44. Stutler, D.; Roesner, L.; Schmidt, M. Urban Runoff and Diffuse Pollution. In *Non Point Pollution and Urban Stormwater Management*; Novotny, V., Ed.; Technomic Publishing Company: Lancaster, PA, USA, 1995; Volume 9.
45. Griffiths, J.K. Human cryptosporidiosis: Epidemiology, transmission, clinical disease, treatment, and diagnosis. *Adv. Parasitol.* **1998**, *40*, 37–85.
46. Sischo, W.M.; Atwill, E.R.; Lanyon, L.E.; George, J. *Cryptosporidia* on dairy farms and the role these farms may have in contaminating surface water supplies in the northeastern united states. *Prev. Vet. Med.* **2000**, *43*, 253–267.
47. Kistemann, T.; Classen, T.; Koch, C.; Dangendorf, F.; Fischeder, R.; Gebel, J.; Vacata, V.; Exner, M. Microbial load of drinking water reservoir tributaries during extreme rainfall and runoff. *Appl. Environ. Microbiol.* **2002**, *68*, 2188–2197.
48. Chui, K.K.; Jagai, J.S.; Griffiths, J.K.; Naumova, E.N. Hospitalizations due to non-specific gastrointestinal diseases: A search for etiological clues. *Am. J. Public Health* **2011**, *101*, 2082–2086.
49. Majowicz, S.E.; Edge, V.L.; Fazil, A.; McNab, W.B.; Dore, K.A.; Sockett, P.N.; Flint, J.A.; Middleton, D.; McEwen, S.A.; Wilson, J.B. Estimating the under-reporting rate for infectious gastrointestinal illness in Ontario. *Can. J. Public Health* **2005**, *96*, 178–181.
50. Cook, S.M.; Glass, R.I.; LeBaron, C.W.; Ho, M.S. Global seasonality of rotavirus infections. *Bull. World Health Organ.* **1990**, *68*, 171–177.
51. Hale, A.; Mattick, K.; Lewis, D.; Estes, M.; Jiang, X.; Green, J.; Eglin, R.; Brown, D. Distinct epidemiological patterns of norwalk-like virus infection. *J. Med. Virol.* **2000**, *62*, 99–103.
52. Glass, R.I.; Bresee, J.; Jiang, B.; Gentsch, J.; Ando, T.; Fankhauser, R.; Noel, J.; Parashar, U.; Rosen, B.; Monroe, S.S. Gastroenteritis viruses: An overview. *Novartis Found Symp.* **2001**, *238*, 5–19, discussion 19-25.
53. Environmental Protection Agency (EPA). *Clean Water Act Section 303*; EPA: Washington, DC, USA, 2008. Available online: http://www.epa.gov/waterscience/standards/ rules/ 303.htm (accessed on 1 March 2009).
54. Rotkin-Ellman, M.; Quirindongo, M.; Sass, J.; Solomon, G. *Deepest Cuts: Repairing Health Monitoring Programs Slashed under the Bush Administration*; Natural Resources Defense Council: New York, NY, USA, 2008.

CHAPTER 7

Biohydrogen Production from Wastewaters

Periyasamy Sivagurunathan[1*], Gopalakrishnan Kumar[1], Arivalagan Pugazhendhi[2], Guangyin Zhen[3], Takuro Kobayashi[1] and Kaiqin Xu[1, 4*]

1 Center for Materials Cycles and Waste Management Research, National Institute for Environmental Studies, Tsukuba, Japan

2 Department of Environmental Engineering, Daegu University, Gyeongsan, Gyeongbuk, Republic of Korea

3 School of Ecological and Environmental Sciences, East China Normal University, Shanghai, China

4 School of Environmental Science and Engineering, Shanghai Jiao Tong University, Shanghai, China

ABSTRACT

Biohydrogen production technology is an emerging field for the advanced wastewater treatment with cogeneration of energy. Besides, hydrogen is an excellent candidate with high energy value (122 kJ/g) than other known carbon-based fuels with no adverse effects to the environment as it releases only water vapor as the by-products during the combustion. Biohydrogen production technology can be assisted through two major pathways: (a) light-dependent reaction (biophotolysis and photofermentation) and (b) light-independent reaction (dark fermentation and microbial electrohydrogenesis cells). The light-dependent reaction can be catalyzed by photosynthetic bacteria, whereas the dark fermentation catalyzed by the heterotrophic bacterial group of facultative and obligate anaerobes. The wastewaters are a rich source of organic nutrients which supports the growth of hydrogen producers along with the disposal of waste and energy recovery. In the present chapter, the recent advancements on biohydrogen production technology from wastewaters with respect to the (a) inoculum development, (b) process optimization, (c) scale-up and (d) the challenges and perspectives toward the improvement of this emerging technology for the wastewater treatment.

Keywords: biohydrogen, dark fermentation, wastewater

1. AN OVERVIEW OF BIOHYDROGEN PRODUCTION

The growing demand of the energy for daily life purposes urged us to seek an alternative and renewable energy carrier with less emission of the pollutants. Hydrogen is an essential and promising candidate for replacing the fossil fuels depletion and greenhouse gas emission reduction. When burning, it releases only water vapor as a by-product with no adverse harmful gases such as NO_x and SiO_2, and hence, it is considered as clean and carbon-free energy carrier. The energy content of hydrogen is 122 kJ/g, which is 2.75-fold greater than the existing hydrocarbon fuels makes an ideal energy carrier for various industrial, transportation and power generations.

Different types of hydrogen production are available such as fossil fuel by hydrocarbon reforming, coal gasification and partial oxidation which requires high temperature and pressure. The biologically adopted hydrogen production methods can be classified as (i) biophotolysis of water using algae/cyanobacteria, (ii) photodecomposition of organic compounds using photosynthetic bacteria, (iii) dark fermentative hydrogen production using strict anaerobic or facultative bacteria and (iv) microbial fuel cells (MFC). Each biological production method had distinct advantages and limitations. For example, the green algae/cyanobacteria decomposes the water into gas (H_2) and liquid (H_2O) in the presence of sunlight by photosynthesis pathway, whereas the slow growth of the algal cells and an inhibition of hydrogenase enzyme with the presence of traces of oxygen limit their application in large scale extent. The photosynthetic bacteria and dark fermentation bacteria share a similar metabolism for the breakdown of organic compounds for their energy and the liberation of energy [1, 2]. The photosynthetic bacteria use organic acids as a substrate and prone to the ammonium and oxygen toxicity, making it as unsuitable for commercial hydrogen production. In contrast, the dark fermentation degrades wide range of organic waste from complex lignocellulose, food waste and industrial wastewater to simpler monomers (sucrose, glucose). However, the chemical oxygen demand (COD) removal efficiency of the dark fermentation is relatively lower 33%, as it requires further treatment before discharge into the system. Moreover, the biomass growth rate and hydrogen production rate of the dark fermentation are comparatively higher than the other hydrogen production methods and make it as attractive candidate for industrial and commercial biohydrogen production [3]. Recently, the auxiliary methods for the hydrogen production from hydrogen effluent have been emerged through microbial fuel cell (MFC) or bioelectrochemical systems (BES) technology.

2. HYDROGEN-PRODUCING MICROORGANISMS

Table 1 displayed the microbial strains helpful for biohydrogen production through dark fermentation [4]. Hydrogen production during fermentation involves either facultative anaerobic bacteria or strict anaerobic bacteria. Facultative anaerobes are capable of growing in the absence of oxygen. The most common hydrogen-*producing* facultative anaerobes are *Klebsiella pneumoniae* [5], *Escherichia coli* [6], *Enterobacter aerogenes* [7], *Rhodospirillum rubrum, Methanobacterium formiccium* [4]. Chookaew et al. [5] reported that *Klebsiella* sp. TR17 is able to produce biohydrogen from crude glycerol in an up-flow anaerobic sludge blanket (UASB) reactor with highest HPR of 242.15 mmol H_2/L/d and HY of 44.27 mmol H_2/g glycerol. Besides, the *Klebsiellapneumoniae* produce valuable by-products such as 1,3-propanediol and 2,3-butanediol [8]. Reungsang et al. [7] reported that the immobilized *E. aerogenes* ATCC 13048 produced major soluble metabolite products (SMPs), such as ethanol, 1,3-propanediol (1,3-PD), formic acid and acetic acid.

2.1. Facultative anaerobes

Facultative anaerobes play important roles in H_2 production by biological routes, as it can grow in the presence of oxygen, higher biomass growth rate and utilization of wide range of organic wastes. The widely studied facultative anaerobic model for hydrogen production is *E. coli* and *E.aerogenes*. Facultative anaerobes convert pyruvate to acetyl-coA and formate with the catalysis of pyruvate formate-lyase complex and then release H_2 with formate hydrogen lyase. The maximum theoretical hydrogen yield is 2 mol of H_2 per mole of glucose. The glucose metabolic pathway yields succinate, lactate, acetate, ethanol and formate, as fermentation end-products. *Enterobacter* sp. have been widely used in various reactor configuration from batch to continuous mode operation. Several attempts like coculture of the facultative anaerobes with strict anaerobes have been assessed to improve the biohydrogen production. The coculture has advantages over pure culture due to the less maintenance, technical feasibility and faster substrate utilization rate. Sivagurunathan et al. [9] demonstrated that the addition of enriched mixed culture with *Enterobacter cloacae* enhanced the hydrogen production rate of 2.25 L/L-d from beverage wastewater. In another report [6], immobilization of *E. coli* cells using sodium alginate increased the hydrogen production efficiency from fructose (1.17 mol/mol hexose) and beverage wastewater (1.65 mol/mol hexose), respectively.

Table 1. Hydrogen production using pure cultures WW, wastewater; CMS, condensed molasses soluble

Wastewater type	Inoculum source	Hydrogen yield (HY) (mol/mol hexose added)	References
Distillery effluent	*Enterobacter cloacae*	165.3 mL/g COD	[32]
Cassava WW	*Clostridium acetobutylicum*	2.41 mol/mol glu	[33]
Rice mill WW	*Enterobacter aerogens*	1.74 mol/mol sugar	[34]
Rice mill WW	*Citrobacter ferundii*	1.40 mol/mol sugar	[34]
Rice mill WW	*Enterobacter aerogens* RM08	1.97 mol/mol	[35]
CMS	*Clostridium tyrobutyricum*	0.7 mmol H_2/g COD	[36]
CMS	*Clostridium pasteurianum*	1.1 mmol H_2/g COD	[36]
CMS	*Clostridium sporosphaeroides*	0.9 mmol H_2/g COD	[36]

2.2. Mixed consortia

The mixed consortia can be derived from a variety of different natural sources, such as sewage sludge, anaerobically digested sludge, compost, animal manure and contaminated soil (Table 2). Mixed culture contains different types of bacteria; it also contains methanogens or hydrogen-consuming bacteria. Mixing also determines the local shear stress that the flow applies to microorganisms. Mixed culture can be obtained from aerobic or anaerobic sludge in wastewater treatment plants or compost piles or any other source of bacteria. Currently, researchers mainly focused two routes for microbial fermentative hydrogen production: one utilizes pure microbial strains and the other employs a mixed microbial consortium. Generally, the hydrogen-producing efficiency and hydrogen yield of pure bacteria are lower than mixed consortia. Several investigators have focused on hydrogen production by microbial fermentation using a mixed microbial consortium, because of low-cost organic substrates, high hydrogen yields and operated in non-sterile conditions.

Table 2. Hydrogen production using mixed consortia.

Wastewater type	Inoculum source	Hydrogen yield (HY) (mol/mol hexose added)	References
BWW	EMC-sewage sludge + pig slurry	1.95 mol/mol glu	[37]
BWW	EMC + *E. coli* XL1 blue	260 mL/g COD	[11]
Sugar beet juice	Anaerobic sludge	2.0 mol/mol glu	[38]
Distillery WW	Anaerobic sludge	10.95 mmol/g COD	[39]
Dairy WW	Anaerobic sludge	15.33 mmol/g COD	[40]
Cheese processing WW	Mixed cultures	10.2 mM/g COD	[41]
Organic WW	Soil	2.32 mol/mol	[42]
Herbal WW	Slaughter house sludge	165 mL/g COD	[43]
CMS	Anaerobic sludge	1.5 mol/mol	[44]
Brewery WW	Anaerobic sludge	1.21 mol/mol	[45]
GWW	Anaerobic sludge	0.75 mol/mol	[46]

[i] - WW, wastewater; BWW, beverage wastewater; CMS, condensed soluble molasses; GWW, glycerine wastewater.

3. PROCESS OPTIMIZATION FOR SCALE-UP

Biohydrogen production is an emerging research area in the sustainable biofuel production via anaerobic fermentation technology. Though the hydrogen production from biological routes seems attractive over other commercial process, the operational conditions are essential to optimize in order to attain the maximum achievable hydrogen production rates and yields. A few important parameters on these aspects are as follows:

a) Inoculum pretreatment

b) pH

c) Nutrient availability

d) Hydraulic retention time

Biohydrogen production through mixed consortia is a complex bioprocess where the inoculum source, substrate type, environmental factors (pH, temperature and substrate concentration), nutrient availability and HRT can influence the metabolic reactions of hydrogen producers. Optimizing these factors is a paramount importance for enhancing the hydrogen production efficiency from organic wastes.

3.1. Inoculum pretreatment

The active acidogenic hydrogen-producing biocatalyst role is crucial, notably in a complex mixed culture microenvironment. In general, the hampering hydrogen yield from mixed consortia was observed due to (i) the competition of hydrogen-consuming microbes and (ii) diversion of the metabolic flux toward non-favorable hydrogen by-products. The hydrogen consumers, such as lactic acid bacteria, methanogenes and sulfur-reducing bacteria, not only act as a competitor for the hydrogen producers but also synthesize various by-products, which affect the growth of hydrogen producers. For instance, the release of proteinaceous toxin (bacteriocins) by lactate-producing bacteria acts as a suppressing factor for hydrogen production and microbial growth [10]. Thus, when the mixed culture is used as an inoculum source, pretreatment step acts as an important role in determining the efficiency of the hydrogen production from mixed consortia. Table 3 showed the various pretreatment methods for enriching the hydrogen producers. The pretreatment step promotes the selective enrichment of hydrogen producers with a suppression of the hydrogenotrophic methanogenes and other hydrogen consumers. The suppression of the hydrogen consumers by pretreatment process allows the mixed consortia to produce the hydrogen as a major product. The fundamental basics relied with the pretreatment method are the physiological difference of the microorganisms. The spore-forming hydrogen producers survive under the harsh pretreatment conditions, whereas the vegetative cells ruptured/killed during the pretreatment. Various pretreatment methods, such as heat shock, acid shock, alkali shock, chemical agents, load shock and oxygen shock, have been assessed for enriching the hydrogen producers from mixed consortia. Each pretreatment step has a significant impact on the suppression of the microbial populations and also the distribution of the microbial metabolism.

Table 3. Inoculum pretreatment method for enriching hydrogen production mixed consortia.

Substrate	Inoculum source	Pretreatment method	Hydrogen yield (HY)	References
Deoiled jatropha waste	Anaerobic digester sludge	Heat shock	20 mL H_2/g VS	[11]
Glucose	Anaerobic sludge	Acid shock	0.80 mol/mol	[12]
Sucrose	Anaerobic digester sludge	Base shock	3.06 mol/mol	[13]
Glucose	Anaerobic granular sludge	Chloroform	1.55 mol/mol	[14]
Desugared molasses	Digested manure	Load shock	237 mL H_2/g-sugar	[16]
Glucose	Anaerobic sludge	Repeated aeration	1.96 mol/mol	[17]
Glucose	Anaerobic sludge	Gamma irradiation	2.15 mol/mol	[47]

Among the various pretreatment methods, the heat-shock [11] pretreatment has been widely accepted as a suitable method for preparing the hydrogen-producing seed inocula, due to the relatively simple method for the suppression of the hydrogen consumers and selective enrichment of the sporulating hydrogen-producing bacteria such as *Clostridium* sp. The acid-shock [12] and base-shock [13] pretreatments suppress the methanogenic activity by the narrow selective growth pH range of the methanogenes (6–7.5), whereas the *Clostridium* populations survive in the harsh condition due to the spore-forming capability. The chemical shock methods such as chloroform [14] and 2-bromoethanesulfonic acid (BESA)

[15] have a complex structure, analog to the methanogenic coenzyme, and it acts as a inhibitor for the methanogenes. This method facilitated the suppression of the methanogenes, whereas the other non-spore-forming hydrogen producers such as *Enterobacter* sp. can also survive with the presence of *Clostridium* sp., thus enhancing the substrate utilization and hydrogen yield. The load-shock [16] treatment is directed by the exposure of the inoculum to a higher substrate concentration, and it leads to the surge in the pH with an accumulation of organic acids and inhibits the methanogenic populations.

Ren et al. [17] demonstrated that application of various pretreatment methods, such as acid, alkaline, heat-shock and repeated aeration, can greatly affect the metabolic pathway and the microbial community distribution pattern. The dominant butyric acid-mediated hydrogen metabolism was observed with heat-shock and alkaline treatment, and mixed-type fermentation pathway was observed with the acid pretreatment, whereas the ethanol-type pathway was observed with repeated aeration treatment with a maximum hydrogen yield of 1.96 mol/mol glucose. The microbial community characterized by denaturing gradient gel electrophoresis (DGGE) revealed that the changes in the composition of the microbial dynamics affect the hydrogen yield. The strain *Ethanoligenens harbinens* was detected under repeated aeration condition with an ethanol-mediated pathway, and the hydrogen-consuming propionic acid bacterium *Propionibacterium propionicus* was detected in acid treatment with low hydrogen productivity. The heat-shock-mediated mixed culture was dominated with *Clostridium* sp. which represents the butyric-acid-type metabolic pathway. Based on the evidence, the appropriate pretreatment method is essential for enriching the hydrogen-producing bacterial populations and enhanced hydrogen production.

3.2. PH

pH is the key driven parameter affecting the cellular metabolism of hydrogen-producing bacterial populations, since the prevalent end products of the bacterial metabolism vary with the changes in the medium pH. Based on the pH and the major end products formation, three metabolic pathways have been proposed (a) ethanol type (EtOH) (Eq. 1), (b) butyric type (HBu) (Eq. 2) and (c) propionic type (HPr) (Eq. 3). The former, HBu type, involved in the hydrogen-generating reactions, whereas the latter, HPr type, involved in the hydrogen-scavenging reactions. Hence, the elimination of the propionate formation is an essential step for the enhancement of hydrogen production.

$$C_6H_{12}O_6 + 2H_2O \rightarrow 2CH_3CH_2OH + 2HCO_3^- + 2H_2$$

$$\Delta G^0 = -235.0\ \text{kJ/mol} \tag{1}$$

$$C_6H_{12}O_6 \rightarrow CH_3CH_2CH_2COOH + 2CO_2 + 2H_2$$

$$\Delta G'_O = -254.0\ \text{kJ/mol} \tag{2}$$

$$C_6H_{12}O_6 + 2H_2 \rightarrow 2CH_3CH_2COOH + 2H_2O$$

$$\Delta G'_O = -279.4\ \text{kJ/mol} \tag{3}$$

pH affects the physiological conditions of the bacterial growth, metabolism and ions transport. Optimizing the pH is considering a key factor influenced the redox environment and thdirection of electron flow toward the hydrogen formation. The experimental reports demonstrated that the optimal pH for the bacterial growth does not result in the elevated hydrogen production performances [3]. For the dark fermentative hydrogen fermentation, the optimal pH for efficient hydrogen production lied between 5.5 and 6.5 for various wastewaters and pure substrates [18]. In addition, the acidic pH induces the pyruvate transformation to volatile fatty acids (VFA) with concomitant hydrogen production, whereas the neutral pH facilitated the methanogenic pathway. Maintaining the acidogenic (5.5–6.5) pH is essential for controlling the methanogenic populations and efficient hydrogen production.

3.3. Nutrients

The inorganic nutrient supplements, such as nitrogen (N), phosphorus (P) and iron (Fe), along with carbon (C) source, are important for microbial growth and improvement in the hydrogen production. The nutrient at proper concentration is beneficial for hydrogen production. For instance, Lin and Lay [19] explained that at a carbon/nitrogen (C/N) ratio of 47, the hydrogen yield from sucrose was 1.9 times higher than the control with a value of 4.8 mol/mol substrate. In a pure culture thermotolerant *Kelbsiella* sp., the maximum hydrogen yield of 0.28 mol/mol glycerol was observed with 11.21 g/L glycerol, 2.84 g/L KH_2PO_4 and 5.66 g/L NH_4Cl, respectively [20].

Wang et al. [21] mentioned that the hydrogen production efficiency of glucose (313.3 mL/g glucose) was improved with low supplementation of nitrate 0.1 g/L; however, increased concentration of nitrate over 0.1 g/L significantly affected the hydrogen yield and the substrate consumption rate. The drop in hydrogen production is attributed by the inhibition of nitrogenize activity by surplus ammonium ions [22, 23].

The iron (Fe) is an important element essential for the hydrogenase activity, which directs the metabolic pathway by stimulating the active site for the ferredoxin (Fd). The addition of iron supplement was shown to improve the hydrogen production. Gadhe et al. [24] demonstrated the effects of nano-sized iron and nickel oxide nanoparticles by using dairy wastewater as a substrate, and it showed that an enhancement in hydrogen yield of 17.2 mmol/g COD is due to the enhanced activity of the ferredoxin oxidoreductase, ferredoxin and hydrogenase enzymes.

Moreover, the optimal value for the Fe^{2+} concentration is varied with the type of substrates used. For instance, the optimal concentration reported by Liu and Shen [25] was 10 mg/L from starch, whereas palm oil mill effluent showed an optimal value of 257 mg/L [26].

3.4. Hydraulic retention times

The hydraulic retention time (HRT) is one of the key process control parameters influencing the continuous hydrogen production. HRT enables the better process control of the microorganisms that can regulate the metabolic pathway favorable for efficient hydrogen production. The long HRT permits the growth of hydrogen consumers mainly archaea, which is unsuitable for hydrogen production, whereas too low HRT leads to the washout of active biomass and deterioration of the reactor performances. The optimization of HRT is a paramount importance for the scale-up, long-term and sustainable hydrogen production.

HRT controls the organic loading rate (OLR), substrate degradation and reaction kinetics. The organic wastes required long HRT, whereas the simple organics required short HRT [2]. The reported optimum HRT value for the wastewater ranges from 0.5 to 24 h. For example, the short HRT (0.5 h) provided the maximum hydrogen production rate of 14 L/L-d from condensed soluble wastewater [27], whereas the long HRT (24 h) is required for efficient conversion of olive mill wastewater with a HPR of 7.0 L/L-d [28].

The process parameters discussed above significantly influenced the hydrogen production; hence, careful assessment of each individual factor is important for stable hydrogen production.

4. BIOREACTOR DESIGN CONSIDERATIONS FOR CONTINUOUS HYDROGEN PRODUCTION

Bioreactor configuration is a notable factor in dark fermentative hydrogen production, as it influences the contact between the organic waste and hydrogen producers, substrate utilization, biomass dilution rate, etc. According to the feeding regime, the biohydrogen production can be conducted in batch, semi-continuous and continuous mode (Table 4). The batch mode operation is relatively simple and easier to control. Hence, the batch mode hydrogen reactors have been widely used to determine the feasibility of the organic waste feedstock and to optimize the environmental parameters such as pH, temperature, substrate concentration. In semi-continuous mode operation, the organic substrate was operated in a sequencing batch which includes feeding, reaction, settle and decant stages [29].

Table 4. Bioreactor types used in hydrogen production.

Substrate	Inoculum source	Reactor mode	HPR (L/L/d)	References
Palm oil mill effluent	Anaerobic digester sludge	ASBR	6.7	[29]
Condensed molasses	Anaerobic sludge	CSTR	14.04	[27]
Beverage WW	Enriched mixed cultures	CSTR	37.56	[30]
Tofu processing WW	Anaerobic digester sludge	MBR	19.86	[48]
Desugared molasses	Anaerobic sludge	UASB	5.6	[49]
Olive mill WW	Anaerobic sludge	PBR	7.0	[28]
Beverage WW	Enriched mixed cultures	ICBR	55.4	[31]
Beverage WW	Anaerobic digester sludge	PBR	88.7	[50]

[i] - WW, wastewater; ASBR, anaerobic sequencing batch reactor; CSTR, continuously stirred tank reactor; MBR, membrane bioreactor; PBR, packed bed reactor; ICBR, immobilized cell bioreactor.

The sequencing batch operation is recommended for a viscous substrate like a POME and solid organic biomass like food waste and lignocellulosic biomass, where the physical contact between the substrate and microorganisms is limited, and this reactor mode operation enables the better hydrolysis rate, avoids clogging in the pipes and retains the effective biomass concentration. In continuous mode operation, the continuous supply of nutrients and the removal of the pollutants occur simultaneously with the aid of peristaltic pumps.

Although various reactor models assessed, the continuous mode operation is preferred for bench-scale and commercial-scale applications. The widely investigated model for continuous mode operation is the CSTR type, wherein the substrates and feedstocks are well mixed inside the reactor with the aid of the mechanical rotor; however, the biomass washout usually occurred at lower HRT [27, 30]. In some cases, the biofilm formed inside the CSTR is resistance to the biomass washout and thereby enhancing the hydrogen production performances. Chu et al. [27] investigated the CSTR reactor model by using condensed soluble molasses as a substrate with suspended and immobilized cells as inoculum source. The hydrogen production from immobilized cell was relatively lower with a maximum HPR of 7.6 L/L/d; however, the suspended cells operation provided the maxim HPR of 14.04 L/L/d, respectively. The observed variation is attributed by the washout of the active biomass in immobilized cells system (9.8 g volatile suspended solids (VSS)/L), poor mass transfer between the microbes and substrates and the increased lactic acid formation. On the other hand, the suspended cell system formed a hydrogen-producing granule (HPG) inside the reactor, and thus, it retains the active biomass (12.30 g VSS/L) and less formation of the lactic acid. Sivagurunathan et al. [30] demonstrated that the hydrogen production from ICBR [31] was higher (55 L/L/d) than the suspended cells CSTR (37.56 L/L/d) operation. The superior performance of the ICBR is due to the formation of granular biomass at short HRT of 3 h with the presence of *Selenomonas* sp. and further maturation of granules with the presence of active hydrogen-producing *Clostridium* Sp. The *Selenomonas* sp. act as a bio-glue for the development of granules. Moreover, the energy content analysis of the beverage wastewater with immobilized cells system analysis showed that it has the capability of reducing the CO_2 reduction efficiency of 2832 ton CO_2 equivalent/year.

5. CONCLUSION

Biohydrogen production from industrial wastewaters seems to be appropriate and environmental benign option for future sustainable hydrogen economy with simultaneous energy recovery and waste disposal. Various studies revealed the hydrogen production potential of wastewaters. Among them, sugar-rich wastewaters are the promising substrate for high-efficient hydrogen production rates and yields, due to their easier degradation rate and higher substrate concentration. Other key challenges that rely on dark fermentative hydrogen production from organic wastes are the low substrate conversion efficiency, moderate-to-low hydrogen yield and residual organics in the effluents. In general, biohydrogen production is a primary step for wastewater treatment, in which a maximum 4 mol/mol glucose representing 33% of COD removal efficiency; nearly 70–80% of the residual organics remain untreated with the hydrogen-producing effluent, thus seeks further disposal of the effluent in the wastewater streams. The post-residual effluent has to be integrated with various two-step processes, such as methane production, photofermentation, microbial electrolysis cells, bioplastics production and microalgae cultivation, for maximizing the energy recovery.

ACKNOWLEDGEMENTS

The author Dr. Periyasamy Sivagurunathan greatly acknowledged the financial assistance from Japan Society for Promotion of Science: JSPS ID: P15370 for this study.

REFERENCES

1. Singh L, Wahid ZA. Methods for enhancing bio-hydrogen production from biologicalprocess: A review. Journal of Industrial and Engineering Chemistry. 2015; 21:70–80.
2. Sivagurunathan P, Kumar G, Bakonyi P, Kim S-H, Kobayashi T, Xu KQ, et al. A criticalreview on issues and overcoming strategies for the enhancement of dark fermentativehydrogen production in continuous systems. International Journal of Hydrogen Energy.2016;41:3820–36.
3. Ghimire A, Frunzo L, Pirozzi F, Trably E, Escudie R, Lens PNL, et al. A review on darkfermentative biohydrogen production from organic biomass: Process parameters anduse of byproducts. Applied Energy. 2015;144:73–95.
4. Elsharnouby O, Hafez H, Nakhla G, El Naggar MH. A critical literature review onbiohydrogen production by pure cultures. International Journal of Hydrogen Energy.2013;38:4945–66.
5. Chookaew T, O-Thong S, Prasertsan P. Biohydrogen production from crude glycerolby immobilized *Klebsiella* sp. TR17 in a UASB reactor and bacterial quantification undernonsterile conditions. International Journal of Hydrogen Energy. 2014;39:9580–7.
6. Sivagurunathan P, Kumar G, Lin C-Y. Hydrogen and ethanol fermentation of variouscarbon sources by immobilized *Escherichia coli* (XL1-Blue). International Journal ofHydrogen Energy. 2014;39:6881–8.

7. Reungsang A, Sittijunda S, Angelidaki I. Simultaneous production of hydrogen andethanol from waste glycerol by *Enterobacter aerogenes* KKU-S1. International Journal ofHydrogen Energy. 2013;38:1813–25.
8. Chookaew T, O-Thong S, Prasertsan P. Fermentative production of hydrogen andsoluble metabolites from crude glycerol of biodiesel plant by the newly isolated thermotolerant *Klebsiella pneumoniae* TR17. International Journal of Hydrogen Energy.2012;37:13314–22.
9. Sivagurunathan P, Gopalakrishnan K, Lin CY. Enhancement of fermentative hydrogenproduction from beverage wastewater via bioaugmentation and statistical optimization.Current Biochemical Engineering. 2014;1:92–8.
10. Gomes SD, Fuess LT, Mañunga T, Feitosa de Lima Gomes PC, Zaiat M. Bacteriocins oflactic acid bacteria as a hindering factor for biohydrogen production from cassava flourwastewater in a continuous multiple tube reactor. International Journal of HydrogenEnergy. 2016;41:8120–31.
11. Kumar G, Bakonyi P, Sivagurunathan P, Nemestóthy N, Bélafi-Bakó K, Lin C-Y.Improved microbial conversion of de-oiled Jatropha waste into biohydrogen via inoculumpretreatment: process optimization by experimental design approach. BiofuelResearch Journal. 2015;2:209–2014.
12. Wang J, Wan W. Comparison of different pretreatment methods for enriching hydrogenproducingbacteria from digested sludge. International Journal of Hydrogen Energy.2008;33:2934–41.
13. Zhu H, Béland M. Evaluation of alternative methods of preparing hydrogen producingseeds from digested wastewater sludge. International Journal of Hydrogen Energy.2006;31:1980–8.
14. Hu B, Chen S. Pretreatment of methanogenic granules for immobilized hydrogen fermentation.International Journal of Hydrogen Energy. 2007;32:3266–73.
15. Chang S, Li JZ, Liu F. Evaluation of different pretreatment methods for preparinghydrogenproducing seed inocula from waste activated sludge. Renewable Energy.2011;36:1517–22.
16. Kongjan P, O-Thong S, Angelidaki I. Biohydrogen production from desugared molasses(DM) using thermophilic mixed cultures immobilized on heat treated anaerobic sludgegranules. International Journal of Hydrogen Energy. 2011;36:14261–9.
17. Ren NQ, Guo WQ, Wang XJ, Xiang WS, Liu BF, Wang XZ, et al. Effects of differentpretreatment methods on fermentation types and dominant bacteria for hydrogen production.International Journal of Hydrogen Energy. 2008;33:4318–24.
18. Lin CY, Lay CH, Sen B, Chu CY, Kumar G, Chen CC, et al. Fermentative hydrogenproduction from wastewaters: A review and prognosis. International Journal ofHydrogen Energy. 2012;37:15632–42.
19. Lin CY, Lay CH. Effects of carbonate and phosphate concentrations on hydrogen productionusing anaerobic sewage sludge microflora. International Journal of HydrogenEnergy. 2004;29:275–81.
20. Chookaew T, O-Thong S, Prasertsan P. Statistical optimization of medium componentsaffecting simultaneous fermentative hydrogen and ethanol production from crude glycerolby thermotolerant *Klebsiella* sp. TR17. International Journal of Hydrogen Energy.2014;39:751–60.
21. Wang B, Wan W, Wang J. Effects of nitrate concentration on biological hydrogen productionby mixed cultures. Frontiers of Environmental Science & Engineering in China.2009;3:380–6.
22. Wang B, Wan W, Wang J. Effect of ammonia concentration on fermentative hydrogenproduction by mixed cultures. Bioresource Technology. 2009;100:1211–3.
23. Redwood MD, Macaskie LE. A two-stage, two-organism process for biohydrogen fromglucose. International Journal of Hydrogen Energy. 2006;31:1514–21.
24. Gadhe A, Sonawane SS, Varma MN. Enhancement effect of hematite and nickelnanoparticles on biohydrogen production from dairy wastewater. International Journalof Hydrogen Energy. 2015;40:4502–11.

25. Liu G, Shen J. Effects of culture and medium conditions on hydrogen production fromstarch using anaerobic bacteria. Journal of Bioscience and Bioengineering. 2004;98:251–6.
26. OThong S, Prasertsan P, Intrasungkha N, Dhamwichukorn S, Birkeland N-K.Optimization of simultaneous thermophilic fermentative hydrogen production andCOD reduction from palm oil mill effluent by Thermoanaerobacteriumrich sludge.International Journal of Hydrogen Energy. 2008;33:1221–31.
27. Chu CY, Wu SY, Hsieh PC, Lin CY. Biohydrogen production from immobilized cellsand suspended sludge systems with condensed molasses fermentation solubles.International Journal of Hydrogen Energy. 2011;36:14078–85.
28. Scoma A, Bertin L, Fava F. Effect of hydraulic retention time on biohydrogen and volatilefatty acids production during acidogenic digestion of dephenolized olive mill wastewaters.Biomass and Bioenergy. 2013;48:51–8.
29. Badiei M, Jahim JM, Anuar N, Sheikh Abdullah SR. Effect of hydraulic retention time onbiohydrogen production from palm oil mill effluent in anaerobic sequencing batch reactor.International Journal of Hydrogen Energy. 2011;36:5912–9.
30. Sivagurunathan P, Lin C-Y. Enhanced biohydrogen production from beverage wastewater:process performance during various hydraulic retention times and their microbialinsights. RSC Advances. 2016;6:4160–9.
31. Sivagurunathan P, Sen B, Lin C-Y. High-rate fermentative hydrogen production frombeverage wastewater. Applied Energy. 2015;147:1–9.
32. Mishra P, Das D. Biohydrogen production from Enterobacter cloacae IIT-BT 08 usingdistillery effluent. International Journal of Hydrogen Energy. 2014;39:7496–507.
33. Cappelletti BM, Reginatto V, Amante ER, Antônio RV. Fermentative production ofhydrogen from cassava processing wastewater by *Clostridium acetobutylicum*. RenewableEnergy. 2011;36:3367–72.
34. Ramprakash B, Muthukumar K. Comparative study on the production of biohydrogenfrom rice mill wastewater. International Journal of Hydrogen Energy. 2014;39:14613–21.
35. Ramprakash B, Muthukumar K. Comparative study on the performance of various pretreatmentand hydrolysis methods for the production of biohydrogen using *Enterobacteraerogenes* RM 08 from rice mill wastewater. International Journal of Hydrogen Energy.2015;40:9106–12.
36. Hsiao CL, Chang JJ, Wu JH, Chin WC, Wen FS, Huang CC, et al. Clostridium strancocultures for biohydrogen production enhancement from condensed molasses fermentationsolubles. International Journal of Hydrogen Energy. 2009;34:7173–81.
37. Sivagurunathan P, Sen B, Lin C-Y. Batch fermentative hydrogen production by enrichedmixed culture: Combination strategy and their microbial composition. Journal ofBioscience and Bioengineering. 2014;117:222–8.
38. Dhar BR, Elbeshbishy E, Hafez H, Lee H-S. Hydrogen production from sugar beet juiceusing an integrated biohydrogen process of dark fermentation and microbial electrolysiscell. Bioresource Technology. 2015;198:223–30.
39. Gadhe A, Sonawane SS, Varma MN. Evaluation of ultrasonication as a treatmentstrategy for enhancement of biohydrogen production from complex distillerywastewater and process optimization. International Journal of Hydrogen Energy.2014;39:10041–50.
40. Gadhe A, Sonawane SS, Varma MN. Enhanced biohydrogen production from dark fermentationof complex dairy wastewater by sonolysis. International Journal of HydrogenEnergy. 2015;40:9942–51.
41. Yang P, Zhang R, McGarvey JA, Benemann JR. Biohydrogen production from cheeseprocessing wastewater by anaerobic fermentation using mixed microbial communities.International Journal of Hydrogen Energy. 2007;32:4761–71.
42. Sharma Y, Li B. Optimizing hydrogen production from organic wastewater treatmentin batch reactors through experimental and kinetic analysis. International Journal ofHydrogen Energy. 2009;34:6171–80.
43. Sivaramakrishna D, Sreekanth D, Sivaramakrishnan M, Sathish Kumar B, HimabinduV, Narasu ML. Effect of system optimizing conditions on biohydrogen production

fromherbal wastewater by slaughterhouse sludge. International Journal of Hydrogen Energy.2014;39:7526–33.
44. Wu JH, Lin CY. Biohydrogen production by mesophilic fermentation of food wastewater.Water Science and Technology. 2004;49:223–8.
45. Shi XY, Jin DW, Sun QY, Li WW. Optimization of conditions for hydrogen productionfrom brewery wastewater by anaerobic sludge using desirability function approach.Renewable Energy. 2010;35:1493–8.
46. Fernandes BS, Peixoto G, Albrecht FR, Saavedra del Aguila NK, Zaiat M. Potential toproduce biohydrogen from various wastewaters. Energy for Sustainable Development.2010;14:143–8.
47. Yin Y, Hu J, Wang J. Enriching hydrogen-producing bacteria from digested sludgeby different pretreatment methods. International Journal of Hydrogen Energy.2014; 39:13550–6.
48. Kim MS, Lee DY, Kim DH. Continuous hydrogen production from tofu processingwaste using anaerobic mixed microflora under thermophilic conditions. InternationalJournal of Hydrogen Energy. 2011;36:8712–8.
49. Kongjan P, O-Thong S, Angelidaki I. Hydrogen and methane production from desugaredmolasses using a two-stage thermophilic anaerobic process. Engineering in LifeSciences. 2013;13:118–25.
50. Chu CY, Hastuti ZD, Dewi EL, Purwanto WW, Priyanto U. Enhancing strategy onrenewable hydrogen production in a continuous bioreactor with packed biofilter fromsugary wastewater. International Journal of Hydrogen Energy. 2016;41:4404–12.

CHAPTER 8

Wastewater treatment by local microalgae strains for CO2 sequestration and biofuel production

Abeera A. Ansari[1,2], Asif Hussain Khoja[1,3], Azra Nawar[1], Muneeb Qayyum[1] and Ehsan Ali[1,4]

1 US-PAK Centre for Advance Studies in Energy (CAS-EN), National University of Sciences and Technology (NUST), Sector H-12, Islamabad 44000, Pakistan

2 Department of Civil and Environmental Engineering, College of Engineering, University of Massachusetts (UMass), Amherst, MA 01002, USA

3 Chemical Reaction Engineering Group (CREG), Faculty of Chemical and Energy Engineering, Universiti Teknologi Malaysia (UTM), Skudai, 81310 Johor Bahru, Malaysia

4 University of Agriculture, Agriculture University Road, Faisalabad 38000, Pakistan

ABSTRACT

Currently, the scientific community is keenly working on environmental-friendly processes for the production of clean energy and sustainable development. The study was conducted to cultivate microalgae in raw institutional wastewater for water treatment, enriched production of biomass and CO_2 sequestration. The strains which were used in this study are *Scenedesmus* sp. and *Chlorella* sp. which were isolated from Kallar Kahar Lake, Pakistan. Both strains were cultivated in synthetic growth medium (Bold's Basal Medium) to enhance biomass production. Afterward, microalgae cultures were inoculated in wastewater sample in mixotrophic mode under ambient conditions. The impurities in wastewater were successfully removed from the original sample by the 7th day of operation. COD 95%, nitrate 99.7% and phosphate 80.5% were removed by applying *Scenedesmus* sp. Meanwhile, *Chlorella* sp. reduced 84.86% COD, 98.2% nitrate and 70% phosphate, respectively. Interestingly, sulfates were removed from wastewater completely by both strains. Besides being useful in wastewater remediation, these microalgae strains were subsequently harvested for lipid extraction and potential biofuel production was determined. Therefore, the applied method is an environmentally safe, cost-effective and alternative technology for wastewater treatment. Furthermore, the achieved biomass through this process can be used for the production of biofuels.

Keywords: Wastewater treatment, Microalgae, COD Nutrients, Biofuel

1. INTRODUCTION

The demand for energy and clean water is increasing day by day due to the rapid increase in population, industrialization and urbanization. The basic source of the energy is fossil fuels, which are exhaustible and not easily replenished on a human time scale. The consumption of fossil fuels is a major cause of emissions of harmful gases and a distinct thread to climate change. To meet the increasing energy demands as well as keeping in view the environmental protection, there is a need to find out eco-friendly renewable energy sources for sustainable development. Among the renewable sources, biomass production and utilization for biofuel generation is the most attractive technology. Biofuel such as biodiesel is a biodegradable and nontoxic source of fuels. It is being produced from variety of biomass including soybean, sunflower, coconut, canola, palm oil and waste oil; however, its production on large scale is not sustainable (Mostafa et al. 2012). The algae are one of the strong contenders for producing biodiesel. The algae have been grown successfully in different synthetic mediums as well as in different types of wastewater; hence, the water can also be treated in different stages via algae cultivation. The idea of algae cultivation in wastewater is a hybrid approach for the treatment of wastewater, CO_2 utilization and production of biomass for biofuels.

Wastewater treatment is a multistage process to purify wastewater before it enters into the natural water body. The purpose of treatment is removal or at least maximize the reduction of organic matter, solids, nutrients or other pollutants through algae cultivation (Buljan et al. 2011). Currently in

Pakistan, there is a trend of discharging untreated wastewater (having agricultural, municipal or industrial origin) directly into the nearby water bodies, which significantly alters their physical, chemical and biological nature. There are several issues; in particular, eutrophication, oxygen depletion, increase in turbidity, habitat and aquatic life loss are associated with excessive nutrient loading, which is courtesy of untreated wastewater disposal. More recently, there are some industries that have established wastewater treatment plants on site, which mostly work on the activated sludge method due to its reliability (Irfan 2009). Traditional wastewater treatment methods can be broadly categorized into physical, chemical and biological, such as sedimentation, chlorination and activated sludge. The removal efficiency of most of them are high; however, these conventional methods are not considered as sustainable, thereby presenting a need for a cost-effective and sustainable method.

According to the literature, microalgae is a strong contender for biofuel generation as it can be grown successfully in different synthetic mediums as well as different types of wastewater. Wastewater can also be treated by microalgae cultivation. Previous studies have proved that microalgae utilizes nutrients from wastewater for biomass production which can be used for further production of several useful products such as biodiesel, ethanol, methanol, biomolecules, hydrogen, fertilizer and carotenoids. Hence, treating wastewater by microalgae is a zero waste, environmentally friendly approach and cost-effective, as useful by-products can potentially offset the cost. Microalgae is, hence, a bioremediation agent which is currently being deployed in many facilities. However, recent studies have shown that microalgae potential in wastewater treatment is much wider than its current use. This has been supported by a report published by the National Renewable Energy Laboratory (NREL) which stated that under a controlled environment, microalgae are capable of producing 40% surplus oil for biodiesel per unit area of land as compared to crops such as palm seed or sunflower. Microalgae are being cultivated for the generation of biomass for biofuels; however, microalgae require nutrients to grow and with growing water scarcity issues there is a need for an economical growth medium. With that being said, wastewater can be the best option to cultivate microalgae for biomass, since wastewater contains nutrients and water which can be reclaimed after treatment with microalgae (Woertz 2007). Microalgae growth in wastewater has been indicated as a cost-effective process. Literature depicts that cultivation of microalgae in a controlled environment requires low initial capital, operational cost and low technical costs. It also states that low land-cost areas enriched with sunlight and warm temperature will be advantageous for microalgae-based wastewater system (Woertz 2007; Yun et al. 1997).

Apart from water treatment, microalgae are also major contributors to CO_2 sequestration. Emission of CO_2 is the main cause of global warming.

Microalgae use CO_2, NO_2 and SO_2 as nutrients for their growth. CO_2 in flue gas also did not have any inhibitory effect on the algal growth (Yun et al. 1997).

In this study, the isolated algae strains (*Scenedesmus* sp. and *Chlorella* sp.) were cultivated in raw institutional wastewater for simultaneous wastewater treatment, CO_2 sequestration and biomass harvesting. The aim of the work was to determine the treated water quality, CO_2 intake and the lipid contents in biomass. Furthermore, FTIR analysis was carried out on algal biomass to justify its potential of being a feedstock for biofuel (biodiesel) prospects.

2. MATERIALS AND METHODS

2.1 Synthetic medium preparation

Bold's Basal Medium (BBM) was used as a synthetic growth medium which is ideally used for culturing a wide variety of freshwater algae strains. Having a pH of 6.6, it contained the following chemical composition according to Leslie and Summerell (Leslie and Summerell 2006): K_2HPO_4 (75 mg/L), $MgSO_4$ (75 mg/L), NaCl (25 mg/L), $CaCl_2$ (25 mg/L), $NaNO_3$(250 mg/L), KH_2PO_4 (105 mg/L), $FeCl_3$ (0.194 g/L), $MnCl_2$ (0.082 g/L), $CoCl_2$ (0.16 g/L), $Na_2MoO_4 \cdot 2H_2O$ (0.008 g/L), and $ZnCl_2$ (0.005 g/L).

2.2 Isolation of local microalgae strains

Local microalgae strains were obtained from Kallar Kahar Lake, located in Jhelum, Pakistan, and were isolated and preserved in the Biofuel Lab at Centre for Advance Studies in Energy (CAS-EN), NUST (Pakistan). These strains were revived prior to isolation in BBM at ambient temperature (20 ± 3 °C) with constant CO_2 supply (0.04% v/v) at an average flow rate of 0.3 L/min using an aquarium air pump. Illumination was provided via natural sunlight, and the pH was maintained at 7.0. After the revival, serial dilution and agar plating techniques were performed until unialgal cultures were obtained (CSIRO 2013). The resulting isolated strains of *Scenedesmus* sp. and *Chlorella* sp. were identified under light microscope (40×, 100×) on the basis of their morphological and other cellular features by comparing with previous studies (Ponnuswamy et al. 2013; Van Vuuren 2006). These strains were then enriched in 10 ml BBM to increase cell concentration. It should be noted that, before the experiments, the BBM medium was first purged with nitrogen gas for a duration of 3 min to remove the dissolved gases and ensure that the initial dissolved CO_2 concentration is zero. This is a significant point and is important for CO_2 sequestration.

2.3 Inoculum preparation and maintenance

Isolated green microalgae were inoculated at 10% ($V_{inoculum}/V_{media}$) in 250 ml Erlenmeyer flasks containing 100 ml BBM media. These culture flasks were incubated on a shaker (150 rpm) under ambient conditions of (20 ± 3 °C) temperature and natural sunlight at alternating light–dark photoperiod (14 h:10 h) for 20–30 days (Wang et al. 2010). Furthermore subculturing was performed by inoculating 50 µl of algae cultures onto test tubes containing solidified BBM slants with 1.5% (w/v) of agar powder. These test tubes were also incubated at the above conditions for 2–3 weeks during which growth appeared on slants (Prabakaran and Ravindran 2012).

2.4 Characteristics of wastewater

Raw institutional wastewater from the primary septic tank located in Environmental Sciences and Engineering Department, NUST, Pakistan, was used. The representative wastewater sample was grayish black in color and was collected by taking measured volumes of wastewater after every hour and mixing them together. Different physiochemical parameters were evaluated using standard protocols described by the American Public Health Association (APHA) (Eaton 2005). Among the parameters, nitrate nitrogen (NO_3-N) was determined using UV–visible spectrophotometry, while phosphate phosphorus (PO_4^{-3}-P) and sulfates (SO_4^{-2}) were quantified using spectrophotometer Hach DR/2010. COD was determined using closed reflux method, while turbidity, DO and pH were evaluated using turbidity, DO and pH portable meters, respectively.

2.5 Cultivation in wastewater

10% ($V_{inoculum}/V_{media}$) microalgae cultures were inoculated in reagent bottles containing 1L wastewater samples, respectively. Just like the BBM media, the wastewater samples were first purged with nitrogen gas for 3 min to remove absorbed gases prior to inoculation. The cultures were illuminated under natural sunlight (14 h:10 h) (light–dark photoperiod) at an ambient temperature range of (24 ± 4 °C). CO_2 was provided continuously by aquarium air pumps and pH of all cultures was maintained at 7.0–8.0. The incubation period was about 12 days during which microalgae growth parameters as well as wastewater parameters were analyzed.

2.6 Analysis of microalgae growth

Samples were taken from the inoculated wastewater on a daily basis for cell count (CC), which was measured using a hemocytometer (Castellanos 2013), whereas dry biomass weight and chlorophyll a were determined after every 3 days. Chlorophyll a (mg/L) was determined spectrophotometrically after extraction by 90% acetone (Arar 1997). Dry biomass weight (g/L) was calculated by filtering 10 ml microalgae culture samples through pre-heated, pre-weighed glass microfiber filters (Whatman GF/C, 47 mm). After filtration, the filters containing microalgae suspension were dried at 103 °C to a constant weight, cooled in a desiccator and weighed on an electronic balance (Irving and Allen 2011).

2.7 Growth kinetic parameters

Dry biomass values (g/L) from the exponential growth phase were used for conducting the growth kinetics study. The specific growth rate (μ, day^{-1}) was calculated using Eq. (1), where $X_2$and X_1 are the dry biomass weight (g/L) at time t_2 and t_1, respectively.

$$\mu = \frac{\ln X_2 - \ln X_1}{t_2 - t_1} \quad (1)$$

The maximum specific growth rate (μ_{max}, day^{-1}) was determined from the different μ values calculated, while the maximum biomass obtained was designated as X_{max} (g/L) (de Morais and Costa 2007). Cell doubling time (t_d, day) was estimated using Eq. (2), while productivity (*P*, g/L/day) was calculated using Eq. (3) (Sankar et al. 2011):

$$t_d(d) = \frac{ln2}{\mu_{max}} \quad (2)$$

$$P = \frac{X_t - X_0}{t_t - t_0} \quad (3)$$

where 'X_t' is the biomass concentration at time '*t*', while '$X_{o'}$' is the initial biomass concentration at inoculation time t_o. P_{max} (g/L/day) was designated to the maximum productivity. Carbon dioxide uptake rate (Pco_2) (g/L/day) was measured using Eq. (4) (Elvira-antonio and Ruí 2013), whereas '*P*' is the productivity calculated above in Eq. (3).

$$Pco_2 = 1.88 \times P \quad (4)$$

2.8 Lipid analysis

The protocol by Bligh and Dyer (1959) was used to extract lipids from microalgae biomass. Microalgae culture was centrifuged (4000*g*, 10 min) to obtain wet pellet which was mixed in distilled water, methanol and chloroform (4:5:10 v/v) and left overnight on a shaker at moderate speed. The following day, distilled water and chloroform (1:1 v/v) were introduced into the sample and allowed to mix for 5–6 h.

Finally the sample was centrifuged (4000*g*, 10 min) to produce a biphasic layer in which the bottom layer contained lipids dissolved in chloroform. This layer was extracted using micropipette and dried at 50 °C for 2 h in an oven to obtain the lipid content.

2.9 FTIR analysis

Fourier transforms infrared (FTIR) spectroscopy analysis was conducted on dried algal biomass at room temperature using Perkin Elmer Spectrum 100 FTIR spectrophotometer (Ponnuswamy et al. 2013). The algal pellet was obtained via centrifugation (4000*g*, 10 min) and was placed in an oven at 103 °C to dry to constant weight.

The obtained dried algal powder was then blended with potassium bromide (KBr) powder and pressed into tablets before analysis in the wavelength region of 4000–500 cm^{-1}.

3. RESULTS AND DISCUSSION

3.1 Analysis of raw wastewater

The physiochemical analysis of raw wastewater revealed the presence of color in the sample, which was due to the presence of high level turbidity and other dissolved and undissolved substances that may have also contributed to the odor (Iram et al. 2013). The chemical oxygen demand (COD) level was observed to be slightly above the National Environmental Quality Standards (NEQs) of 150 mg/L (Shams 2000).

The overall nutrient load suggested that the wastewater was in low-medium strength range as reported in (Gogate and Pandit 2004). Table 1 shows the initial physiochemical analysis of the collected raw wastewater.

Table 1 Raw wastewater parameters before treatment

Parameter	Units	Initial values
COD	mg/L	185 ± 10.7
DO	mg/L	3.51 ± 0.2
Nitrate-nitrogen	mg/L	9.1 ± 0.42
Phosphate-phosphorus	mg/L	18 ± 0.6
Sulfates	mg/L	52 ± 1.8
Turbidity	NTU	64 ± 5
pH		7.4 ± 0.4
Odor		Pungent
Color		Brownish gray
Temperature	°C	28.06 ± 3

3.2 Analysis of algae growth in wastewater

The isolated microalgae strains demonstrated no lag phase, as good adaptability was observed in the wastewater medium since the start. Both strains showed a similar growth trend till 6 days, after which *Chlorella* sp. displayed a decline while *Scenedesmus* sp. continued to exhibit increased exponential growth till the 7th day. After 7 days, both strains entered the decline phase as algae cells continued to die off. Growth curves of *Chlorella* sp. and *Scenedesmus* sp. in terms of CC (million cells per ml) vs time are shown in Fig. 1.

In this study, the *Scenedesmus* sp. achieved the maximum production of chlorophyll a (19.04 mg/L), while *Chlorella* sp. showed the second best results with 13.1 mg/L in wastewater culture on the 8th day. Afterward, both strains showed decline in chlorophyll a content to 8.59 and 8.88 mg/L, respectively, on the 12th day as microalgae cells entered death phase.

According to Young et al. (1996), chlorophyll a production shows similar growth trend to cell concentration, i.e., the highest concentration of chlorophyll a will be obtained at the highest cell concentration or at the end of the exponential phase of growth and can be observed in Fig. 2.

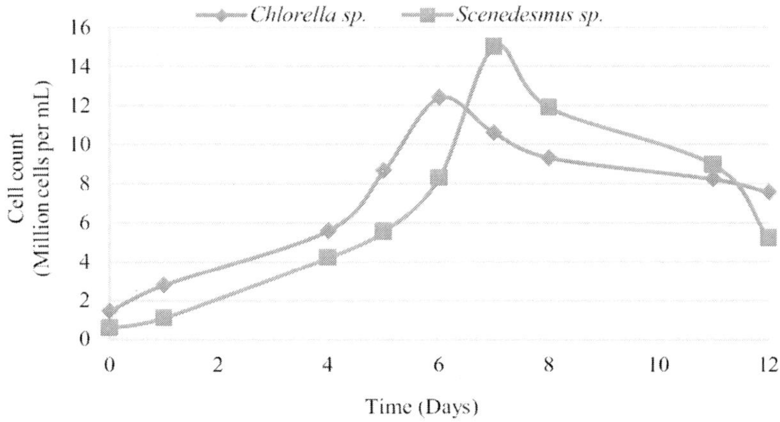

Figure. 1 Growth (CC) curves of isolated microalgae in a wastewater sample

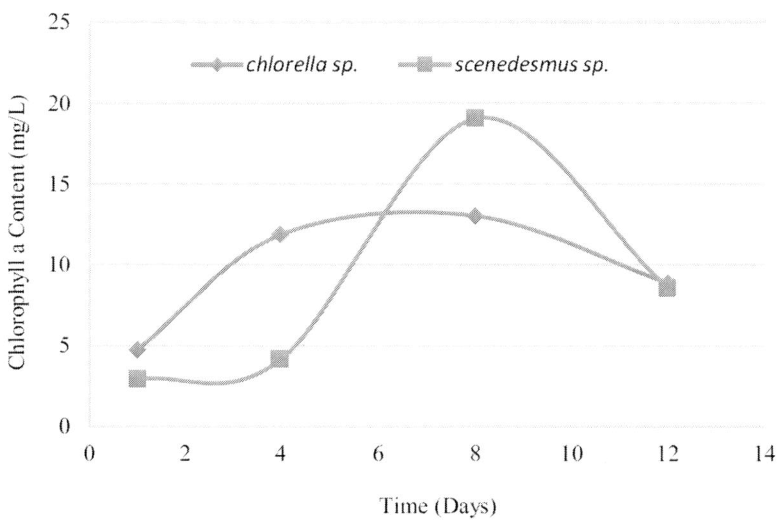

Figure. 2 Chlorophyll a content of isolated microalgae in a wastewater sample

Biomass yield is an expression of organic production, as the dry weight of the organic mass produced over a period of time; it is also used to express microalgae growth. The highest dry weight was achieved using *Scenedesmus* sp. (0.445 g/L) as compared to *Chlorella* sp. (0.39 g/L) in wastewater medium on the 8th day. Afterward, both strains showed drastic reduction till the 12th day, representing nutrient deficiency in the media. Furthermore, the data showed linear biomass growth, which reflected linear carbon dioxide (CO_2) sequestration as algae uses carbon dioxide from air to carry out photosynthesis to produce oxygen and glucose. Greater CO_2 uptake demonstrates greater biomass production as an indicator of microalgae growth. The CO_2 uptake rates of the selected microalgae strains are presented in Table 2 along with overall growth kinetics and dry weight for both strains (see Fig. 3).

Table 2 Growth kinetic study of both strains in wastewater sample

Parameters	Units	***Chlorella* sp.**	***Scenedesmus* sp.**
Growth kinetics			
μ_{max}	(d^{-1})	0.297	0.33
t_d	(d)	2.33	2.1
X_{max}	(g/L)	0.39	0.445
P_{max}	(g/L/day)	0.0507	0.0587
PCO_2	(g/L/day)	0.0953	0.1103

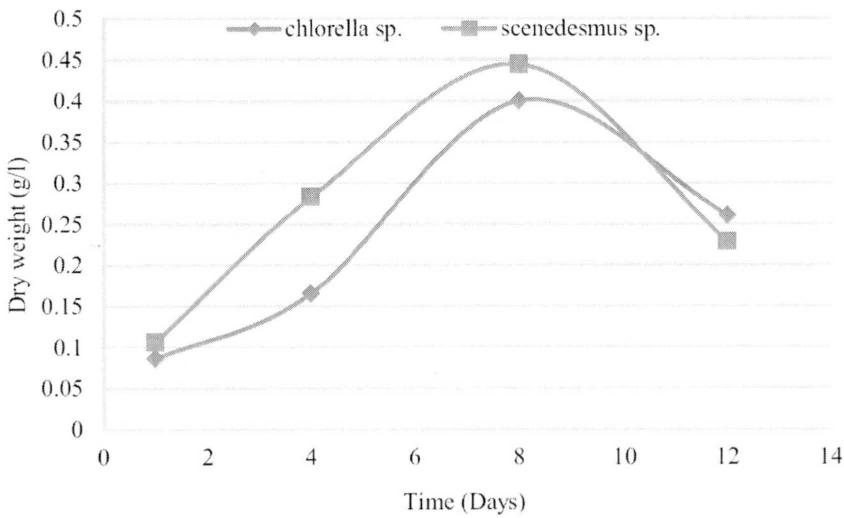

Figure. 3 Dry weight of isolated microalgae in a wastewater sample

From the growth studies, it is evident that the growth of *Scenedesmus* sp. is better than that of *Chlorella* sp. in our wastewater sample. It was also reflected in the overall growth kinetics (Table 2).

3.3 FTIR analysis and lipid content of isolated microalgae strains

FTIR Spectroscopy was employed to determine the presence of vibrationally active functional groups (including O–H, N–H, C=O, C–H, CH_2, C–O–C and >P = O) in the isolated microalgae strains (Venkatesan et al. 2012). *Chlorella* sp. and *Scenedesmus* sp. both showed strong peaks in the range of 2809–3012 and 3300–3560 cm^{-1}, indicating the presence of lipid–carbohydrate and protein–carbohydrate functional groups (Table 3). This, along with peaks in the range of 1000–1200 cm^{-1}, strongly indicated the presence of carbohydrates in *Chlorella* sp. depicting higher carbohydrate potential as compared to *Scenedesmus* sp. Both *Chlorella sp.* and *Scenedesmus* sp. showed strong peaks at protein amide I band. This band is primarily due to C=O stretching vibrations in amide complexes (Dean et al. 2010). The presence of alcoholic (1260–1350 cm^{-1}), ester (1250–1300 cm^{-1}) and nucleic acid (1191–1356 cm^{-1}) functional groups was observed in *Chlorella* sp., while the functional group of antioxidant enzymes (1030–1120 cm^{-1}) was observed in both algal strains. Overall, from FTIR analysis, it was observed that *Chlorella* sp. had higher lipid and carbohydrate content along with nucleic acid in comparison to *Scenedesmus* sp., making it a better biofuel feedstock, if optimum cultivation and later harvesting conditions are provided.

Table 3 Tentative FTIR spectra band analysis of the selected algal strains

Main peak (cm^{-1})		**Functional groups**	**Frequency range (cm^{-1})**
***Chlorella* sp.**	***Scenedesmus* sp.**		
3436	3423	(O–H) stretching vibration existence of carbohydrate	3300–3560
		(N–H) stretching vibration existence of secondary amines (lipids, protein)	
2909	2958	Lipid–carbohydrate mainly Vas (CH_2) and Vs (CH_2) stretching	2809–3012
		(C–H) Aliphatic stretching	

		vibration	
1650	1685	Protein amide I (C=O) stretching vibration (esters, amino acids)	1583–1750
1451	Not detected	(CH_2) bending vibration CH_2–CO existence of carbonyl compounds	1405–1477
		lipid as (CH_2) bending of methyl	
1223	Not detected	Nucleic acid, Vas > P = 0 stretching associated with phosphorus compounds	1191–1356
		C–O stretching, O–H bending vibration existence of alcohol	
		C–O asymmetric C–O–C stretching existence of esters	
1100	1045	Carbohydrate V (–O–C) of polysaccharides. Nucleic acid	1030–1120
		Symmetric C–H stretching vibration indicating antioxidant enzyme presence	
		SO_3 stretching vibration, ionic sulfonates	

After the FTIR analysis, the next step was to validate and quantify the lipid content potential of the microalgae strains via experimentation. For this purpose, the lipid content was extracted from the harvested biomass at the exponential growth phase. The results validated the observations of FTIR analysis by indicating that *Chlorella* sp. have higher lipid content than *Scenedesmus* sp. in both growth media (Fig. 4). The total lipid percentages were reported to be high, as 45% of pure cultures of *Scenedesmus* sp. and *Chlorella* sp. were grown in BBM (Thompson 1996). The experiments were conducted in ambient conditions without optimization, i.e., under lower illumination, temperature and without pure CO_2 supplement. The resulting lipid contents of the pure strain cultures grown in BBM were much lower than those previously reported (Prabakaran and Ravindran 2013; Sharma et al. 2016). On the other hand, strains cultivated in institutional wastewater demonstrated higher lipid content under similar ambient conditions. This might be due to the wastewater sample being nitrogen deficient with respect to BBM. When the nitrogen is exhausted from medium culture, the cell nitrogen is used in enzymes and essential cellular structures and a portion of the carbon dioxide content is converted to lipid and carbohydrates, giving higher lipid yields (Kamalanathan et al. 2016). Overall, the results suggest that the strains have the potential of producing higher lipid content in the wastewater treatment application if the process is optimized.

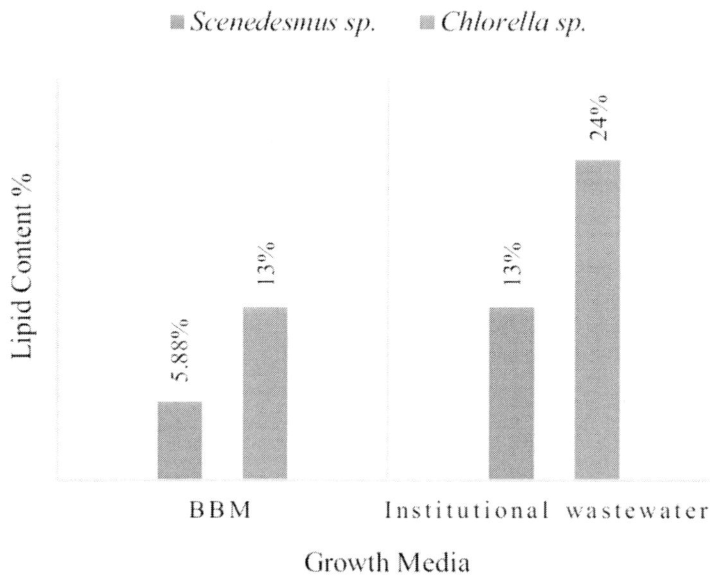

Figure. 4 Lipid content of isolated microalgae strains

3.4 Analysis of treated wastewater sample

The effect of microalgae cultivation on NO_3-N, SO_4^{-2} and PO_4^{-3}-P nutrient stripping was observed. The results indicated a significant reduction in the above nutrients during the course of experimentation. PO_4^{-3}-P was removed up to 70 and 80.5% from wastewater on the 7th day by *Chlorella* sp. and *Scenedesmus* sp. respectively. It depicted effective uptake of phosphorus to be accumulated and synthesized as polyphosphates in their bodies. Nearly, 50% removal was observed after 3 days by both strains as seen in Fig. 5.

The removal rates for NO_3-N (the major inorganic nitrogen form in the influent) in 7 days were 98.2 and 99.7% for *Chlorella* sp. and *Scenedesmus* sp., respectively. The results from the experiment showed up to 47% reduction by *Chlorella* sp. in 4 days, while *Scenedesmus* sp. achieved 50%

removal by the 5th day (Fig. 6). According to Wang et al. (2010), the relative constancy of nitrate uptake by algae might be due to the saturation of assimilators to amino grouping production for entry into nitrogenous metabolism. This homeostasis of nitrogen assimilation enables it to maximize growth, which supports our findings.

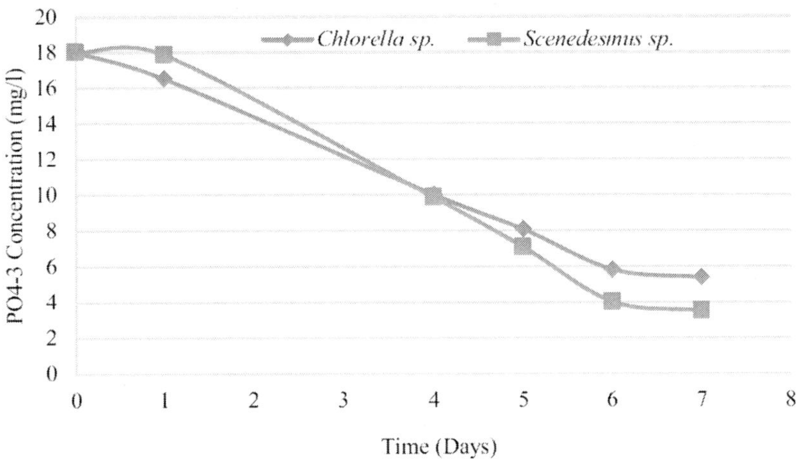

Figure. 5 Phosphate-phosphorous reduction by isolated microalgae strains

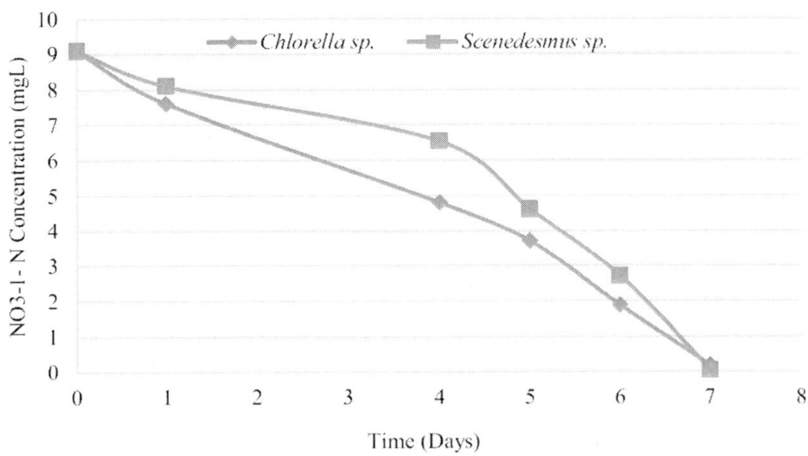

Figure. 6 Nitrate-nitrogen reduction by isolated microalgae

SO_4^{-2} was removed completely within 7 days by both strains. Being the only form of sulfur source in wastewater medium, eukaryotic microalgae consumed SO_4^{-2} to manufacture S-amino acids, sulfo-lipids and other S-containing metabolites during its growth phase (Prieto and Vega 1992).

Besides nutrients, both strains displayed a significant reduction in COD and turbidity levels. The efficiency for turbidity removal was above 90% for both strains. COD ,which represents the organic matter present in wastewater that can be oxidized chemically, was reduced up to 84.86 and 95% by *Chlorella* sp. and *Scenedesmus* sp. on the 7th day of cultivation, respectively (Fig. 7).

This COD removal was attributed to the attached and suspended growth of the algal biomass on the aeration tube as well as in the mixture. During cultivation, algal cells visibly increased, turning the wastewater color from blackish gray to green, which when filtered produced clear water. The overall reduction in wastewater pollutants within 7 days of incubation is listed in Table 4.

DO increased significantly from 3.51 mg/L to 8.65 and 8.5 mg/L in wastewater by *Chlorella* sp. and *Scenedesmus* sp., thereby improving the overall water quality and making it acceptable for discharging or reuse purpose according to the National Environmental Quality Standards (NEQs).

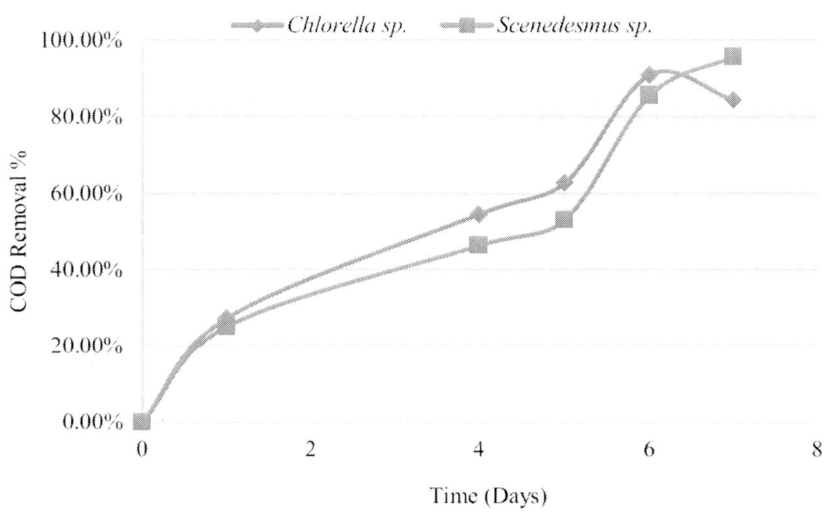

Figure. 7 Percentage removal of COD vs time

Table 4 Percentage removal (maximum) of pollutants by the 7th day of incubation

Parameters		***Chlorella* sp.**		***Scenedesmus* sp.**		**NEQs limit**
Name	**Units**	**Final**	**% Removal**	**Final**	**% Removal**	
PO_4^{-3}–P	mg/L	5.3 ± 0.56	70.5	3.5 ± 0.56	80.5	NA
NO_3-N	mg/L	0.165 ± 0.005	98.2	0.024 ± 0.02	99.7	NA
SO_4^{-2}	mg/L	BDL[a]	100	BDL	100	1
COD	mg/L	28 ± 5.6	84.86	8 ± 2.5	95	150
Turbidity	NTU	4 ± 0.5	93.73	3.5 ± 0.8	94.93	NA

[a]*BDL* below detection level

4. CONCLUSION

Discharged municipal, industrial and agricultural wastewaters contain many organic and inorganic contaminants which stimulate adverse impacts on human health and aquatic life. Treatment of such wastewaters by microalgae is an attractive green option as it can sequester carbon dioxide, reduce nutrient load and generate biomass as feedstock for biofuel production. In this study, *Chlorella* sp. and *Scenedesmus* sp. were isolated and grown on raw institutional wastewater for simultaneous nutrient removal and biomass/lipid production. Results indicated significant improvement in water quality within a detention period of 7 days. This treated wastewater can be reused or discharged into the environment, as excessive nutrients and COD were reduced to levels within the limits of the National Environmental Quality Standards (NEQs). Both strains showed amazing growth trends and significant biomass yield. The harvested biomass was checked for the biofuel potential via FTIR analysis, whose results regarding lipid content were validated by performing lipid extraction and quantification. The outcome demonstrated the potential of these local strains as suitable candidates for biodiesel production to combat rising fuel issues, especially *Chlorella* sp. if favorable environmental conditions are provided. Furthermore, the algal residue left after oil extraction can be utilized for fertilizer or other biofuel production, thereby presenting possibilities for an economical treatment system. With that being said, more research should be conducted to determine the extent of scale-up for this technology and an energy/economics analysis to understand the level of sustainability of the system.

ACKNOWLEDGEMENTS

The authors are grateful to the Biofuel laboratory, Centre for Advanced Studies in Energy (CAS-EN) NUST, for providing funds for this study. The authors are also thankful to the Institute of Environmental Sciences and Engineering (IESE), NUST, Islamabad, Pakistan, for providing laboratory facilities. Without their support, this research could not have progressed well.

REFERENCES

1. Arar EJ (1997) In Vitro determination of chlorophylls a, b, c 1 + c 2 and pheopigments in marine and freshwater algae by visible spectrophotometry adapted by National Exposure Research Laboratory Office of Research and Development U. S. Environmental Protection Agen," no. September, pp 1–26
2. Bligh EG, Dyer WJ (1959) Bligh-Dyer method. Can J Biochem Physiol 37(8):911–917
3. Buljan FSJ, Kral I, Clonfero G (2011) Introduction to treatment of tannery effluents. UNIDO, Vienna
4. Castellanos CS (2013) Batch and continuous studies of *Chlorella vulgaris* in photo-bioreactors. In: Electronic thesis and dissertation repository, The University of Western Ontario, paper 1113
5. CSIRO (2013) Microalgal isolation techniques. [Online]. http://www.marine.csiro.au/. Accessed 15 Sep 2015
6. de Morais M, Costa J (2007) Isolation and selection of microalgae from coal fired thermoelectric power plant for biofixation of carbon dioxide. Energy Convers Manag 48(7):2169–2173
7. Dean AP, Sigee DC, Estrada B, Pittman JK (2010) Using FTIR spectroscopy for rapid determination of lipid accumulation in response to nitrogen limitation in freshwater microalgae. Bioresour Technol 101(12):4499–4507
8. Eaton AD (2005) Standard methods for the examination of water and wastewater, 21st edn. American Public Health Association, Washington, D.C
9. Elvira-antonio N, Ruí A (2013) Effect of nitrogen content and CO_2 consumption rate by adding sodium carbonate in the lipid content of *Chlorella vulgaris* and *Neochloris oleoabundans*. Int J Environ Prot 3(2):13–19
10. Gogate PR, Pandit AB (2004) A review of imperative technologies for wastewater treatment II: hybrid methods. Adv Environ Res 8(3):553–597
11. Iram S, Kanwal S, Ahmad I, Tabassam T, Suthar V, Mahmood-Ul-Hassan M (2013) Assessment of physicochemical parameters of wastewater samples. Environ Monit Assess 185(3):2503–2515
12. Irfan M (2009) Wastewater treatment in textile, tanneries and electroplating industries especially by activated sludge method—a technical report. J Pakistan Inst Chem Eng XXXVII:4–5
13. Irving TE, Allen DG (2011) Species and material considerations in the formation and development of microalgal biofilms. Appl Microbiol Biotechnol 92(2):283–294

14. Kamalanathan M, Pierangelini M, Shearman LA, Gleadow R, Beardall J (2016) Impacts of nitrogen and phosphorus starvation on the physiology of *Chlamydomonas reinhardtii*. J Appl Phycol 28(3):1509–1520
15. Leslie F, Summerell A (2006) Media—recipes and preparation. Fusarium Laboratory Manual, Hoboken, New Jersey
16. Mostafa SSM, Shalaby EA, Mahmoud GI (2012) Cultivating microalgae in domestic wastewater for biodiesel production. Notulae Sci Biologicae 4(1):56–65
17. Ponnuswamy I, Madhavan S, Shabudeen S (2013) Isolation and characterization of green microalgae for carbon sequestration, waste water treatment and bio-fuel production. Int J Bio-Sci Bio-Technol 5(2):17–26
18. Prabakaran P, Ravindran A (2012) Scenedesmus as a potential source of biodiesel among selected microalgae. Curr Sci 102(4):616–619
19. Prabakaran P, Ravindran D (2013) Selection of microalgae for accumulation of lipid production. Caribb J Sci Technol 1:131–137
20. Prieto JL, Vega JM (1992) Sulfate uptake in *Chlamydomonas reinhardtii*. Phyton 32:1989–1992
21. Sankar V, Daniel DK, Krastanov A (2011) Carbon dioxide fixation by *Chlorella minutissima* batch cultures in a stirred tank bioreactor. Biotechnol Biotechnol Equip 25(3):2468–2476
22. Sharma AK, Sahoo PK, Singhal S, Patel A (2016) Impact of various media and organic carbon sources on biofuel production potential from *Chlorella* spp. 3 Biotech 6(2):116
23. Shams K (2000) Review on NEQS. In: The gazette of Pakistan, vol 57, no. M, pp 189–194
24. Thompson GA (1996) Lipids and membrane function in green algae. Biochem Biophys 1306:17–45
25. Van Vuuren SJ (2006) Easy identification of the most common freshwater algae: a guide for the identification of microscopic algae in South African freshwaters. Resource quality services (RQS), Pretoria, South Africa
26. Venkatesan S, Pugazhendy K, Sangeetha D, Vasantharaja C, Prabakaran S, Meenambal M (2012) Fourier transform infrared (FT-IR) spectoroscopic analysis of spirulina. Int J Pharm Biol Arch 3(4):969–972
27. Wang L, Min M, Li Y, Chen P, Chen Y, Liu Y, Wang Y, Ruan R (2010) Cultivation of green algae *Chlorella* sp. in different wastewaters from municipal wastewater treatment plant. Appl Biochem Biotechnol 162(4):1174–1186
28. Woertz IC (2007) Lipid productivity of algae grown on dairy wastewater as a possible feedstock for biodiesel. California Polytechnic University, San Luis Obispo
29. Young AJ, Harker M, Tsavalos AJ (1996) Reference. J Ferment Bioeng 82(2):113–118
30. Yun YS, Lee SB, Park JM, Lee CI, Yang JW (1997) Carbon dioxide fixation by algal cultivation using wastewater nutrients. J Chem Technol Biotechnol 69(4):451–455

CHAPTER 9

Spreading of Antibiotic Resistance with Wastewater

Sadik Dincer and Esra Sunduz Yigittekin

Biology Department, Science and Letter Faculty, Cukurova University, Adana, Turkey

ABSTRACT

The recent statistics show that the world's population is rapidly increasing. This increase negatively affects the water resources and it increases the water demand progressively. Along with the increase in the world's population, the insensible use of water resources, pollution, and drought lead to the increasing reduction of water resources. Due to these factors, all countries, primarily developed countries, have started looking for new water resources. This search has been extended to extraterrestrial water. However, the existing technology and opportunities direct countries toward the purification of wastewater rather than searching for new water resources. For the reasons outlined above, purification and recycling of wastewater become important. In addition to the natural resistance of microorganisms against antibiotics, a resistance also arises because of the unconscious and overuse of antibiotics. This resistance spreads through wastewater progressively. Antibiotic resistance shows an increase according to the scientific data. In order to prevent the resistance, it is of capital importance to treat the wastewater in which the domestic pollution burden is high. In this study, the role of domestic wastewater in the occurrence and spread of antibiotic resistance will be revealed.

Keywords: antibiotics, spreading antibiotic resistance, water, wastewater, domestic wastewater

1. INTRODUCTION

Water contains millions of microscopic living beings within itself. The plenty amount of water is accessible on our planet for living beings to maintain their vital activities.

Along with 14 billion m^3 of water, 97.5% of it is salty water, 2.6% of it is freshwater, and 0.8% of the total amount of water is present as freshwater in the state of constant vaporization, precipitation, and flow. Water scarcity is indicated as one of the main problems of the twenty-first century in the whole world, and for this reason, lives of many people depend on the right usage of water. People need water primarily for civic, industrial, and agricultural areas. However, water is regarded as a limited source. For the fact that water resources become insufficient and decrease in quality creates serious concerns. The population increase, urbanization, agricultural practices, and industrialization increase the water demand. Wastewater treatment is built for the purpose of reducing the pollution by removing pathogens, nutrients, and biodegradable substances, and protecting public health and the environment. Furthermore, with the increase in water demand, the recycling of wastewater has been brought into question [1].

Wastewaters are divided into two groups such as domestic and industrial. Domestic wastewater can originate from house, workplace, and hospital because of its content. The complete treatment of this type of wastewater is impossible even if it goes through many stages. This situation causes many problems. The emergence and spread of antibiotic-resistant bacteria are the leading reasons for this problem and they make humans and animals sick.

According to American Centers for Disease Control and Prevention (CDC), antibiotic resistance constitutes one of the most important health problems of a country. In America, it is estimated that 2 million people become sick and 700,000 people die worldwide because of resistant bacteria. In a report of 2013, CDC indicated that the usage of antibiotics in the production of food animals causes the emergence of resistant *Campylobacter* that is contagious to humans. Resistance genes can be transferred between zoonotic bacteria types, among the bacteria species, through food chain and contact with feces of ill animals and contaminated environment [2].

Antimicrobial resistance causes many problems in humans and animals in the case of the spread through wastewater, spread wastewater treatment output's being low, and the usage of these waters in agricultural practices and irrigation fields, the emergence of many antibiotic-resistant microorganisms.

2. SIGNIFICANCE OF WATER

Water is an essential substance which is necessary for vital activities such as nourishment, circulation, respiration, excretion, and reproduction to occur in every period of human life. At the same time, water itself is a habitat as one of the basic elements in nature while forming a habitat. The presence of water in a habitat and its quality are extremely important for life [3].

As the most important one of all natural sources for all living beings, water is a habitat and it contains millions of microscopic living beings. It constitutes approximately three-quarters of the Earth [4].

Water is crucial for the life of living beings as the most common natural resource on the Earth. Seventy-five percent of the Earth's surface, seventy percent of the human body, and seventy-eight percent of the blood consist of water [5].

Ninety-seven percent of the water body on the Earth consists of oceans and seas, two percent of lakes, rivers, and underground waters, and one percent of glaciers and snows. Water has been used during the development of civilizations for many purposes such as personal hygiene, agricultural irrigation, industrial production, and electric power production [6].

Water submerges more than 70% of the Earth as fresh and salty water and these environments are defined as aquatic environments [7]. Salty water constitutes more than 96% of the water on the Earth. More than 68% of the present freshwater is found in ice and glaciers. In this way, it is considered that water stays in stock. Thirty percent of freshwater consists of underground water. Two-thirds of underground waters are located deeper than 800 m. Surface freshwater sources such as rivers and lakes constitute 93,100 km^3 (22,300 cubic miles), which is 1/700 of 1% of all water on the Earth [8].

The minerals, salts, and sulfates contained in the water are very important along with its other characteristics. The presence of these substances in a certain amount in water is essential for life while their presence in small or greater quantities continually affects life in a negative way. At the same time, water is a habitat. The pollution of this environment creates danger for life [9, 10].

An increase in industrialization, urbanization, and population that started at the beginning of the twentieth century, and an increase in the use of natural resources have caused the emergence of the problems called the environmental pollution jeopardizing human life. An increase in the variety and amount of solid and liquid wastes that are disposed into the environment causes air and water pollution [6].

Water is a component which is significant for the life cycle of all living beings on the Earth. While three-quarters of the Earth is covered by water, two-thirds of the human body is covered by water. This rate significantly affects all living beings while being important for both the Earth and human beings. Water has many important roles from systems in the living organism to cellular functions. Even a small decrease in the amount of water can endanger life.

Since the existence of the Earth, all civilizations have settled in the places around or close to water. This shows us that water is a functional substance which is completely life-oriented. The decrease in the amount of water arises from both the environmental pollution and unconscious consumption. Because of this decrease, countries are in search of new sources and wars break out.

3. WATER CONSUMPTION

Population increase on the Earth leads to the decrease of water and pollution of clean and potable water. This will cause water scarcity in the future. Rivers and lakes constitute most of the water that people use daily. The pollution of these water sources will create water shortage. The amount of water that meets our needs is 0.25% of all water sources on the Earth [11].

About 97.39% of 1384.10^9 km^3 water on the Earth is found in the oceans and seas. The remaining 2.01% consist of glaciers and 0.60% consists of underground water, lakes, and rivers. This situation shows that the available freshwater supply constitutes quite a small amount of all water sources on the Earth [12].

The world's population that is approximately 6 billion is able to use 54% of the renewable surface and underground water supplies. It is considered that this rate will increase to 70% with the population's increase as the conditions of use remain the same. At the same time, it is estimated that 90% of the present freshwater sources will be used with the increase of life standards and the increase of water usage per person. For other living beings, there will be 10% of the available water supply. It is indicated that there will not be enough water for environmental and ecological functions because of the population increase and unconscious use of water resources [13].

Water resources are also used in a sectorial aspect besides meeting daily needs. The usage of water is classified as agricultural, industrial, and domestic sectors [14].

Sixty-nine percent of the freshwater resources on the Earth are used for agriculture, twenty-three percent are used for industry, and eight percent are used for domestic purposes. These rates differ from continent to continent. For instance, while the rates of agricultural, industrial, and domestic usage of water in Africa are 88, 5, and 7%, respectively, these rates in Europe are 33, 54 and 13%, respectively [15].

Water consumption in the world has increased 10-fold since 1900. In the studies conducted, it is determined that water consumption will increase 17% in agriculture, 20% in industry, and 70% in domestic consumption in 2015. Moreover, it is told that 20% of 6 billion world population is deprived of clean water resources. The water amount per person decreased to 7300 m^3 in 2000 while it was 16,800 m^3 in 1950 [16–18]

While it is estimated that world population will be 8 billion, it is considered that water consumption per person will decrease to 4800 m^3 in 2025. This decrease in consumption will arise from water resources shortage. Furthermore, the present available water resources will be polluted in 2025 and then water will not be provided [19]. It is estimated that the curve of the increase in water demand and the curve of the decrease of clean water resources will intersect in 2023.

Recent studies show that population growth will increase the consumption of water. However, this situation is inversely proportional to the number of water resources. Due to the decrease of clean and available water resources, the quantity of water per person has decreased. The reason for these situations is water scarcity which arises from the unconscious use and pollution.

4. WATER POLLUTION

Water pollution has a negative effect on public health and ecology because of the degradation of water quality and natural balance. Water pollutants contain surplus metal, some radioactive isotopes, nitrogen, phosphorus, sodium, and other beneficial and necessary elements along with especially some faecal originated pathogenic bacteria, parasites, and viruses which can be human or animal originated.

Mixing of any organic, inorganic, radioactive, or biologic substance that inhibits or disturbs the usage of water resources by impairing their quality into water is called water pollution [20, 21].

The reasons of water pollution are particularly domestic, industrial, agricultural, physical, chemical, radioactive, and microbial pollution.

Domestic wastewater contains organic and inorganic substances that are suspended, colloidal, and dissolved. Domestic wastewater consists of organic foods such as too much carbon, nitrogen, and phosphorus and highly concentrated microorganisms [22]. With the increase in urbanization, the flow of domestic waste into water through sewerage system also increased. In particular, detergents which are used in washing machines, oils poured out into lavabo, and the dispersion of wastes that should be accumulated in dustbins and recycled into the environment cause water pollution [23]. The characteristics of industrial wastewater differ from industry to industry [22].

Apart from domestic and industrial wastewaters that are discharged into water sources without being treated, the unconscious fertilization and unconscious usage of agricultural pesticides are also the reasons for pollution. These pollutants become crucial with negative effects on the water resources regarded as inadequate according to the world average, environmental, and public health and in terms of economy [6]. In the fields close to water, the incorrect ploughing mixes into water through the wind and causes pollution in water [3]. An increase in the usage of synthetic manure and pesticides in agriculture and industry and chemical substances that are used in the industry create a risk of water pollution [23].

Industrial establishments cause physical pollution. Power plants, steel, paper, car, plastic, and packing factories, which are big industrial establishments, throw environmentally hazardous solid and liquid substances. These substances are mostly toxic like arsenic, phenol, cyanide, chromium, and cadmium [24].

The chemical pollution of water began to cause critical health problems. It is estimated that in the future one of the most important water pollution problems will be the pollution caused by chemicals. Main metals causing chemical pollution are copper, zinc, mercury, nitrate, and phosphate [23].

Radioactive pollution in water can result from research agencies, hospitals, and some industrial fields. Radioactivity increases because of testing nuclear weapons. Therefore, rain water is getting dirty and, as a consequence, surface water is exposed to radioactive pollution [25].

Water might also be polluted by some pathogenic bacteria, parasites, and viruses that can originate from humans and animals [6]. Microorganisms that cause water pollution in terms of hygiene generally originate from diseases or human and animal excrements and urine being a porter [26]. An increase in bacteria population leads to bacterial pollution as a result of the decomposition of organic substances that are accumulated in the sea and inland waters or mixing of various pollutants apart from sewerage [25]. In stored waters, there is a lot of bacterial communities like the members of *Pseudomanas* sp., *Micrococcus* sp., *Achromobacter* sp., *Streptomyces* sp., and especially *Enterobacteriaceae.* The members of *Enterobacteriaceae* do not reproduce in clean water and their natural habitat is not potable water. Coliform bacteria are important in terms of human health and as an indicator of water pollution [24]. Their presence indicates that stool is mixed up with water through sewage directly or indirectly in one or more phases starting with the raw material to the transfer of water [27].

When it comes to water pollution, microbial pollution comes to mind first, even though there are many reasons of pollution. The most important of them is the pollution that arises from domestic and industrial wastes. Since domestic wastewater contains sewage waters and detergents, it also causes the indirect microbial and chemical pollution.

5. WASTEWATERS AND CLASSIFICATION

Wastewaters are formed as a result of the pollution of water used in households and industrial establishments [28].

For the waters that are disposed by being used in households or industry, "wastewater" definition is used. Wastewaters demonstrate biological, chemical, and physical pollution. While biological pollution consists of bacteria, fungi, parasites, and virus particles, and chemical pollution consists of toxic substances, decomposed organic substances, and phosphor, physical pollution consists of color, scent, foaming, temperature increase, and suspended matters. Heavy metals contain colorants that belong to the group of chemical pollutants and include industrial wastes and some pesticides [29].

5.1. Classification of wastewaters

Wastewaters are classified into two groups as domestic and industrial.

5.1.1. Domestic wastewaters

Wastewaters that originate from the dirty water from households and workplaces and do not include the industrial content of factories are called domestic wastewaters. Although their pollution rate is low they contain a high level of oily compounds, proteins, particles, chemical oxygen demand (COD), and detergents. For this reason, domestic wastewaters have a complex structure (**Table 1**) [30].

Domestic wastewaters are the waters that contain dirty looking and colorful soluble and insoluble matters from food wastes, kitchen lavabos, bathrooms, washing, and dishing machines and the matters that have organic and inorganic content and 99% of water [31].

Table 1. Physical, chemical, and biological components of domestic wastewater [32].

Physical properties	Chemical components			Biological components
	Organics	Inorganics	Gases	
Solid matters	Carbohydrates	pH	Methane	Living cells
Heat	Oil and grease	Nitrogen	Oxygen	Plants
Color	Pesticides	Phosphorus	Hydrogen	Single cells
Smell	Phenols	Alkalinity	Sulfur	Viruses
	Proteins	Chlorides		
	Surface active agent	Heavy metals		
		Sulfur		
		Toxic components		

Domestic wastewaters contain suspended, colloidal, and dissolved organic and inorganic substances. As well as this pollution arises from sewerages and detergents, it can also originate primarily from households and business enterprises. Moreover, domestic wastewaters contain pathogenic microorganisms such as bacteria, helminth, protozoa, and viruses. This situation increases the pollution rate of waters and indicates that water treatment is absolute. The indicator of the treatment's necessity is that some bacteria include R-plasmid. Since R-plasmid ensures antibiotic resistance to bacteria, untreated domestic wastewaters cause antibiotic-resistant bacteria to infect people and animals and create disease.

One of the most significant wastewaters that belong to domestic wastewaters is hospital-acquired wastewaters.

5.1.1.1. Hospital-acquired wastewaters

Hospital wastewaters contain micro and macro pollutants that come from various sources such as operating rooms, laboratories, investigation units, polyclinics, and drug use. The most important macro pollutants are bacteria and viruses while the most important micro pollutants are antibiotics, heavy metals (Hg, Pt, Gd, etc.), hormones and detergents/antiseptics. While microbiologic quality is determined for the usage of water, faecal pollution is identified by biologic and chemical indicators. In the content of biological indicators, there are total coliforms, faecal coliforms, faecal streptococcus, and *Clostridium perfringens*. Total coliforms are in the form of aerobic and facultative anaerobic, asporogenic and Gram-negative bacteria. Faecal coliforms, the marker of the pollution of water in which faecal coliforms and total coliform bacteria are found and which indicates the presence of pathogenic bacteria with human or animal excrements represent the presence of pathogenic bacteria and limited virus contamination. In hospital wastewaters, antibiotics such as ciprofloxacin, erythromycin, and sulfamethoxazole are found in high numbers in accredited adsorbable organic halide (AOX) and paracetamol. In municipal sewage, antibiotics such as ofloxacin and erythromycin are found in high numbers in AOX, paracetamol, and ibuprofen which is an analgesic [33].

In order for drugs to be stored longer and be easier to take, they must be quite durable and of high mobility quality in the liquid-phase while being produced. For this reason, active substances in drugs and biotransformation products lead to various factors by accumulating in the ecosystem. A lot of drugs such as antibacterial drugs, antibiotics, antifebrile, anodyne, synthetic steroids, cholesterol medicines, beta blockers and cytostatic drugs are the drugs detected in the ecosystem by studies performed [34].

Various drugs are used for various purposes during the treatment, protection, and development of human and animal diseases. These drugs cannot completely metabolize and they are removed from the body as they are or as a by-product in the form of ordure, urine, sweat, etc. [35].

In order for living beings to be treated, protected against microorganisms and infections and become resistant, many drugs should be taken. After the functions of these drugs in the body are over, they are removed through liver and kidneys. Medicine taken reaches the maximum level in the blood and when it starts to decrease the excretion also begins. While the excretion periods of drugs such as painkillers and antibiotics out of body are different, antibiotics are not removed for a long time. Drugs are removed from the body as urine, ordure, or metabolized product. In this way, they mix into wastewaters through the sewer system.

Drugs mix into wastewaters not only through excretion. With the disposal of unused drugs in households and hospitals, they also mix into wastewaters through the sewage.

The medicines found in wastewaters cannot be completely refined through the refinement. One of the biggest reasons for this situation is that hospital wastewaters directly mix with domestic waters without pretreatment. This affects primarily potable waters, underground waters, lakes, and rivers in a negative way.

Medicine remnants that mix into the potable water as a result of the inadequate refinement of domestic wastewaters negatively affect living beings in many ways. This effect arises especially from antibiotics.

Antibiotics that enter the body through water cause pathogenic microorganisms to become resistant.

Due to the negative outcomes on living beings, hospital wastewaters should be refined before they are transferred to domestic wastewaters.

5.1.2. Industrial wastewaters

Pollution in the environment that originates from unavailable or economically unvalued wastes in the industrial system is called industrial pollution. The accumulation of permanent and toxic organic substances in industrial wastewaters creates serious problems. The facts that these wastewaters are not discharged into the receiving environment, pollutants are not biodegraded, and they have a toxic influence upon living beings create many troubles [36].

Industrial wastewaters comprise of various resources such as refrigerant waters, process wastewaters, and domestic qualified wastewaters. Because of this content, the refinement of industrial wastewaters becomes crucial [11].

Since industrial wastewaters contain heavy metal content, they create the most crucial environmental problem of the present day. Wastewaters containing heavy metals are the waters that are generally acidic and have a low biochemical oxygen demand (BOD) value. Aquatic life is affected by mixing of wastewaters into the receiving environment. Because of this situation, expensive refinement systems are needed in order to use water resources as potable water sources. Heavy metals contained in wastewaters make the mud impossible to use for agricultural purposes by affecting the refinement efficiency of domestic wastewaters. For this reason, the discharge of industrial wastewaters with heavy metal content into the sewer system has an important role [37].

5.2. Wastewater treatment

The treatment of domestic wastewaters takes place in three stages as mechanic, biologic, and chemical.

Physical treatment covers the refinement of solid matters in wastewaters. This treatment stage comprises of four units as grid/sieve/grinder, sand catcher/oil slinger, preliminary settling, and flotation [38].

The biologic treatment contains the stage in which organic matters contained in wastewaters are refined. This treatment happens along with the decrease in the organic matter amount by using and decomposing of organic substances as a nutrition substance by microorganisms. Domestic wastewater generally decreases nutrition and organic substances such as nitrogen and phosphorus contained in it. Biologic treatment helps microorganisms such as fungi, algae, protozoa, and metazoans and organisms belonging to bacteria and archaea. The most used processes in biologic treatment are activated sludge processes, air-conditioned lagoons, trickling filters, revolving biodiscs, and stabilization pools. Basic operations of this treatment are nitrification, denitrification, dephosphorization, waste stabilization and eliminating organics which are measured especially as BOD_5 and COD in wastewater [39, 40, 38].

Micropollutants that are common in water resources cannot be effectively eliminated with the present treatment systems and environmental impacts in the receiving environment. In particular, antibiotics and pharmaceutics are released into the environment after their production and consumption so they create a threat in the receiving environment. Conventional treatment processes, primarily biologic treatment systems, remain insufficient in the elimination of antibiotics. In order to remove antibiotics that are resistant to biodegradation, advanced oxidation processes with a high oxidation potential should be used [41].

In wastewater treatment establishments, antibiotics are generally removed from the environment by biodegradation and sorption with activated sludge. Antibiotic-resistant bacteria spread in nature thanks to the removing methods of activated sludge containing antibiotic-resistant organisms such as agricultural practices or burying into the pit [42].

Antibiotics are used to help the growth of animals along with the treatment of human and animal diseases. Antibiotics that enter the body are removed without being metabolized at the rates reaching 90%. For this reason, the main source of antibiotic pollution in nature is the antibiotics in human and animal faeces. In recent studies, it is determined that antibiotics are found in animal faeces and domestic wastewater sewage sludge besides various compartments. By regarding physical and chemical properties, antibiotics can reach sediments, soil, and underground water. It is determined that conventional treatment methods remain insufficient in the removal of low concentrated antibiotics in water. The high concentration of antibiotics in the environment causes the degradation of ecological balance by creating a toxic effect on microorganisms, and their low concentration causes pathogenic and nonpathogenic bacteria to gain antibiotic resistance. For this reason, in order to remove antibiotic pollution, alternative treatment methods are necessary [43].

Gao et al. [44] determined 14 antibiotics in total in the wastewater, and 18 antibiotics in the activated sludge in their study. In the activated sludge, floroquinons, and ofloxacin were determined at the highest rate. Wastewater treatment establishments cannot remove antibiotics completely and the removal rate ranges from 34 to 72%. The amount of antibiotics in water is determined to be higher in winter months in comparison with spring and fall months. At the same time, antibiotic remnants have an adverse effect on very different organisms in nature (they encourage reproduction). Because of the low treatment effect, wastewater treatment establishments are the major source of antibiotics in aquatic environments.

According to Li et al. [45], the removal activity of target antibiotics from water changed between 32 and 78% through conventional treatment. With the advanced treatment methods, the removal rate of target antibiotics became 85–100% and pollution probability of antibiotics decreased. In addition to this, in the risk assessment, the effects of ofloxacin and erythromycin on microorganisms in water are investigated by refining it. The majority of antibiotics cannot be absorbed or metabolized in the body. Moreover, the large part pass into the sewage system through urine and faeces and it comprises a significant part of the antibiotic source in nature.

In the study conducted by Zhang et al. [46], the elimination mechanism of three β-lactam, two fluoroquinolones, and two macrolide antibiotics was investigated in the wastewater treatment establishment, which has four different treatment methods among six wastewater treatment establishments in China, Dalian. In this study, fluoroquinolones and macrolide antibiotics were determined as dominant antibiotics at the exit of wastewater treatment establishment and in coastal waters. It is revealed that β-lactams are removed through biodegradation, for fluoroquinolones pretreatment is more effective than biologic treatment, and macrolide concentration increases dramatically after biological treatment. The reason for this is that macrolides that are covered by faeces particles are revealed [46].

Xu et al. [47] examined antibiotics and their resistant genes in a water treatment establishment in Beijing, China and the situation of the river into which water was discharged in their study. A total of 13 antibiotic resistance genes (ARGs) were examined. Sul-arg was found at the highest rate among all antibiotic resistance genes (ARGs). ARG quantity in the wastewater treatment establishment is higher than in the river. According to the correlation analysis, there is a positive relationship between tetracyclines and tetargs in water. This correlation could not be performed between sul-args and sulfonamides. A negative relationship was observed between the concentration of quinolone genes and enrofloxacin. When ARG abundance of the waters that are treated in the treatment establishment is examined, treatment establishment causes resistant genes to increase. Results show that treatment establishments have a function of a warehouse for resistance genes. As a result, treated water needs advanced treatment before it is sent to the natural aquatic environment. In the study, three antibiotic groups were studied as tetracycline, sulfonamides, and quinones that are known for their permanence in the aquatic environment.

Tetracyclines are removed at the rate of 87.9% in sludge elimination establishments. In the elimination of tetracycline, biodegradation, and adsorption have an important role. In sludge elimination establishments, teta, tetm, tetw, and teto genes are the ones that are mostly found [47].

6. ANTIBIOTICS

Antibiotics are bioactive substances that kill or stunt the growth of the microorganism and have a high effect on synthetic or biological origin [48].

Antibiotics that are naturally obtained from plants and their extracts and are used for medical purposes have been brought into use as a result of Paul Ehrlich's studies in 1908. Paul Ehrlich revealed some chemical substances that are harmful to some bacteria and are less harmful to the host cells by investigating them [49].

Antibiotics are produced in nature by bacteria or fungi. The production of antibiotics by these living beings and their release into the environment result from their food competition with other species. Therefore, they produce antibiotics in their environment which extinguish other microorganisms or inhibit their growth. Antibiotics do not affect fungi, viruses, and protozoa since they are active only in bacterial infections. At the present time, antibiotics are produced synthetically. The microorganisms the production of which has provided the invention of antibiotics are fungi [50].

As antibiotics can be broad-spectrum affecting numerous bacteria, they can also be narrow-spectrum affecting limited bacteria. Furthermore, antibiotics with bactericide effect have an effect on bacteria by killing bacteria and antibiotics with bacteriostatic effect have an effect on bacteria by stopping their reproduction [51].

Although it has not been a long time since antibiotics have come into use, a rapid increase has been observed in their development. However, many problems have occurred during and after the consumption of these drugs. One of the main problems among them is bacterial resistance that develops against antibiotics.

There are a lot of reasons that bacteria develop resistance to antibiotics. The most important among them is that antibiotics are used without need and unconsciously.

Drugs that are used most frequently and in an excessive amount in the world are antibiotics. This usage also covers the unnecessary and unconscious use besides the proper use for treatment. The use for wrong purposes, misuse, and unnecessary use of antibiotics lead to bacterial resistance. For this reason, information about the usage of antibiotics should be given and the excessive and unnecessary usage should be prevented.

6.1. Classification of antibiotics

Antibiotics are separated into two groups according to their effect on microorganisms:

Classification according to antibiotic potencies

1) Bacteriostatic: This type of antibiotics prevents the development and reproduction of bacteria without killing the cells.
2) Bactericide: This type of antibiotics destroys bacterial cells by causing heavy damage.

6.2. Mechanisms of action of antibiotics

6.2.1. The ones that inhibit cell wall synthesis

Bacteria are prokaryote microorganisms. They do not have real nucleus but they have cell walls. Cell walls protect bacteria from the external environment and antimicrobials. Cell wall contains pores 1–2 nm in diameter that is convenient to the transition of substances found in the external environment and nonselective. In short, they are not semipermeable. The transition of antimicrobials depends on the structure of the cell wall and molecular size of the drug.

Human cells have no cell wall. Thus, antibiotics (Penicillins and Beta-lactams) in this group cannot spoil the adhesion of human cells. These antibiotics affect either by adhering to Penicillin-Binding Proteins (PBP) or by spoiling the synthesis of cell wall without adhering to PBP (**Table 2**) [51].

6.2.2. The ones that inhibit the protein synthesis

These chemotherapeutic drugs are generally broad-spectrum and have a bacteriostatic effect. Tetracyclines which belong to this antibiotic group prevent the adhesion of t-RNA to ribosomes. As human ribosomes (60S + 40S) and bacterial ribosomes (50S + 30S) are structurally different, these antibiotics that show an effect by adhering to ribosomes do not affect human ribosomes and protein synthesis (**Table 2**) [51].

Table 2. Classification of antibiotics [53].

The ones that inhibit the cell wall synthesis	Beta-Lactams:
	Penicillines
	Cephalosporins
	Monobactams (Aztreonam)
	Carbapenems (imipenem, Meropenem)
	Cycloserine
	Ristocetin
	Bacitracin
	Teicoplanin
	Vancomycin
The ones that inhibit cytoplasm membrane permeability	Polymyxins
	Gramicidin Nystatin
	Amphotericin B
	Candicein
	Ketoconazole and other antifungal imidazols
	Fluconazole and other antifungal trizols
	Hexachlorophene
	Cationic detergents
The ones that inhibit ribosome's protein synthesis	Tetracyclines
	Aminoglycosides
	Macrolides
	Amphenicols
	Lincosamides Fucidicasid
The ones that effect bacteria's genetic material break DNA and RNA	Fluoroquinolones
	Rifamycins
	Nalidixicasid
	Metronidazole
	Actinomycins Mitomycins
	Bleomycins
	Acyclovir
	Doxorubicin
	Daunorubicine
	Methotrexate
Bacterial antimetabolites	Sulfonamides
	Sulfones
	PAS
	Isoniazide (INH)
	Ethambutol
	Trimethoprim

6.2.3. The ones that inhibit nucleic acid synthesis

The most important antibiotics which belong to this group are rifampicin and quinones. Rifampicin inhibits the transcription (RNA inhibition dependent on DNA). Quinones inhibit the formation of supercolid (DNA gyrase inhibitors).

Topoisomerases which are used in human DNA and RNA synthesis and enzymes which are used in the nucleic acid synthesis of microorganisms are different. For this reason, these antibiotics do not have a toxic effect on human cells (**Table 2**) [51].

6.2.4. The ones that increase cytoplasmic membrane permeability

These antimicrobials create an effect by splitting the membrane substances in bacteria, inhibiting sterol synthesis in fungi or spoiling the permeability by binding sterols.

The cytoplasmic membrane of human cell bears a resemblance with cytoplasmic membranes of bacteria and fungi. Therefore, these antibiotics can have a toxic effect on human cells when they are used in a systemic way (**Table 2**) [51].

6.2.5. The ones with antimetabolic activity

Antibiotics in this group are generally bacteriostatic. The ones that are broadly known are the drugs such as sulfonamides, sulfons, para-amino salicylic acid (PAS), ethambutols, and isoniazid. Sulfonamides and Sulfons stop the function of PAS and para-amino benzoic acid (**Table 2**) [51].

6.3. Basic antibiotic groups

6.3.1. Beta-lactams

Antibiotics containing beta-lactam circle which is found in the nucleus and is responsible for the antibacterial effect of molecules are called beta-lactam antibiotics. The beta-lactam circle is a saturated circle, which comprises one nitrogen and three carbons. Antibiotics in this group have bactericide effect by influencing the cell wall which consists of the murine of bacteria. Penicillin and Ampicillin are the most known antibiotics in the beta-lactam group [52].

6.3.2. Vancomycin

They have an effect on multiresistant bacteria [54]. These antibiotics inhibit cell wall synthesis by stopping RNA synthesis in bacteria, break the continuity of the peptidoglycan chain and spoil the cytoplasmic membrane structure. Vancomycin which has a narrow antibacterial spectrum affects Gram (+) cokes and *Clostridiums* [51].

6.3.3. Tetracycline

These antibiotics inhibit protein synthesis by adhering to 30S subunit of the microorganism ribosome. Tetracycline which affects both Gram (+) and Gram (-) bacteria is broad-spectrum and has bacteriostatic effect [55, 56]. Tetracycline affects numerous and various bacteria types. It is also effective against *Rickettsia* sp., *Chlamydia* sp., *Spirochaete* sp., *Mycoplasma* sp., *Leptospira* sp., and some protozoa [51].

6.3.4. Aminoglycosides

Aminoglycosides inhibit protein synthesis in ribosomes by adhering to 30S subunit of bacterial ribosomes. Moreover, they cause the misreading of genetic code that m-RNA has. These antibiotics are narrow-spectrum and have bactericide effect. They are effective only in aerobe bacteria as they are dependent on oxygen in the membrane cell [51].

6.3.5. Macrolides

These antibiotics inhibit protein synthesis that is dependent on RNA in bacteria. They provide this effect by preventing the continuity of the peptide chain and adhesion of t-RNA by binding 70S ribosome to 50S subunit. Bacteriostatic macrolides have an intense effect against Gram (+) cokes and bacillus [57].

6.3.6. Chloramphenicol

Chloramphenicol is the first broad-spectrum antibiotic. These antibiotics inhibit peptidyl transferase enzyme by binding bacterial ribosomes to 50S subunit and thus they inhibit protein synthesis in a reversible way. They are sensitive to coke, aerobe, anaerobe Gram (+) bacilli, and most of the Gram () bacteria. Furthermore, these antibiotics inhibit protein synthesis of bacteria in the tissue by transferring into the tissue [55, 58].

6.3.7. Quinolones

Quinolones affect bacteria by inhibiting DNA gyrase. This effect prevents DNA replication and creates bactericide impact. Moreover, the bacteria that are exposed to this antibiotic do not divide and die from stretching abnormally. They are effective in most Gram (-) bacteria and Gram (-) bacteria [59].

6.3.8. Trimethoprim-sulfamethoxazole

It is also known as cotrimoxazole. When Sulfamethoxazole (STX) is a sulfonamide, Trime-thoprim (TMP) is a diaminopiriminid which inhibits bacterial dihydrofolate reductase competitively. They affect many Gram (+) and Gram (-) bacteria by causing unnoticeable synergistic bactericide effect when both drugs are used separately.

As a rule, the maximum synergistic activity of both antibacterial drugs, Trimethoprim (TMP), and Sulfamethoxazole (STX), occurs in bacteria types which are sensitive to both drugs. In the determination of the activity, sensitivity to TMP is more important [60].

6.4. Antibiotic resilience

Antibiotic resilience is simply the ability to resist against any antibiotic which spoils the reproduction function of a microorganism or causes its death. Resistance concerns the microorganism, patient, antibiotic, and environment or all of them. Resistance has no connection with virulence [61].

1) Antibiotic resistance spreads in three ways in bacteria:
2) Transfer of bacteria between people
3) Transfer of resistant genes between bacteria (generally through plasmids)
4) Transfer of resistant genes between genetic elements in bacteria [60]

The resistance that microorganisms show against antibiotics is classified in two groups as natural (phenotypic) and acquired (genotypic).

6.4.1. Natural resistance

Natural resistance is the situation that occurs when the microorganism cannot carry the structure affected by the drug as its quality or it cannot reach the target due to the structure of the drug. This resistance is not hereditary besides it is the key feature of bacteria and it is not related to the use of drugs.

For instance, microorganisms such as L-forms of bacteria and *Mycoplasma* that have no membrane have a natural resistance to antibiotics such as penicillin which inhibit the cell wall synthesis. Another example is that vancomycin cannot affect Gram (-) bacteria due to the fact that it cannot pass from adventitia [62].

6.4.2. Acquired resistance

Depending on the change in bacteria's genetic characteristic, it is the resistance which occurs as a result of taking DNA series that have resistance gene from another bacterium through transformation, transduction, or conjugation as it can be through mutations in a plasmid, chromosome, or transposon DNA. Furthermore, these bacteria can gain resistance against antibiotics to which they have been sensitive before [63]. Genetic originated resistance is examined in two groups as chromosomal and extrachromosomal.

6.4.2.1. Chromosomal resistance

It occurs as a result of mutations which happen spontaneously in the bacterial chromosome. Spontaneous mutations arise from some physical or chemical factors. Consequently, structural changes occur in the bacterial cell. In this situation, changes can happen in the drug's target in the cell or permeability of the cell to the drug can decrease [62].

6.4.2.2. Extrachromosomal resistance

Bacteria have extrachromosomal resistance plasmids that are called extrachromosomal elements, transposons that are active elements found on the chromosomes and bring chromosomes new antibiotic resilience, integrons, and antibiotics.

6.4.2.2.1.+ plasmids

The structures that can be inside bacteria or outside the chromosomes in the DNA structure, bring some qualities to these bacteria and keep these qualities under control genetically are called plasmids.

Plasmids can have virulence factors besides resistance genes against antimicrobics and heavy metals. Plasmids which have resistance genes are called R-plasmids. R-plasmids transfer the resistant gene package by passing into other bacteria through transformation, transduction, and conjugation. Thus, they provide the spread of resistance [64].

6.4.2.2.2. Transposons

Transposons are the structures which can settle in different places in the bacterial chromosome or can be transferred from chromosome to plasmid, from plasmid to plasmid, from plasmid to DNA or bacteriophage. These structures are DNA series found over the replicon like a chromosome, plasmid or bacteriophage as they cannot replicate by themselves. They have an active role in the spread of the multiple drug resistant isolates of transposons by revealing them in a short time [62, 65].

6.4.2.2.3. Integrons

Integrons are active DNA elements which have the ability to capture genes, which codify antibiotic-resistant genes in enteric bacteria, with specific recombination. These genes that are captured by integrons are called gene tapes. Gene tapes are active genetic elements which comprise of only one gene and recombination zone which is free, little-alkali and called the 59-base element. As well as these gene tapes may not present in integrons at all, there can be 100 of them [66].

6.4.3. Cross-resistance

It is the situation when some microorganisms are resistant both to some drugs and at the same time to other drugs that have a similar mechanism. This resilience can be seen between structurally similar drugs like erythromycin and kanamycin as it can be seen between completely different drugs like erythromycin and lincomycin [62].

7. ANTIBIOTIC RESISTANCE IN AQUACULTURE AND AGRICULTURE

Antibiotic concentrations below curative doses cause antibiotic resistance in many patient groups especially in critically ill patients [67].

The emergence of antibiotic-resistant bacteria is seen as an important health problem. For, thousands of patients die because of resistant bacteria. All efforts are concentrated on the decrease of existing antibiotic-resistant bacteria and antibiotic usage [2].

Rapidly developing antibiotic-resistant bacteria force public health services and health centers. American Centers for Disease Control and Prevention and Food and Agricultural Organization stated that antibiotic resilience has seriousness over the world. According to the predictions, 700,000 people die because of antibiotic resistance in a year. With the changes in temperature and rain regime, climate-sensitive bacteria and diseases will increase and spread to new regions, consequently, the situation will worsen [68].

Determination frequency and antibiotic concentrations are generally higher in January and May. The reason for this is that low-flow and low-temperature conditions cause antibiotics to be trapped by sediments. Antibiotic quantities vary per region. The highest quantities are found in estuaries and places where sewage is disposed. Antibiotic usage is more than 100,000–200,000 tonnes over the world and more than 25,000 tonnes in China. 80–90% of these antibiotics are released into nature through human urine and faeces. Pharmacologically active compounds in animal manure are used as a fertilizer in agriculture, and in conclusion, these compounds are accumulated in soil or mix into surface or underground waters [69].

Antibiotics are used as an environmental pollutant, in the treatment of diseases in a broad sense, in the protection and treatment of diseases in veterinary, and as growth promotive in aquaculture and agriculture [42].

Veterinary drugs are used for the protection and treatment of animal diseases and are one of the important components of environmental pollution as a result of intensive agricultural and aquaculture actions. Veterinary drugs are among the potentials of chemical pollutants and they have a biological effect in low concentrations like other drugs. While the annual usage of veterinary antibiotics in the United States reaches 11,000 tonnes, China follows it by 6000 tonnes. These quantities contain not only drugs with therapeutic purposes but also antibiotics which are used to promote production. In Europe, France leads these rates with 1064 tonnes, Holland follows it with 514 tonnes, and England with 403 tonnes. The most used antibiotics are tetracyclines, sulfonamides, β-lactams, and macrolides. The presence of veterinary antibiotics in nature causes the emergence of antibiotic-resistant bacteria and nontarget microorganisms are affected by drinking potable water that contains antibiotic remnants or by consumption of animal or herbal foods that contain antibiotics. Mixing of veterinary drugs into nature may cause the development of single, multiple, and cross-resistance in pathogens, commensals, and nonpathogens. Most of the veterinary drugs are feebly absorbed in the animal intestine. The remaining large quantity is removed with faeces. A small combination of these drugs removed undergoes a change, conjugates with polar molecules or remains the same. Consequently, these drugs can be detected in natural environments such as animal manure, soil, surface, and underground water resources. The major source of veterinary drugs in nature is biological remnants and the usage of dirty animal faeces in fertilization [70].

The usage of wastewaters for agricultural and other purposes by treating them provides many advantages such as the formation of alternative water resources, prevention of the pollution of surface and underground waters, and reduction of fertilizer usage. However, along with its advantages, it also has negative effects on public health and the environment. In order to minimize these effects, risks that origin from pathogens and chemicals that emerge from the wastewater usage should be evaluated well [71].

Waters that are polluted in many ways are treated by many methods with the progress of technology. The usage of wastewaters as irrigation waters by putting them through pretreatment or delivering into the land is one of these methods. Causing soil pollution by water pollution occurs in this method. Wastewaters from various resources pollute the soil and they have various effects on soil pollution [72].

Domestic wastewaters can be used in forests, pastures, lawns by being pretreated. The removal of wastewaters by using them in the irrigation of lands in this way creates serious health problems. Moreover, bacteria and pollutants in wastewaters are harmful to human health by being absorbed by the soil and reaching underground waters when the buffering effect decreases [72].

In the sector of aquaculture, antibiotics are intensively used to treat fish and protect it from diseases. Antibiotics that are applied to fish cause fish pathogens and zoonotic fish bacteria to gain resistance to antibiotics. Zoonotic fish bacteria which develop antibacterial resistance create danger for people and cause infections that are hard to treat [73].

The misuse of antibiotics affects human health directly or indirectly and complicates the treatment of fish diseases. Its direct effect is that fish bacteria and zoonotic fish bacteria gain resistance. These strains which are resistant create refractory infections when they infect people. The indirect antibiotic resilience occurs with the transfer of resistance plasmids in bacteria to human pathogens. In this way, human pathogens that gain resistance create resistant infections in people. Also in the studies conducted, it is revealed that multiple antibiotic resistance genes are transferred from fish pathogens to human pathogens [73].

In August 2011, 20 antibiotics that were taken from 20 different samples' regions taken from sediment and aquatic organisms in Dalian coastline were examined. Tetracyclines are dominant antibiotics in sea water. Sulfonamides are dominant antibiotics in sediment and aquatic organisms. Industrial aquaculture is the most significant reason for the pollution of coasts in developed and developing countries because of the intensive antibiotic usage. Antibiotic usage in China comprises the quarter of antibiotic usage over the world [74].

The state of 37 antibiotics was examined on 6 aquaculture farms around Hailing Island. Sulfamethoxazole, salinomycin, and trimethoprim were detected at the highest rate in water; ox tetracycline was detected at the highest rate in shrimp larva pools, enrofloxacin was detected at the highest rate in feed samples, and erythromycin was detected at the highest rate in sediment [75].

Wastewater usage in agriculture and land irrigations can be described as wastewater recycling. This usage brings many problems even if it is very economical. The usage of wastewater in agricultural activities by pretreating it is not enough to eliminate these results. Especially, domestic wastewaters constitute an enormous danger because of their content. Sewage and hospital wastewaters are the reasons for this danger.

Antibiotics cause infections and antibiotic resistance in people besides the fact that they decrease the productivity in agriculture with the bacterial and parasite microorganisms they contain.

In aquaculture, which is a method used in fish farming, it is possible that bacterial and parasite infections occur. Therefore, antibiotics are used for the treatment and protection from infections. Antibiotic usage complicates treatment as well as it creates antibacterial resistance in fish and people.

8. SUGGESTION

Urbanization, an increase in industry and population, increases the water demand with each passing day. The most important need is water and nutrition's existence depends on water. For this reason, the amount of water, as well as its presence, is important for living beings.

The increase of water usage, its unconscious use, the involvement in pollutant activities, and not taking precautions against pollution have a negative influence on water amount. In this situation, people's awareness should be increased, and pollution pretending precautions should be taken. Besides these situations, water reutilization can be provided by treatment.

The reutilization of wastewater by treatment increases water amount and creates some sources for the use of living beings. This brings positive situations as well as many negative situations along with it.

Wastewater causes pathogenic factors in living beings because of its content as well as it causes antibiotic-resistant bacteria and the spread of pathogen microorganisms.

There are many ways for antibiotic resistance to occur and spread. The leading factors among them are the excessive and unconscious usage of antibiotics, the usage of broad-spectrum antibiotics, its accumulation in sewers through taking it to the body and urinating it, especially giving hospital sewer's accumulation to treatment facilities without its pretreatment.

The result of the insufficient treatment of wastewater treatment facilities is that the amount of antibiotics remains and an increase occurs as well. Due to the chemical structure of antibiotics, it may not come to light before entering the wastewater treatment. Antibiotics have the tendency to hold on to sediments due to their structure. In the stages of wastewater treatment, as a result of the decomposition of sediments, antibiotics come out. In this situation, the problems of not treating antibiotics arise. The usage of these waters for agricultural purposes also causes antibiotic resistance to spread.

Antibiotics are used for the purposes of treating diseases in humans and also for the same purposes in animals. In this situation, this can cause the emergence and spread of antibiotic-resistant bacteria.

To prevent the emergence and spread of antibiotic-resistant bacteria, first the awareness of people of antibiotic use should be raised. The usage of antibiotics in human and animal treatment should be reduced. Other waters that belong to the group of all sewage and wastewater, especially hospital sewage, should be pretreated before being discharged to wastewater treatments with biological treatment. Mechanic, chemical, and thermal treatment processes are included in pretreatment. Many pretreatment processes such as the process of oxidation, thermophilic pretreatment, sludge disintegration, ozonation, photocatalytic pretreatment, physicochemical pretreatment, and ultrasonic method should be used. Since wastewater treatment systems used fall short in some cases, new systems and pretreatment systems for antibiotic treatment should be developed.

REFERENCES

1. Polat A. Su Kaynaklarının Sürdürülebilirliği İçin Arıtılan Atıksuların Yeniden Kullanımı. Turkish Journal of Scientific Reviews. 2013; 6 (1): 58–62. (in Turkish)
2. Centner TJ. Recent government regulations in the United States seek to ensure the effectiveness of antibiotics by limiting their agricultural use. Environment International. 2016; 94: 1–7.
3. Akin M, Akin G. Importance of water, water potential in turkey, water basins and water pollution. Ankara University Faculty of Language, History and Geography Journal. 2007; 47, 2:105–118. 92 Biological Wastewater Treatment and Resource Recovery
4. Acikgoz S. Hydrogeochemistry of streams discharging to Sapanca Lake [thesis]. Kocaeli: Kocaeli University; 2008.
5. Mutluay H, Demirak A. Water Chemistry. 1st ed. Istanbul: Beta; 1996. 140 p.
6. Akturk S. Adana determination of microbial quality in water from Adana-Tufanbeyli road line [thesis]. Adana: Cukurova University; 2009.
7. Oner M. Microbial Ecology. 1st ed. Izmir: Ege University Press; 1987. 282 p.
8. Sunter AT. İcme ve Kullanma Sularinin Aritilmasi ve Dezenfeksiyonu. 6th National Sterilization Disinfection Congress; 1–5 April 2009; Samsun. 425–438 pp. (in Turkish)
9. Curtis, H, Barnes NS. Biology. 5th ed. New York: Worth Publishers; 1989. 987 p.
10. Murray RK, Granner DK, Mayes PA, Rodwell VW. Harpers's Biochemistry. 24th ed. Stamford, CT: Appleton & Lange Publishers; 1996. 868 p.
11. Kamisli S. Determination of heavy metal and antibiotic resistance to gram-negative bacteria isolated from waste water treatment plants [thesis]. Adana: Cukurova University; 2014.
12. Mengu GP, Akkuzu E. Global water crisis and water harvesting techniques. Journal of Adnan Menderes University Agricultural Faculty. 2008; 5(2):75–85.
13. United Nations World Water Assessment Programme. The World Water Development Report 1: Water for People, Water for Life. Paris: Berghahn Books; 2003. 36 p.
14. Muslu VA. Water pricing in the world and Turkey [specialist thesis]. Ankara: Ministry of Forest and Water Management; 2015.

15. UNESCO. Water use. [Internet]. 2008. Available from www.unesco.org/water/iyfw2/water_use.shtml.
16. Atalik A. Küresel isinmanin su kaynaklari ve tarim üzerine etkileri. Bilim ve Ütopya.2006; 139:18–21. (in Turkish)
17. Dagli H. İçmesuyu kalitesi ve insan sağlığına etkileri. Bizim İller, İller Bankası Aylık Yayın Organı. 2005; 3: 16–21. (in Turkish)
18. Haviland WA. Cultural Anthropology. 10th ed. Wadsworth Publishing Company; 2002. 576 p.
19. Anonymous. [Internet]. 2014. www.cevreonline.com/su/dunyada%20suyun% 20 dagilimi. html. Gormez K. Cevre Sorunlari. 3rd ed. Ankara: Nobel Publication Distribution; 2015. 188 p.
20. Ozkan A. Transboundary environmental damages in the Black Sea Basin and regional liability Regime. International Conference on Eurasian Economies; 17–18 September 2013; Russia. pp. 970–975.
21. Arceivala SJ, Balman AH, Balman V. Wastewater Treatment for Pollution Control. Place of Publication not Identified. New Delhi; Ankara: Tata McGraw-Hill. 2002. Spreading of Antibiotic Resistance with Wastewater http://dx.doi.org/10.5772/66188 93
22. Guler C, Cobanoglu Z. Su Kirliliği. 1st ed. Ankara: Çevre Sağlığı Temel Kaynak Dizisi; 1997. 47 p. (in Turkish)
23. Ozarslan A. Determinetion of fecal coliform levels and antibiotics resistance frequency of Adana drinking water [thesis]. Adana: Cukurova University; 2009.
24. Tanyolaç J. Limnology. Ankara: Hatipoğlu Publications; 1993. 235 p.
25. Demirekin H. Sensitivity analysis of environmental problems in Isparta city [thesis]. Isparta: Suleyman Demirel University; 2001.
26. Dincer S, Matyar F, Sonmez N. Seyhan nehrinin fekal kirlilik duzeyi ve fekal koliformlarin antibiyotik hassasiyetleri. 12. Biyoteknoloji Kongresi. 2001; Ayvalık, 252–255. (in Turkish)
27. Samsunlu A. Wastewater Treatment. 3rd ed. Istanbul: Birsen Publisher; 2011. 647 p.
28. San NO. Treatment of wastewater with heavy metal and reactive dye by *Rhodotorula sp.* [thesis]. Ankara: Ankara University; 2007.
29. Gunes E. Anaerobic treatment of domestic wastewaters in upflow anaerobic sludge bed (UASB) reactor at temperate conditions and chemical post-treatment applications [thesis]. Istanbul: Istanbul Technical University; 2008.
30. Dogan M, Saylak M. Su Kimyası. Kayseri: Erciyes University Publications; 2000. 120 p. (in Turkish)
31. Metcalf E. Wastewater Engineering, Treatment and Reuse. 4th ed. New York: McGraw-Hill Education; 2003. 1771 p.
32. Yasar A, Dogan EC, Arslan A. Macro and micro pollutants and treatment options in hospital wastewaters. Erciyes University, Journal of Institute of Science and Technology. 2016; *29*(2): 144–158.
33. Saygi S. Cevre ve insan sagligi yönünden ilaç atiklarinin onemi. Marmara Pharmaceutical Journal. 2012; 16: 82–90. DOI: 10.12991/201216406. (in Turkish)
34. Sonmez G, Isik M. Sulardaki Ilac Kalintilarinin Ileri Oksidasyon Yontemleri Ile Giderimi. Turkish Journal of Scientific Reviews. 2013; *6*(1):68–73. SSN: 1308-0040, E-ISSN: 2146-0132. (in Turkish)
35. Turker C. Treatability of segregated textile effluents containing auxiliary chemicals [thesis]. Istanbul: Istanbul Technical University Institute of Science; 2013.
36. Turkman A, Aslan S, Ege I. Lead removal from wastewaters by natural zeolites. Dokuz Eylul University Faculty of Engıneerıng Journal of Engıneerıng Scıence. 2001; 3:2:13–19.
37. Sinan RK. Estimation of primary treatment and biological treatment effluent parameters by artificial neural networks in domestic wastewater treatment plants [thesis]. Selcuk University; 2010. 94 Biological Wastewater Treatment and Resource Recovery
38. Chojnacka K. Using bisorption to enrich the biomoass of Choleralla vulgaris with microelements to be used as mineral feed supplement. World Journal of Microbiology and Biotechnology.2007; 23:1139–1147.

39. Ozturk I, Timur H, Koskan U. Principals of Wastewater Treatment: Municipal. Industrial Wastewater Treatment and Sludge Control. Ankara, Turkey: Ministry of Environment and Forestry; 2005.
40. Ozkal CB, Pagano SB. Evaluation of antibiotics and antibiotic resistant bacteria removal by photo-catalysis. Nigde University Journal of Engineering Sciences. 2016; 5(1): 1–18.
41. Li W, Shi Y, Gao L, Liu J, Cai Y. Occurrence, distribution and potential affecting factors of antibiotics in sewage sludge of wastewater treatment plants in China. Science of the Total Environment. 2013; 445: 306–313.
42. Yalap KS, Balcioglu IA. Effects of water components on the advanced oxidation of a veterinary antibiotic, oxytetracycline. ITU Journal of Water Pollution Control. 2008; 18: 2–3.
43. Gao L, Shi Y, Li W, Niu H, Liu J, Cai Y. (2012). Occurrence of antibiotics in eight sewage treatment plants in Beijing, China. Chemosphere. 2012; 86(6):665–671.
44. Li W, Shi Y, Gao L, Liu J, Cai Y. Occurrence and removal of antibiotics in a municipal wastewater reclamation plant in Beijing, China. Chemosphere. 2013; 92(4): 435–444.
45. Zhang H, Liu P, Feng Y, Yang F. Fate of antibiotics during wastewater treatment and antibiotic distribution in the effluent-receiving waters of the Yellow Sea, northern China. Marine Pollution Bulletin. 2013; 73(1): 282–290.
46. Xu J, Xu Y, Wang H, Guo C, Qiu H, He Y, Meng W. Occurrence of antibiotics and antibiotic resistance genes in a sewage treatment plant and its effluent-receiving river. Chemosphere. 2015; 119: 1379–1385.
47. Topal M, Senel GU, Topal EIA, Obek E. Antibiotics and usage areas. Journal of Institute of Science and Technology. 2015; 31(3):121–127.
48. Basturk S. Escherichia coli, Klebsiella pneumoniae, Pseudomonas aerugino save Acinetobacter baumannii Suslarında Çesitli Kinolon Grubu Antibiyotiklerin Duyarlılıklarının Arastırılması [thesis]. T.C. Sağlık Bakanlığı Haseki Eğitim Ve Arastırma Hastanesi Enfeksiyon Hastalıkları Ve Klinik Mikrobiyoloji Kliniği İstanbul; 2005. (in Turkish)
49. Hoel D, Williams DN. Antibiotics: past, present and future. Unearthing nature's Magic Bullets. 1997; 101(1):114–118.
50. Gurses F. 2015. Treatability of antibiotic formulation effluents by fenton-like and photo-fenton-like advanced oxidation processes [thesis]. Istanbul: Istanbul Technical University; 2015.
51. Oncul O. Antibiyotikler 1. İ.Ü. Cerrahpasa Tıp Fakültesi Sürekli Tıp Eğitimi Etkinlikleri, Akılcı Antibiyotik Kullanımı ve Eriskinde Toplumdan Edinilmis Enfeksiyonlar Sempozyumu. 2002; 31: 23–28. (in Turkish) Spreading of Antibiotic Resistance with Wastewater http://dx.doi.org/10.5772/66188 95
52. Akkan AG. Antibiyotiklerin sınıflandırılması. I.U. Cerrahpasa Faculty of Medicine Continuing Medical Education Activities Antibiotics Practice Symposium; 2–3 May 1997; Istanbul. pp. 52–62.(in Turkish)
53. Strohl WA, Rouse H, Bruce DF. Lippincott's Illustrated Reviews. Microbiology. Nobel nMedicine Bookstores. 2006; 516:44–47.
54. Alcamo EI. Fundamentals of Microbiology. 6th ed. Sudbury Massachusetts: Jones and Bartlett Publishers; 2001. 832 p.
55. Lefever Kee J, Hayes ER, McCuistion LE. Pharmacology: A Nursing Process Approach. 7th ed. Printed in The United States America: Elsevier Saundres; 2012. 983 p.
56. Allen N. Effects of macrolide antibiotics on ribosome function. In Macrolide Antibiotics. Birkhäuser Verlag, Basel. 2002; 261–280 pp.
57. Braibant M, Gilot P, Content J. The ATP binding cassette (ABC) transport systems of Mycobacterium tuberculosis. FEMS Microbiology. 2000; 24: 449–467
58. Algun U, Arisoy A, Gunduz T, Ozbakkaloglu B. The resistance of Pseudomonas aeruginosa strains to fluoroquınolone group of antibiotics. Indian Journal of Medical Microbiology. 2004; 22(2): 112–114.
59. Rang HP, Dale MM, Ritter JM, Moore PK. 2006. Pharmacology. 5th ed. Edinburgh: Elsevier Churchill Livingstone; 2006. 797 p.

60. Durupinar B. Antibiyotiklere dirençte yeni eğilimler. Klimik Journal. 2001; 14(2): 47–56. (in Turkish)
61. Yuce A. Antimikrobik İlaçlara Direnç Kazanma Mekanizmaları. Klimik Journal. 2001; 14(2): 41–46. (in Turkish)
62. Tanir G, Gol N. Antibiyotik Direnci. Klimik Journal. 1999; 2:12. (in Turkish)
63. Gur D, Tutar I, Vardar Unlu G. (2001). İsepamisinin hastane izolatı Gram-negatif bakterilere karsı in vitro etkisi. Turkish Journal of Hospital Infections. 2001; *5*(1):19. (in Turkish)
64. Aygun G, et al. The antibiotic susceptibility patterns of *acinetobacter baumannii* strains isolated from nosocomial infections in intensive care unit. Journal of ANKEM. 2002; 16(1): 85–88.
65. Roy PH. Integrons: Novel mobile genetic elements mediating antibiotic resistance in *Enterobacteria* and *Pseudomonas*. APUA Newsletter. 1995; 13(3): 1–4.
66. Carlier M, Stove V, Wallis SC, De Waele JJ, Verstraete AG, Lipman J, Roberts JA. Assays for therapeutic drug monitoring of β-lactam antibiotics: A structured review. International Journal of Antimicrobial Agents. 2015; 46(4): 367–375.
67. Centner TJ. Efforts to slacken antibiotic resistance: Labeling meat products from animals raised without antibiotics in the United States. Science of the Total Environment. 2016; 563: 1088–1094. 96 Biological Wastewater Treatment and Resource Recovery
68. Shi H, Yang Y, Liu M, Yan C, Yue H, Zhou J. Occurrence and distribution of antibiotics in the surface sediments of the Yangtze Estuary and nearby coastal areas. Marine Pollution Bulletin. 2014; 83(1): 317–323.
69. Bártíková H, Podlipná R, Skálová L. Veterinary drugs in the environment and their toxicity to plants. Chemosphere. 2016; 144: 2290–2301.
70. Kukul YS, Calıskan Unal AD, Anac S. Wastewater reuse in agriculture and health risks. The Journal of Ege University Faculty of Agriculture. 2007; 44(3):101–116.
71. Tolunay D. Toprak kirlenmesi ve yanlıs arazi kullaniminin yarattigi sorunlar ile cozum onerileri. Journal of the Faculty of Forestry Istanbul University (JFFIU). 1992; 42(1–2):156–168. (in Turkish)
72. Avsever ML, Turk N, Tunaligil S. The increase of antibiotic resistance in aquaculture and its effects on human health. research articles. The Journal of Bornova Veterinary Control and Research.2010; 32(46): 19–23.
73. Na G, Fang X, Cai Y, Ge L, Zong H, Yuan X, Zhang Z. Occun, China. Marine Pollution Bulletin. 69(1): 233–237.
74. Chen H, Liu S, Xu XR, Liu SS, Zhou GJ, Sun KF, Ying GG. Antibiotics in typical marine aquaculture farms surrounding Hailing Island, South China: occurrence, bioaccumulation and human dietary exposure. Marine Pollution Bulletin. 2015; 90(1): 181–187. Spreading of Antibiotic Resistance with Wastewater http://dx.doi.org/10.5772/66188

CHAPTER 10

Biological and Chemical Wastewater Treatment Processes

Mohamed Samer[1]

[1] Cairo University, Faculty of Agriculture, Department of Agricultural Engineering, Egypt

ABSTRACT

This chapter elucidates the technologies of biological and chemical wastewater treatment processes. The presented biological wastewater treatment processes include: (1) bioremediation of wastewater that includes aerobic treatment (oxidation ponds, aeration lagoons, aerobic bioreactors, activated sludge, percolating or trickling filters, biological filters, rotating biological contactors, biological removal of nutrients) and anaerobic treatment (anaerobic bioreactors, anaerobic lagoons); (2) phytoremediation of wastewater that includes constructed wetlands, rhizofiltration, rhizodegradation, phytodegradation, phytoaccumulation, phytotransformation, and hyperaccumulators; and (3) mycoremediation of wastewater. The discussed chemical wastewater treatment processes include chemical precipitation (coagulation, flocculation), ion exchange, neutralization, adsorption, and disinfection (chlorination/dechlorination, ozone, UV light). Additionally, this chapter elucidates and illustrates the wastewater treatment plants in terms of plant sizing, plant layout, plant design, and plant location.

Keywords: Wastewater treatment, biological treatment, chemical treatment, bioremediation, phytoremediation, mycoremediation, vermifiltration, treatment plant

1. INTRODUCTION

The chapter concerns with wastewater treatment engineering, with focus on the biological and chemical treatment processes. It aims at providing a brief and obvious description of the treatment methods, designs, schematics, and specifications. The chapter also answers an important question on how the different processes are interrelated and the correct order of these processes in relation to each other. The main objective of this work was to summarize the work of the eminent scientists in this field in order to provide a clear but concise chapter that can be used as a quick reference for environmental engineers and researchers, and to be effectively implemented in higher education teaching undergraduate and graduate students, as well as extension and outreach.

2. CHAPTER DESCRIPTION AND CONTENTS OVERVIEW

The chapter describes the biological and chemical wastewater treatment processes that include:

a) Bioremediation of wastewater using oxidation ponds, aeration lagoons, anaerobic lagoons, aerobic and anaerobic bioreactors, activated sludge, percolating or trickling filters, biological filters, rotating biological contactors, and biological removal of nutrients;

b) Mycoremediation of wastewater using bioreactors;

c) Phytoremediation of wastewater that includes: constructed wetlands, rhizofiltration, rhizodegradation, phytodegradation, phyto-accumulation, phytotransformation, and hyperaccumulators;

d) Vermifiltration and vermicomposting;

e) Microbial fuel cells for electricity production from wastewater;

f) Chemical wastewater treatment processes that include: chemical precipitation, ion exchange, neutralization, adsorption and disinfection (chlorination/dechlorination, ozone, ultraviolet radiation);

g) Wastewater treatment plants. The chapter elucidates and illustrates the plant sizing, plant layout, plant design, and plant location.

3. OVERVIEW

3.1. Wastewater treatment techniques

Wastewater, or sewage, originates from human and home wastewaters, industrial wastes, animal wastes, rain runoff, and groundwater infiltration. Generally, wastewater is the flow of used water from a neighborhood. The wastewater consists of 99.9% water by weight, where the remaining 0.1% is suspended or dissolved material. This solid material is a mixture of excrements, detergents, food leftovers, grease, oils, salts, plastics, heavy metals, sands, and grits [1, 2].

Types of wastewaters include: municipal wastewater, industrial wastewaters, mixtures of industrial/domestic wastewaters, and agricultural wastewaters. Typical agricultural industries include: dairy processing industries, meat processing factories, juice and beverage industries, slaughterhouses, vegetable processing facilities, rendering plants, and drainage water of irrigation systems.

Subsequent to primary treatment of wastewater, i.e., physical treatment of wastewater, it still contains large amounts of dissolved and colloidal material that must be removed before discharge. The issue is how to transform the dissolved materials or particulate matters that are too little for sedimentation into larger particles to allow the separation processes to eliminate them. This can be accomplished by secondary treatment, i.e., biological treatment. The treatment of wastewater subsequent to the removal of suspended solids by microorganisms such as algae, fungi, or bacteria under aerobic or anaerobic conditions during which organic matter in wastewater is oxidized or incorporated into cells that can be eliminated by removal process or sedimentation is termed biological treatment.

Biological treatment is termed secondary treatment. Chemical treatment, or tertiary treatment, using chemical materials will react with a portion of the undesired chemicals and heavy metals, but a portion of the polluting material will remain unaffected. Additionally, the cost of chemical additives and the environmental problem of disposing large amounts of chemical sludge make this treatment process deficient [1]. Alternatively, the biological treatment must be implemented. This treatment process implements naturally occurring microorganisms to transform the dissolved organic matter into a dense biomass that can be separated from the treated wastewater by the sedimentation process. In fact, the microorganisms utilize the dissolved organic matter as food for themselves, where the generated sludge will be far less for chemical treatment. In practice, therefore, secondary treatment tends to be a biological process with chemical treatment implemented for the removal of toxic compounds.

3.2. Aims of wastewater treatment

The goals of treating the wastewaters are:

a) Transforming the materials available in the wastewater into secure end products that are able to be safely disposed off into domestic water devoid of any negative environmental effects;

b) Protecting public health;

c) Ensuring that wastewaters are efficiently handled on a trustworthy basis without annoyance or offense;

d) Recycling and recovering the valuable components available in wastewaters;

e) Affording feasible treatment processes and disposal techniques;

f) Complying with the legislations, acts and legal standards, and approval conditions of discharge and disposal.

3.3. Biological treatment processes

The secondary treatment can be defined as "treatment of wastewater by a process involving biological treatment with a secondary sedimentation". In other words, the secondary treatment is a biological process. The settled wastewater is introduced into a specially designed bioreactor where under aerobic or anaerobic conditions the organic matter is utilized by microorganisms such as bacteria (aerobically or anaerobically), algae, and fungi (aerobically).

The bioreactor affords appropriate bioenvironmental conditions for the microorganisms to reproduce and use the dissolved organic matter as energy for themselves. Provided that oxygen and food, in the form of settled wastewater, are supplied to the microorganisms, the biological oxidation process of dissolved organic matter will be maintained. The biological process is mostly carried out bacteria that form the basic trophic level (the level of an organism is the position it occupies in a food chain) of the food chain inside the bioreactor. The bioconversion of dissolved organic matter into thick bacterial biomass can fundamentally purify the wastewater. Subsequently, it is crucial to separate the microbial biomass from the treated wastewater though sedimentation. This secondary sedimentation is basically similar to primary sedimentation except that the sludge contains bacterial cells rather than fecal solids. The biological removal of organic matter from settled wastewater is conducted by microorganisms, mainly heterotrophic bacteria but also occasionally fungi. The microorganisms are able to decompose the organic matter through two different biological processes: biological oxidation and biosynthesis [1]. The biological oxidation forms some end-products, such as minerals, that remain in the solution and are discharged with the effluent (Eq. 1). The biosynthesis transforms the colloidal and dissolved organic matter into new cells that form in turn the dense biomass that can be then removed by sedimentation (Eq. 2). Figure 1 summarizes these processes. On the other hand, algal photosynthesis plays an important role in some cases (Figure 2).

Oxidation:

$$\underset{\text{(Organic matter)}}{COHNS} + O_2 + \text{Bacteria} \rightarrow CO_2 + \underset{\text{+Energy}}{NH_3} + \text{Other end products} \quad (1)$$

Biosynthesis:

$$\underset{\text{(Organic matter)}}{COHNS} + O_2 + \text{Bacteria} \rightarrow \underset{\text{(New cells)}}{C_5H_7NO_2} \quad (2)$$

3.3.1. Useful terms

The following terms are the most used in biological treatment processes [2]:

a) DO: Dissolved Oxygen (mg L^{-1})

b) BOD: Biochemical Oxygen Demand (mg L^{-1})

c) BOD_5: BOD (mg L^{-1}), incubation at 15°C for 5 days

d) COD: Chemical Oxygen Demand (mg L^{-1})

e) CBOD: Carbonaceous BOD (mg L^{-1})

f) NBOD: Nitrogenous (mg L^{-1})

g) SOD: Sediment Oxygen Demand (mg L^{-1})

h) TBOD: Total BOD (mg L^{-1})

3.4. Chemical treatment processes

In early wastewater treatment technologies, chemical treatment has preceded biological treatment. Recently, the biological treatment precedes chemical treatment in the treatment process. Chemical treatment is now considered as a tertiary treatment that can be more broadly defined as "treatment of wastewater by a process involving chemical treatment". The mostly implemented chemical treatment processes are: chemical precipitation, neutralization, adsorption, disinfection (chlorine, ozone, ultraviolet light), and ion exchange.

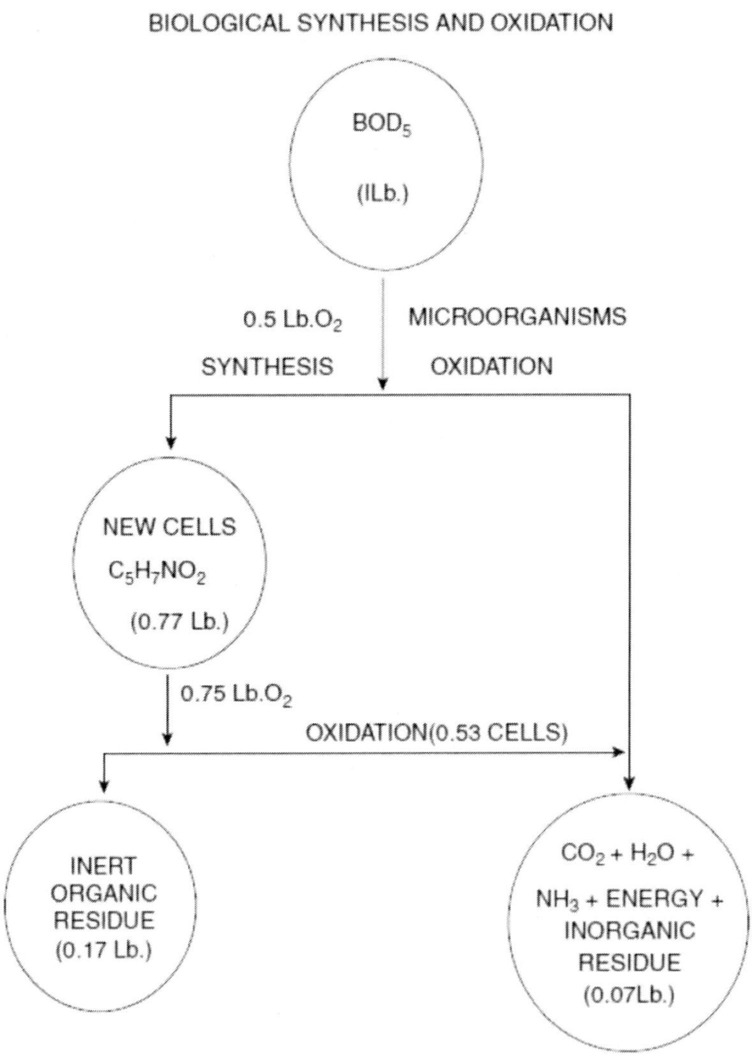

Figure 1. Biological synthesis and oxidation [3].

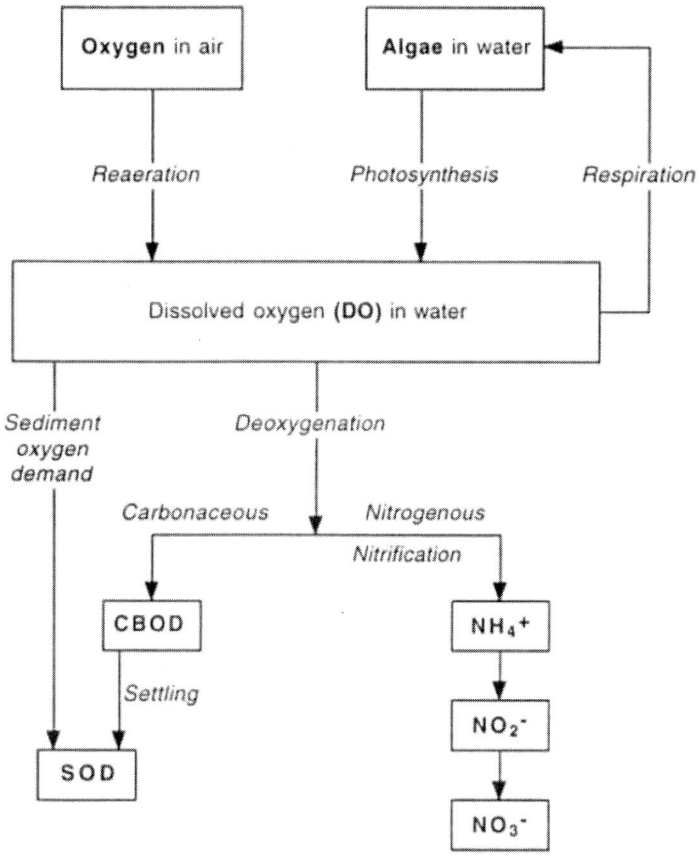

Figure 2. Photosynthesis and oxidation [2].

4. BIOLOGICAL TREATMENT OF WASTEWATER

4.1. Biological growth equation

The biological growth can be described according to the Monod equation:

$$\mu = (\lambda S) / (K_S + S)$$

Where, μ is the specific growth rate coefficient; λ is the maximum growth rate coefficient that occurs at 0.5 μ_{max}; S is the concentration of limiting nutrient, that is BOD and COD; and K_S is the Monod coefficient [3].

Generally, the bacterial growth can be explained by the following simplified figure:

$$\text{Organics+Bacteria+Nutrients+Oxygen}$$

$$\rightarrow \text{NewBacteria} + CO_2 + H_2O + \text{ResidualOrganics} + \text{Inorganics}$$

Several bioenvironmental factors affect the activity of bacteria and the rate of biochemical reactions. The most important factors are: temperature, pH, dissolved oxygen, nutrient concentration, and toxic materials. All these factors can be controlled within a biological treatment system and/or a bioreactor in order to ensure that the microbial growth is maintained under optimum bioenvironmental conditions. The majority of biological treatment systems operate in the mesophilic temperature range, where the optimal temperature ranges from 20°C to 40°C. Aeration tanks and percolating filters operate at the temperature of the wastewater that ranges from 12°C to 25°C; although in percolating filters, the air temperature and the ventilation rate may have a significant effect on heat loss. The higher temperatures increase the biological activity and metabolism, which result in increasing the substrate removal rate. However, the increased metabolism at the higher temperatures may lead to problems of oxygen limitations.

4.2. Bacterial kinetics

The bacterial kinetics can be shown in Figures 3 and 4. The microbial growth curve that shows bacterial density and specific growth rate at the different growth phases is shown in Figure 3. The microbial growth curves that compare the total biomass and the variable biomass are shown in Figure 4.

4.3. Principles of biological treatment

The principles of biological treatment of wastewater were stated by [3]. The following is a summary of the principles:

1) The biological systems are very sensitive for extreme variations in hydraulic loads. Diurnal variations of greater than 250% are problematic because they will create biomass loss in the clarifiers.

2) The growth rate of microorganisms is highly dependent on temperature. A 10°C reduction in wastewater temperature dramatically decreases the biological reaction rates to half.

3) BOD is efficiently treated in the range of 60 to 500 mg L^{-1}. Wastewaters in excess of 500 mg L^{-1}BODs have been treated successfully if sufficient dilution is applied in the treatment process, or if an anaerobic process was implemented as a pretreatment process.

4) The biological treatment is effective in removing up to 95% of the BOD. Large tanks are required in order to eliminate the entire BOD, which is not feasible.

5) The biological treatment systems are unable to handle "shock loads" efficiently. Equalization is necessary if the variation in strength of the wastewater is more than 150% or if that wastewater at its peak concentration is in excess of 1,000 mg L^{-1} BOD.

6) The carbon:nitrogen:phosphorus (C:N:P) ratio of wastewater is usually ideal. The C:N:P ratio of industrial wastewaters should range from 100:20:1 to 100:5:1 for a most advantageous biological process.

7) If the C:N:P ratio of the wastewater is strong in an element in comparison to the other elements, then poor treatment will result. This is especially true if the wastewater is very strong in carbon. The wastewater should also be neither very weak nor very strong in an element; although very weak is acceptable, it is difficult to treat.

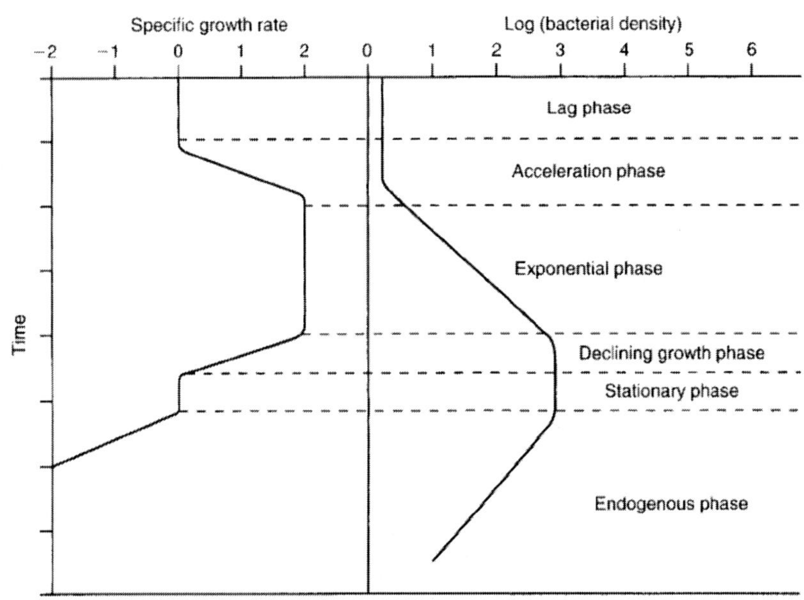

Figure 3. Microbial growth curve [1].

8) Oils and solids cannot be handled in a biological treatment system because they negatively affect the treatment process. These wastes should be pretreated to remove solids and oils.

9) Toxic and biological-resistant materials require special consideration and may require pretreatment before being introduced into a biological treatment system.

10) Although the capacity of the wastewater to utilize oxygen is unlimited, the capacity of any aeration system is limited in terms of oxygen transfer.

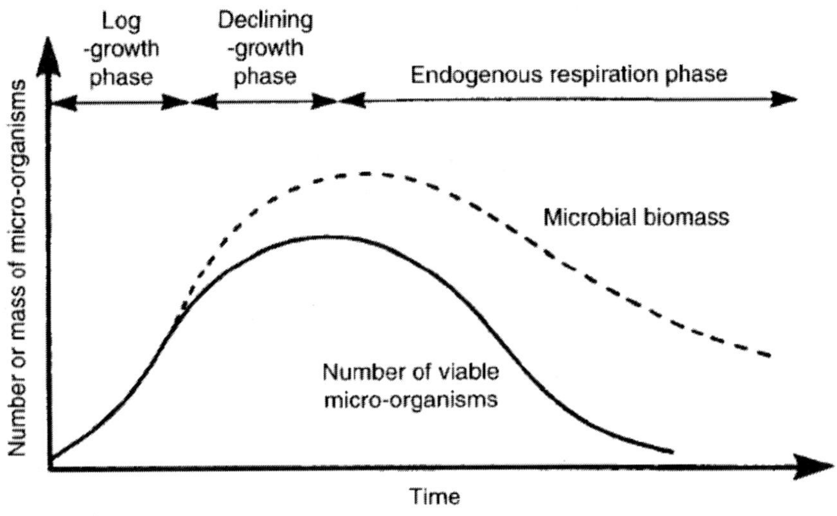

Figure 4. Microbial growth curves [1].

4.4. Bioremediation of wastewater

Bioremediation is a treatment process that involves the implementation of microorganisms to remove pollutants from a contaminated setting. Bioremediation can be defined as "treatment that implements natural organisms to decompose hazardous materials into less toxic or nontoxic materials". Some examples of bioremediation-related technologies are phytoremediation, bioaugmentation, rhizofiltration, and biostimulation. The microorganisms implemented to carry out the bioremediation are called bioremediators. However, some pollutants are not easily removed or decomposed by bioremediation. For example, heavy metals such as lead and cadmium are not eagerly captured by bioremediators. Example of bioremediation: fish bone char has been shown to bioremediate small amounts of cadmium, copper, and zinc.

The bioremediation of wastewater can be achieved by autotrophs or heterotrophs. A heterotroph is an organism that is unable to fix carbon and utilizes organic carbon for its growth. Heterotrophs are divided based on their source of energy. If the heterotroph utilizes light as its source of energy, then it is considered a photoheterotroph. If the heterotroph utilizes organic and/or inorganic compounds as energy sources, it is then considered a chemoheterotroph. Autotrophs, such as plants and algae, that are able to utilize energy from sunlight are called photoautotrophs. Autotrophs that utilize inorganic compounds to produce organic compounds such as carbohydrates, fats, and proteins from inorganic carbon dioxide are called lithoautotrophs. These reduced carbon compounds can be utilized as energy sources by autotrophs and provide the energy in food consumed by heterotrophs. Over 95% of all organisms are heterotrophic.

4.4.1. Aerobic treatment

Aeration has been used to remove trace organic volatile compounds (VOCs) in water. It has also been employed to transfer a substance, such as oxygen, from air or a gas phase into water in a process called "gas adsorption" or "oxidation", i.e., to oxidize iron and/or manganese. Aeration also provides the escape of dissolved gases, such as CO_2 and H_2S. Air stripping has been also utilized effectively to remove NH_3 from wastewater and to remove volatile tastes and other such substances in water [2]. Samer [4] and Samer et al. [5] mentioned that aerobic treatment with biowastes is effective in reducing harmful gaseous emissions as greenhouse gases (CH_4 and N_2O) and ammonia.

4.4.1.1. Oxidation ponds

Oxidation ponds (Figure 5) are aerobic systems where the oxygen required by the heterotrophic bacteria (a heterotroph is an organism that cannot fix carbon and uses organic carbon for growth) is provided not only by transfer from the atmosphere but also by photosynthetic algae. The algae are restricted to the euphotic zone (sunlight zone), which is often only a few centimeters deep. Ponds are constructed to a depth of between 1.2 and 1.8 m to ensure maximum penetration of sunlight, and appear dark green in color due to dense algal development. Samer [6] and Samer et al. [7] illustrated the structures and constructions of the aerobic treatment tanks and the used building materials.

In oxidation ponds, the algae use the inorganic compounds (N, P, CO_2) released by aerobic bacteria for growth using sunlight for energy. They release oxygen into the solution that in turn is utilized by the bacteria, completing the symbiotic cycle. There are two distinct zones in facultative ponds: the upper aerobic zone where bacterial (facultative) activity occurs and a lower anaerobic zone where solids settle out of suspension to form a sludge that is degraded anaerobically.

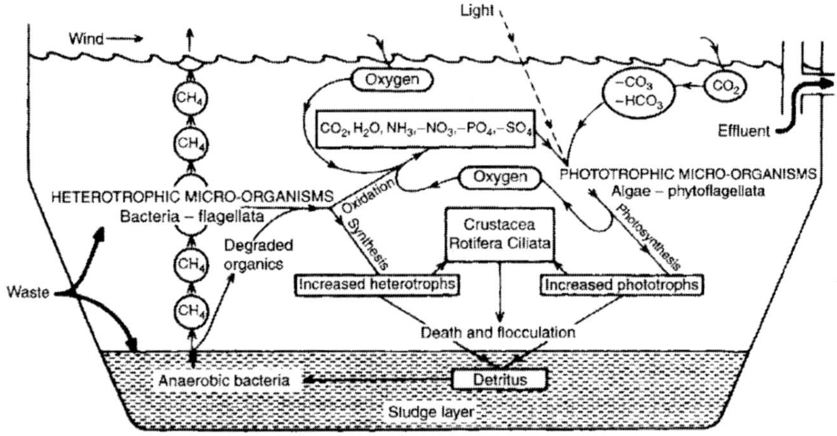

Figure 5. Aerobic system/oxidation pond [1].

4.4.1.2. Aeration lagoons

Aeration lagoons are profound (3–4 m) compared to oxidation ponds, where oxygen is provided by aerators but not by the photosynthetic activity of algae as in the oxidation ponds. The aerators keep the microbial biomass suspended and provide sufficient dissolved oxygen that allows maximal aerobic activity. On the other hand, bubble aeration is commonly used where the bubbles are generated by compressed air pumped through plastic tubing laid through the base of the lagoon. A predominately bacterial biomass develops and, whereas there is neither sedimentation nor sludge return, this procedure counts on adequate mixed liquor formed in the tank/lagoon. Therefore, the aeration lagoons are suitable for strong but degradable wastewater such as wastewaters of food industries. The hydraulic retention time (HRT) ranges from 3 to 8 days based on treatment level, strength, and temperature of the influent. Generally, HRT of about 5 days at 20°C achieves 85% removal of BOD in household wastewater. However, if the temperature falls by 10°C, then the BOD removal will decrease to 65% [1].

4.4.2. Anaerobic treatment

The anaerobic treatments are implemented to treat wastewaters rich in biodegradable organic matter (BOD >500 mg L^{-1}) and for further treatment of sedimentation sludges. Strong organic wastewaters containing large amounts of biodegradable materials are discharged mainly by agricultural and food processing industries. These wastewaters are difficult to be treated aerobically due to the troubles and expenses of fulfillment of the elevated oxygen demand to preserve the aerobic conditions [1]. In contrast, anaerobic degradation occurs in the absence of oxygen. Although the anaerobic treatment is time-consuming, it has a multitude of advantages in treating strong organic wastewaters. These advantages include elevated levels of purification, aptitude to handle high organic loads, generating small amounts of sludges that are usually very stable, and production of methane (inert combustible gas) as end-product.

Anaerobic digestion is a complex multistep process in terms of chemistry and microbiology. Organic materials are degraded into basic constituents, finally to methane gas under the absence of an electron acceptor such as oxygen [8]. The basic metabolic pathway of anaerobic digestion is shown in Figures 6 and 7. To achieve this pathway, the presence of very different and closely dependent microbial population is required.

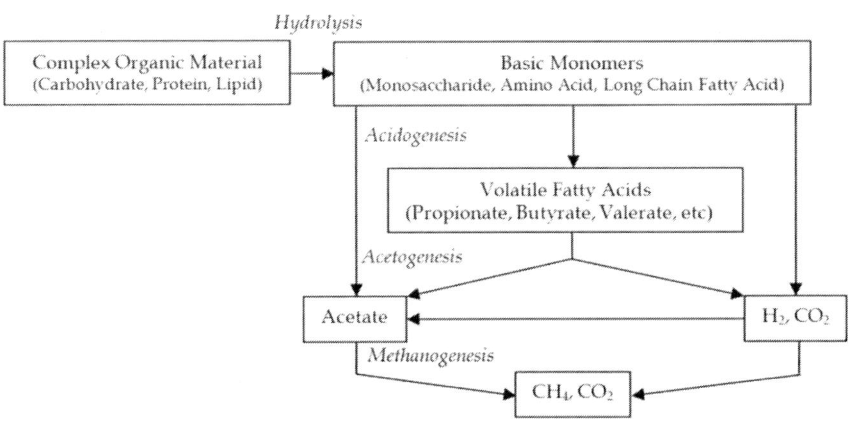

Figure 6. Steps of the anaerobic digestion process [8].

Suitable wastewaters include livestock manure, food processing effluents, petroleum wastes (if the toxicity is controlled), and canning and dyestuff wastes where soluble organic matters are implemented in the treatment. Most anaerobic processes (solids fermentation) occur in two predetermined temperature ranges: mesophilic or thermophilic. The temperature ranges are 30–38°C and 38–50°C, respectively [3]. In contrast to aerobic systems, absolute stabilization of organic matter is not achievable under anaerobic conditions. Therefore, subsequent aerobic treatment of the anaerobic effluents is usually essential. The final waste matter discharged by the anaerobic treatment includes solubilized organic matter that is acquiescent to aerobic treatment demonstrating the possibility of installing collective anaerobic and aerobic units in series [1].

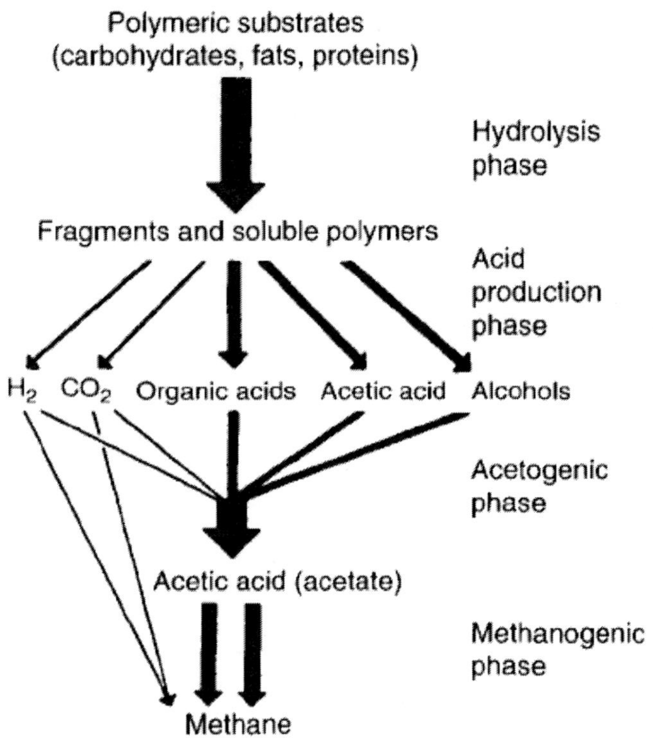

Figure 7. Major steps in anaerobic decomposition [1].

4.4.2.1. Anaerobic digesters

Samer [9] elucidated and illustrated the structures and constructions of the anaerobic digesters and the used building materials. Samer [10] developed an expert system for planning and designing biogas plants. Figures 8 to 13 show different types of anaerobic digesters. While Figures 14 and 15 show some industrial applications. Table 1 shows the advantages and disadvantages of anaerobic treatment compared to aerobic treatment.

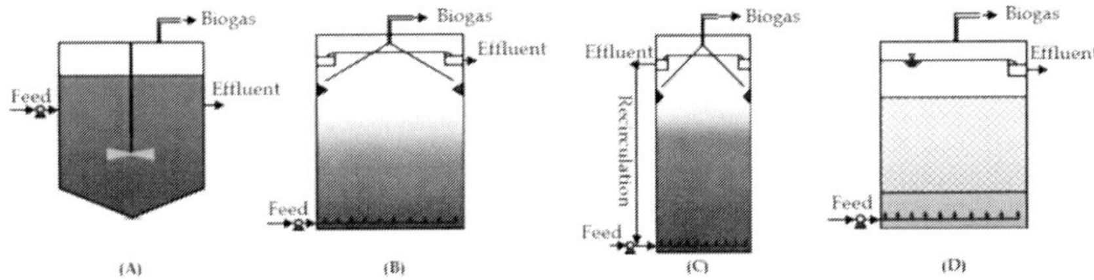

Figure 8. Most commonly used anaerobic reactor types: (A) Completely mixed anaerobic digester, (B) UASB reactor, (C) AFB or EGSB reactor, and (D) Upflow AF [8].

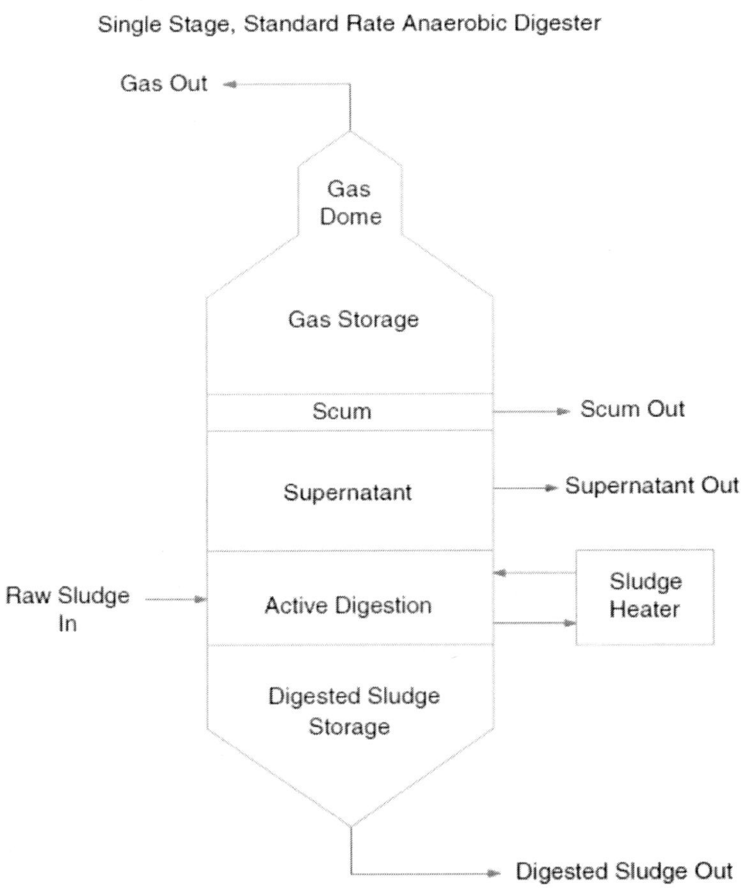

Figure 9. Single-stage conventional anaerobic digester [3].

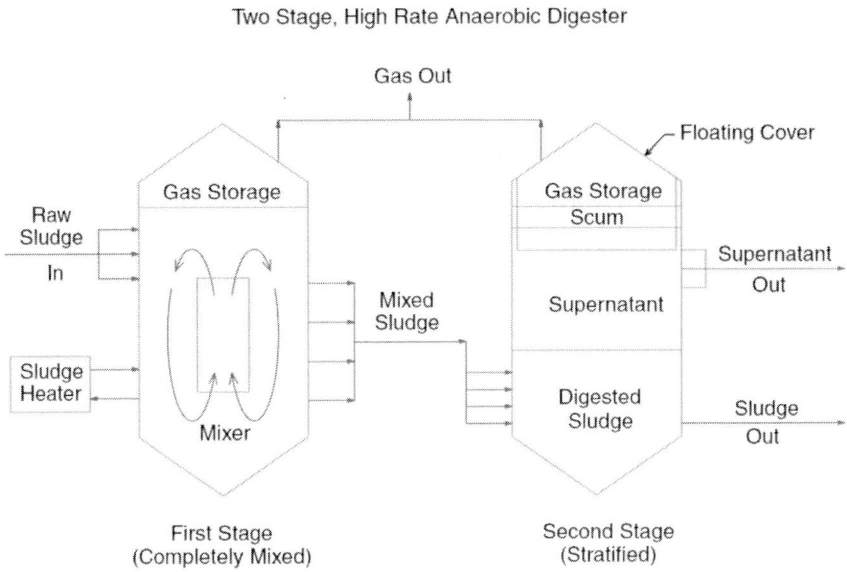

Figure 10. Dual-stage high rate digester [3].

By definition, the anaerobic treatment is conducted without oxygen. It is different from an anoxic process, which is a reduced environment in contrast to an environment without oxygen. Both processes are anoxic, but anaerobic is an environment beyond anoxic where the oxidation reduction potential (ORP) values are highly negative. In the anaerobic process, nitrate is reduced to ammonia and nitrogen gas, and sulfate (SO_3^{2-}) is reduced to hydrogen sulfide (H_2S). Phosphate is also reduced because it is often transformed through the ADP–ATP chain [3].

4.4.2.2. Anaerobic lagoons

An anaerobic lagoon is a deep lagoon, fundamentally without dissolved oxygen, that enforces anaerobic conditions. The anaerobic process occurs in deep ground ponds, and such basins are implemented for anaerobic pretreatment. The anaerobic lagoons are not aerated, heated, or mixed. The depth of an anaerobic lagoon should be typically deeper than 2.5 m, where deeper lagoons are more efficient. Such depths diminish the amount of oxygen diffused from the surface, allowing anaerobic conditions to prevail (U.S. EPA, 2002). Figures 16 to 18 show different types of anaerobic lagoons.

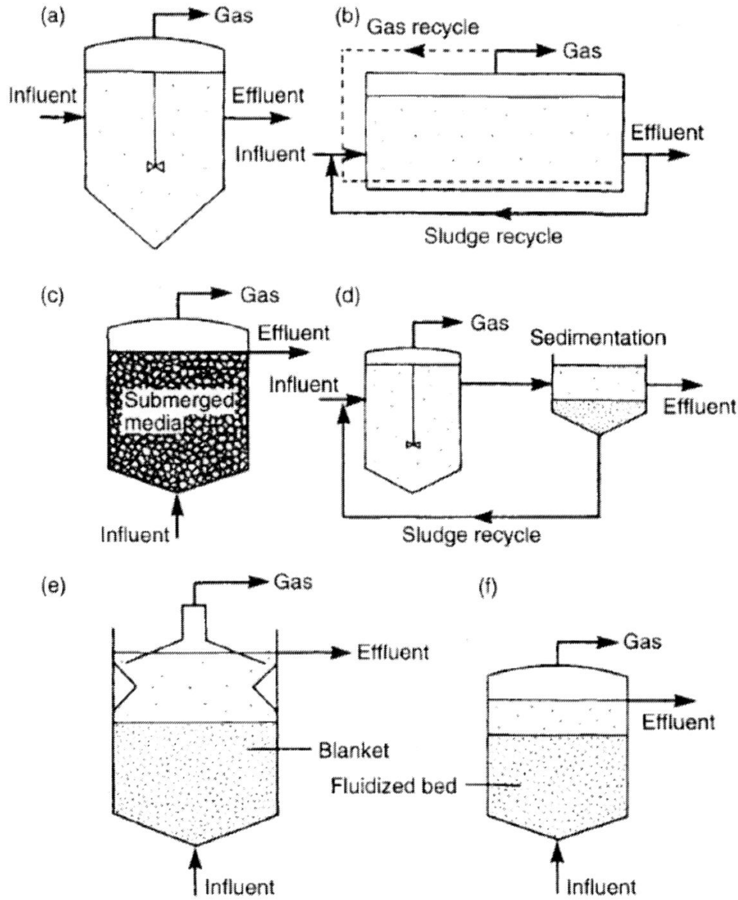

Figure 11. Schematic representation of digester types. Flow-through (A–B) and contact systems (C–F) [1].

4.4.3. Bioreactors

A bioreactor can be defined as "engineered or manufactured apparatus or system that controls the embraced or encompassed bioenvironment". Precisely, the bioreactor is a vessel in which a biochemical process is conducted, where it involves microorganisms (e.g., bacteria, algae, fungi) or biochemical substances (e.g., enzymes) derived from such microorganisms. The treatment can be conducted under either aerobic or anaerobic conditions. The bioreactors are commonly made of stainless steel, usually cylindrical in shape and range in size from liters to cubic meters. The bioreactors are classified as batch, plug, or continuous flow reactors (e.g., continuous stirred-tank bioreactor).

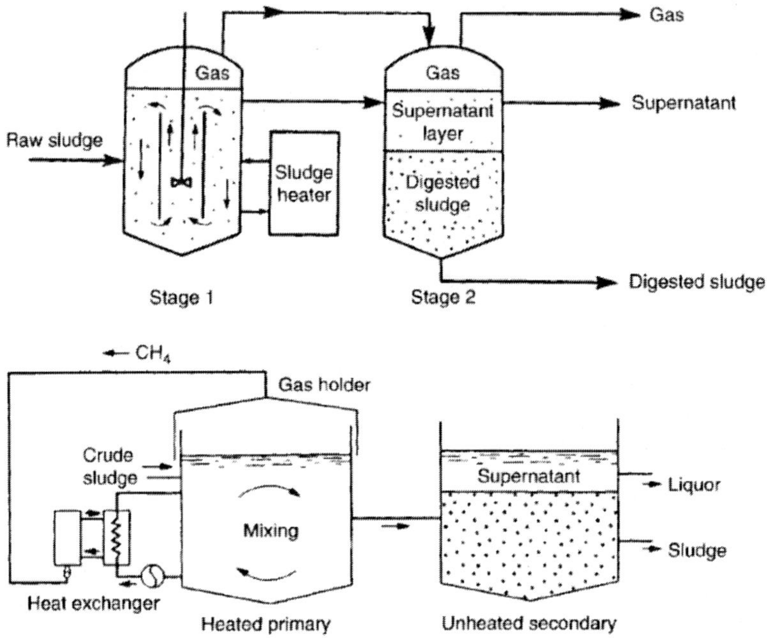

Figure 12. The upper scheme shows a two-stage anaerobic sludge digester, while the lower scheme shows the conventional sludge digestion plant [1].

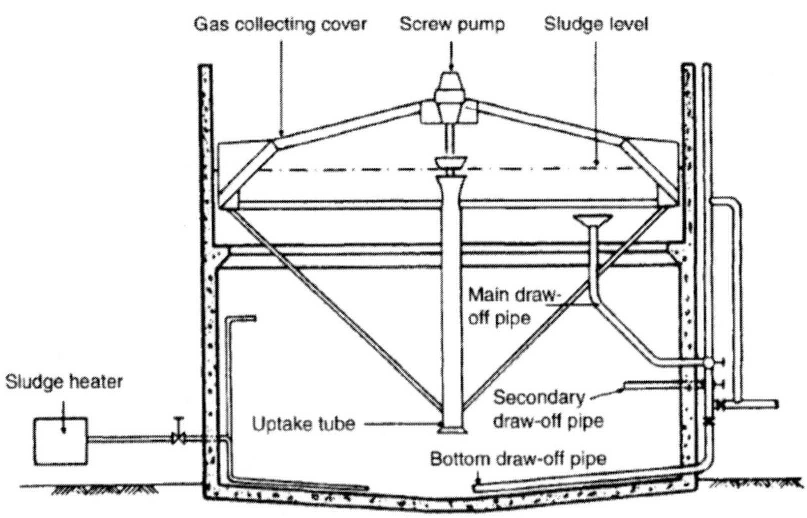

Figure 13. Primary digestion tank with screw mixing pump and external heater [1].

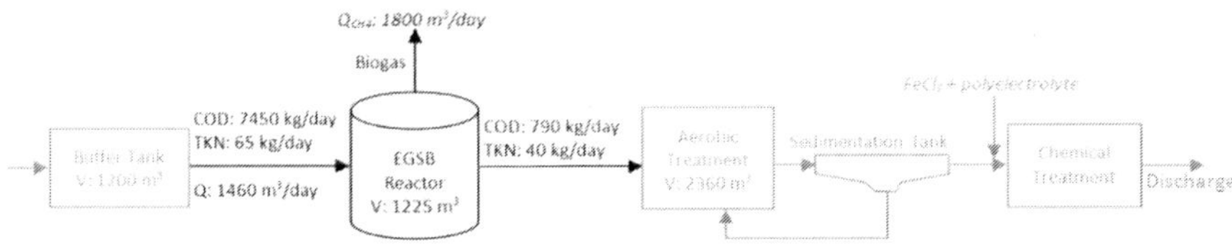

Figure 14. Wastewater treatment plant for corn processing industry [8].

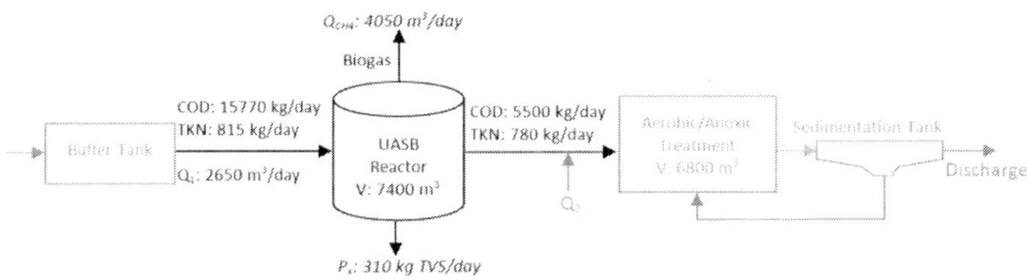

Figure 15. Mass balance study for a wastewater treatment plant of the baker's yeast industry [8].

Table 1. The advantages and disadvantages of anaerobic treatment compared to aerobic treatment [1].

The advantages and disadvantages of anaerobic treatment compared to aerobic treatment

Advantages	*Disadvantages*
Low operational costs	High capital costs Generally require heating
Low sludge production	Low retention times required (>24 h)
Reactors sealed giving no odour or aerosols	Corrosive and malodorous compounds produced during anaerobiosis
Sludge is highly stabilized	Not as effective as aerobic stabilization for pathogen destruction
Methane gas produced as end product	Hydrogen sulphide also produced
Low nutrient requirement due to lower growth rate of anaerobes	Reactor may require additional alkalinity
Can be operated seasonally	Slow growth rate of anaerobes can result in long initial start-up of reactors and recovery periods
Rapid start-up possible after acclimation	Only used as pre-treatment for liquid wastes

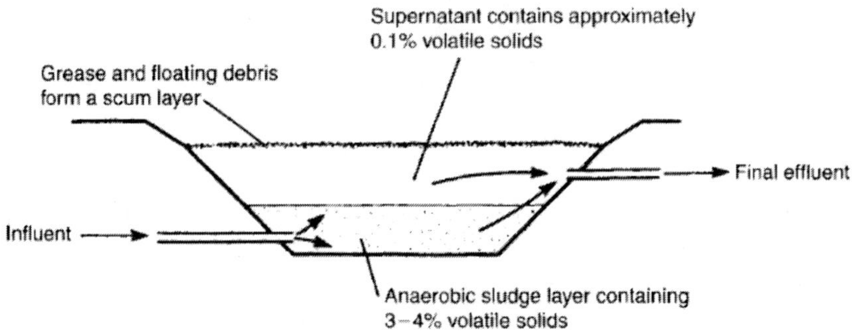

Figure 16. Anaerobic lagoon for strong wastewater treatment, such as meat processing wastewater [1].

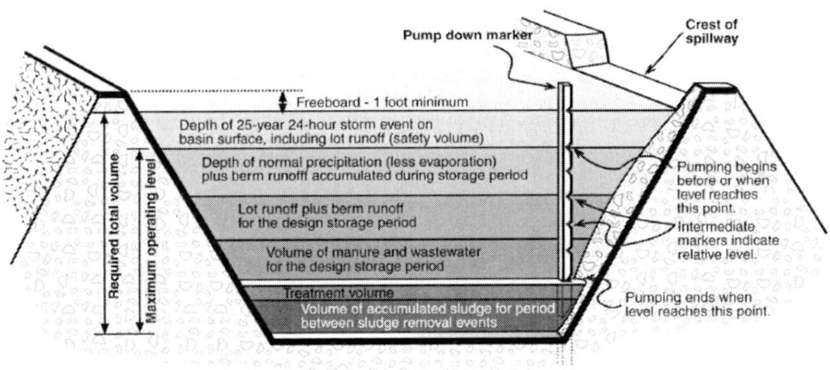

Figure 17. Schematic of volume fractions in anaerobic lagoon design [11].

Mycoremediation is a type of bioremediation where fungi are implemented to break down the contaminants. The term "mycoremediation" refers particularly to the implementation of fungal "mycelia" in bioremediation. The principal role of fungi in the ecological system is the breakdown of pollutants, which is performed by the mycelium. The mycelium, the vegetative part of a fungus, secretes enzymes and acids that biodegrade lignin and cellulose that are the main components of vegetative fibers. Lignin and cellulose are organic compounds composed of long chains of carbon and hydrogen, and therefore they are structurally similar to several organic pollutants. One key issue is specifying the right fungus to break down a determined pollutant. Similarly, mycofiltration is a process that uses fungal mycelia to filter toxic compounds from wastewater. In an experiment, wastewater contaminated with diesel oil was inoculated with mycelia of oyster mushrooms. One month later, more than 93% of many of the polycyclic aromatic hydrocarbons (PAH) had been reduced to non-toxic components in the mycelial-inoculated samples. The natural microbial community participates with the fungi to break down contaminants, eventually into CO_2 and H_2O. Wood-degrading fungi are particularly effective in breaking down aromatic pollutants (toxic components of petroleum), as well as chlorinated compounds (certain persistent pesticides). Figures 19 to 22 show different types and designs of bioreactors.

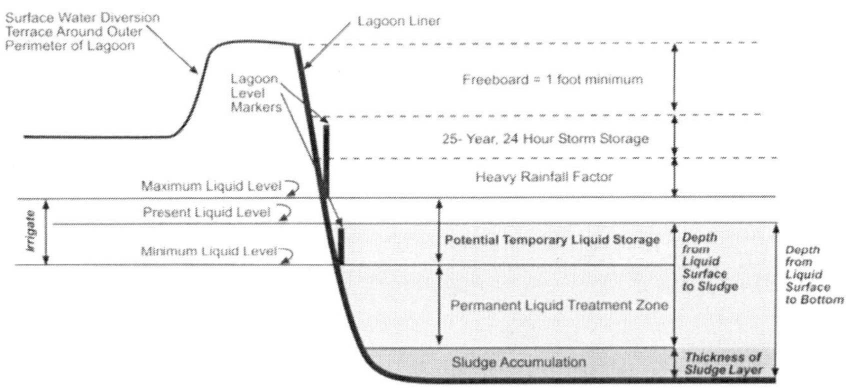

Figure 18. Anaerobic wastewater treatment lagoon [12].

4.4.4. Activated sludge

The activated sludge process is based on a mixture of thick bacterial population suspended in the wastewater under aerobic conditions. With unlimited nutrients and oxygen, high rates of bacterial growth and respiration can be attained, which results in the consumption of the available organic matter to either oxidized end-products (e.g., CO_2, NO_3^-, SO_4^{2-}, and PO_4^{3-}) or biosynthesis of new microorganisms. The activated sludge process is based on five interdependent elements, which are: bioreactor, activated sludge, aeration and mixing system, sedimentation tank, and returned sludge [1]. The biological process using activated sludge is a commonly used method for the treatment of wastewater, where the running costs are inexpensive (Figure 23). However, a huge quantity of surplus sludge is produced in wastewater treatment plants (WWTPs) which is an enormous burden in both economical and environmental aspects. The excess sludge contains a lot of moisture and is not easy to treat. The byproducts of WWTPs are dewatered, dried, and finally burnt into ashes. Some are used in farm lands as compost fertilizer [15]. However, it is suggested that the dried byproducts of WWTPs are fed into the pyrolysis process rather than the burning process.

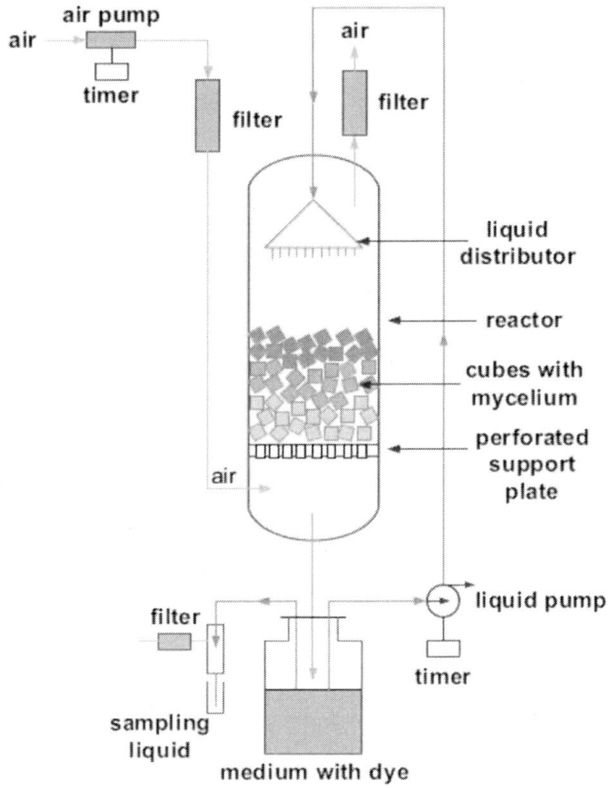

Figure 19. A bioreactor for fungal degradation: trickle bed bioreactor [13].

The sludge volume index (SVI) is an estimation that specifies the tendency of aerated solids, i.e., activated sludge solids, to become dense or concentrated through the thickening process. SVI can be computed as follows: (a) allowing a mixed liquor sample from the aeration tank to sediment in 30 min; (b) determining the concentration of the suspended solids for a sample of the same mixed liquor; (c) SVI is then computed as ratio of the measured wet volume (mL/L) of the settled sludge to the dry weight concentration of MLSS in g/L (Source: Office of Water Programs, Sacramento State, USA).

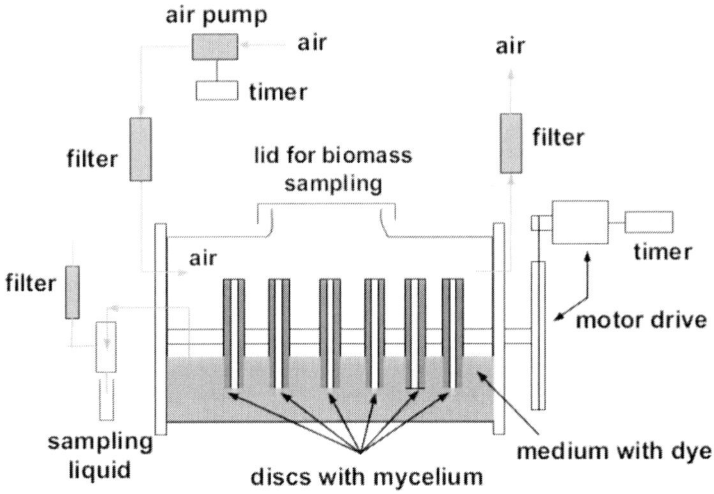

Figure 20. A bioreactor for fungal degradation: rotating disc bioreactor [13].

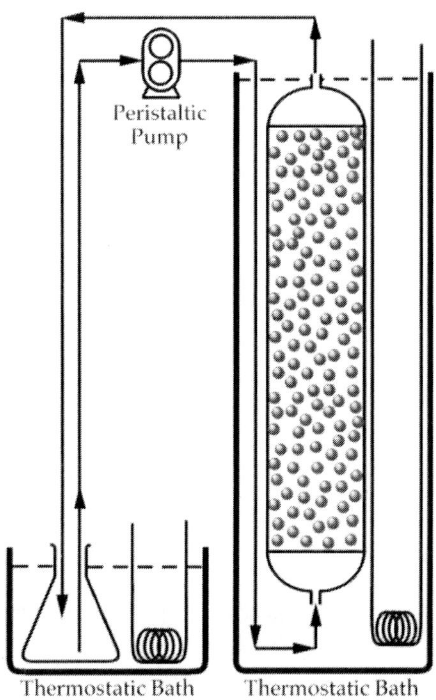

Figure 21. Fluidized bed bioreactor [14].

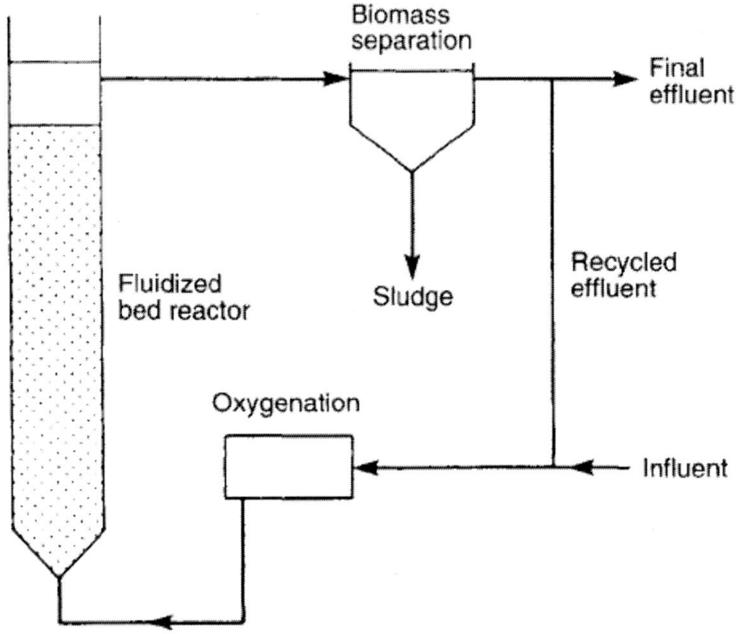

Figure 22. Typical design of fluidized bed reactor system [1].

During the treatment of wastewater in aeration tanks through the activated sludge process (Table 2) there are suspended solids, where the concentration of the suspended solids is termed as mixed liquor suspended solids (MLSS), which is measured in milligrams per liter (mg L^{-1}). Mixed liquor is a mixture of raw wastewater and activated sludge in an aeration tank. MLSS consists mainly of microorganisms and non-biodegradable suspended solids. MLSS is the effective and active portion of the activated sludge process that ensures that there is adequate quantity of viable biomass available to degrade the supplied quantity of organic pollutants at any time. This is termed as Food to Microorganism Ratio (F/M Ratio) or food to mass ratio. If this ratio is kept at the suitable level, then the biomass will be able to consume high quantities of the food, which reduces the loss of residual food in the discharge. In other words, the more the biomass consumes food the lower the BOD will be in the treated effluent. It is important that MLSS eliminates BOD in order to purify the wastewater for further usage and hygiene. Raw sewage is introduced into the wastewater treatment process with a concentration of several hundred mg L^{-1} of BOD. The concentration of BOD in wastewater is reduced to less than 2 mg L^{-1} after being treated with MLSS and other treatment methods, which is considered to be safe water to use.

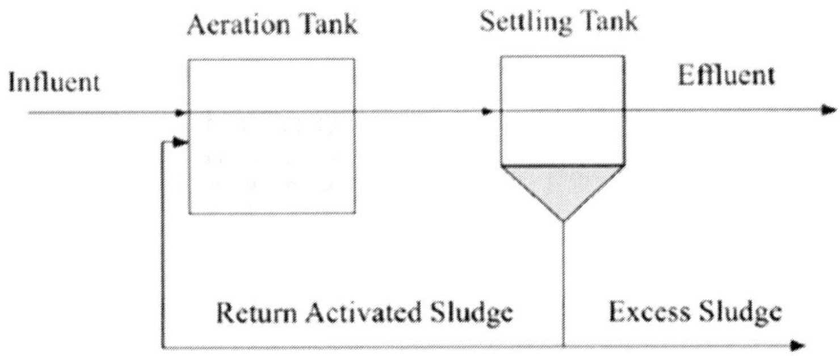

Figure 23. Activated sludge [15].

Table 2. Conventional activated sludge [15].

Specification	Value	Unit
BOD-Sludge Loading	0.40	mg L^{-1}
BOD-Volume Loading	0.20	mg L^{-1}
MLSS	2000	mg L^{-1}
COD of Influent	300	mg L^{-1}
Amount of Influent	4.48	L d^{-1}
Aeration Rate	3.00	L min^{-1}

The biological treatment process is the most commonly implemented method for the treatment of domestic sewage. This method implements bacterial populations that possess superior sedimentation characteristics. The living microorganisms break down the organic matter in the wastewater and consequently purify the wastewater from biological waste [15].

According to [1], the main components of all activated sludge systems are:

1) The bioreactor: it can be a lagoon, tank, or ditch. The main characteristic of a bioreactor is that it contains sufficiently aerated and mixed contents. The bioreactor is also known as the aeration tank.

2) Activated sludge: it is the bacterial biomass inside the bioreactor that consists mostly of bacteria and other flora and microfauna. The sludge is a flocculent suspension of these microorganisms and is usually

termed as the mixed liquor suspended solids (MLSS) that ranges between 2,000 and 5,000 mg L^{-1}.

3) Aeration and mixing system: the aeration and mixing of the activated sludge and the raw influent are necessary. While these processes can be accomplished separately, they are usually conducted using a single system of either surface aeration or diffused air.

4) Sedimentation tank: clarification or settlement of the activated sludge discharged from the aeration tank is essential. This separates the bacterial biomass from the treated wastewater.

5) Returned sludge: the settled activated sludge in the sedimentation tank is returned to the bioreactor to maintain the microbial population at a required concentration to guarantee persistence of treatment process.

Several parameters should be considered while operating activated sludge plants. The most important parameters are: (1) biomass control, (2) plant loading, (3) sludge settleability, and (4) sludge activity. The main operational variable is the aeration, where its major functions are: (1) ensuring a sufficient and continuous supply of dissolved oxygen (DO) for the bacterial population, (2) keeping the bacteria and the biomass suspended, and (3) mixing the influent wastewater with the biomass and removing from the solution the excessive CO_2 resulting from oxidation of organic matter [1].

There are several types of activated sludge processes, e.g., conventional activated sludge plant (Figure 24), complete mix plant (Figure 25), contact stabilization plant (Figure 26), and step aeration plant (Figure 27). Figure 28 shows the food pyramid that represents the feeding relationships within the activated sludge process.

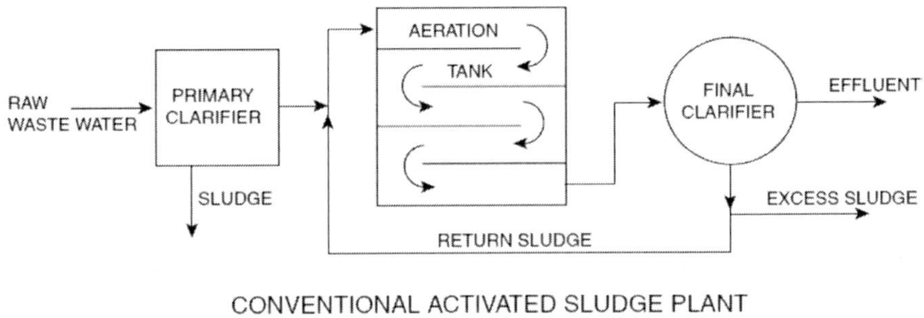

Figure 24. Conventional activated sludge plant [3].

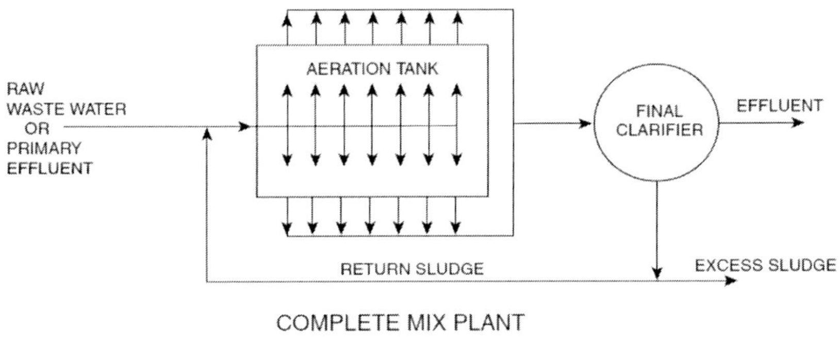

Figure 25. Complete mix plant [3].

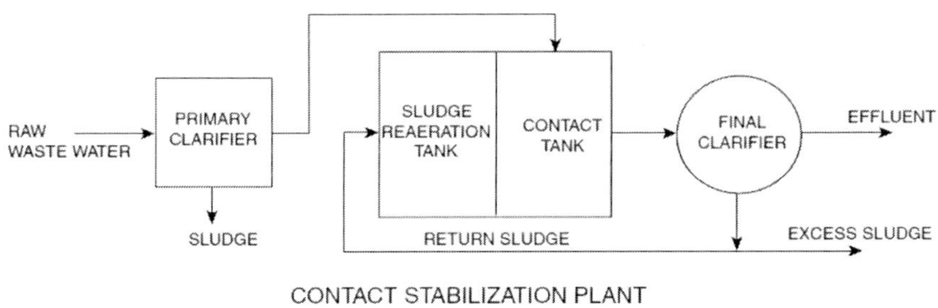

Figure 26. Contact stabilization plant [3].

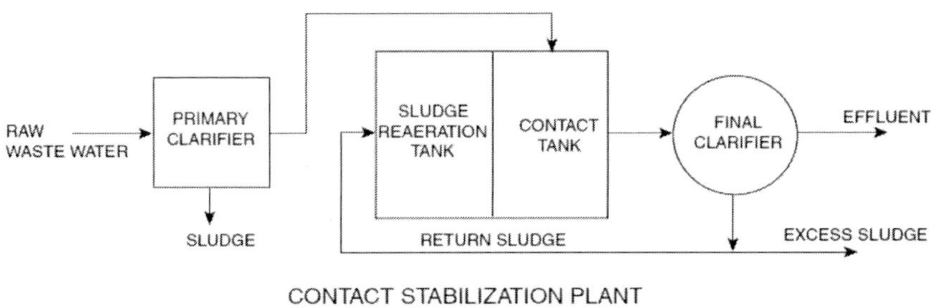

Figure 27. Step aeration plant [3].

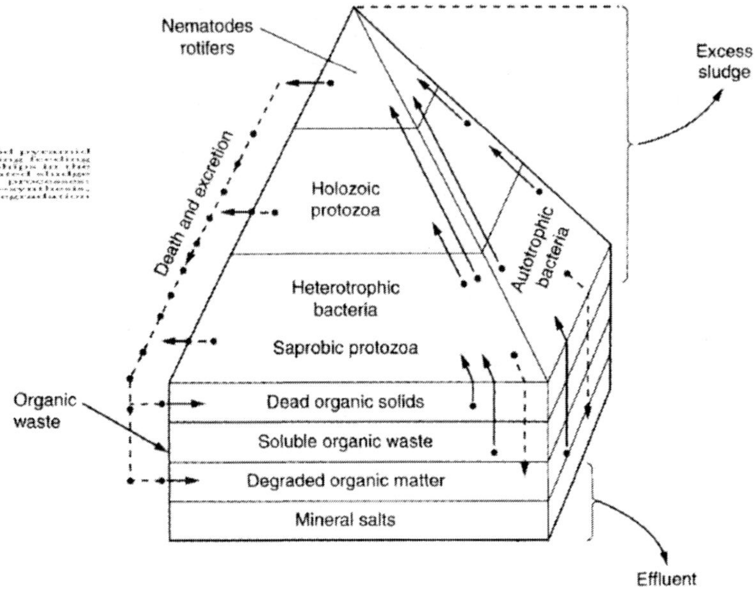

Figure 28. Food pyramid illustrating the feeding relationships within the activated sludge process [1].

4.4.5. Biological filters

The main systems of operation of biological filters are: (a) single filtration, (b) recirculation, (c) ADF, and (d) two-stage filtration with high-rate primary biotower (Figure 29). There are several types of biological filters, for example, submerged aerated filters that are widely known as biological aerated filters (BAFs) and are the commonly implemented design (Figure 30), and the percolating (trickling) filters (Figure 31). The BAFs implement either the sunken granular media with upward (Figure 30a) or download (Figure 30b) flows, or floating granular media with upward flow (Figure 30c), which is the most common design of BAFs. In order to compare the biological filters and the activated sludge systems (Figures 31 and 32), the comparison is based on the oxidation that can be accomplished by three processes:

1) Spreading the wastewater into a thin film of liquid with a large surface area, consequently the required oxygen can be supplied by gaseous diffusion, which is the case of the percolating filters.

2) Aerating the wastewater by pumping air in the form of bubbles or stirring forcefully, which is the case of the activated sludge process.

3) Implementing algae to produce oxygen by photosynthesis, which is the case of the stabilization ponds.

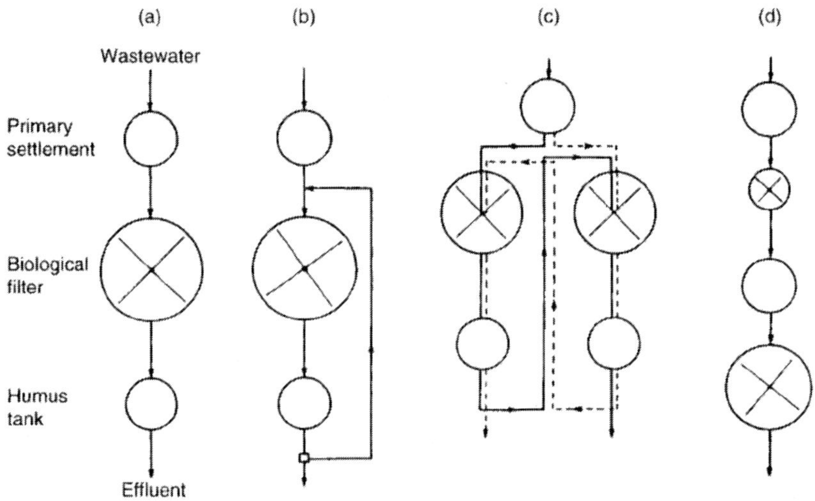

Figure 29. The main systems of operation of biological filters [1].

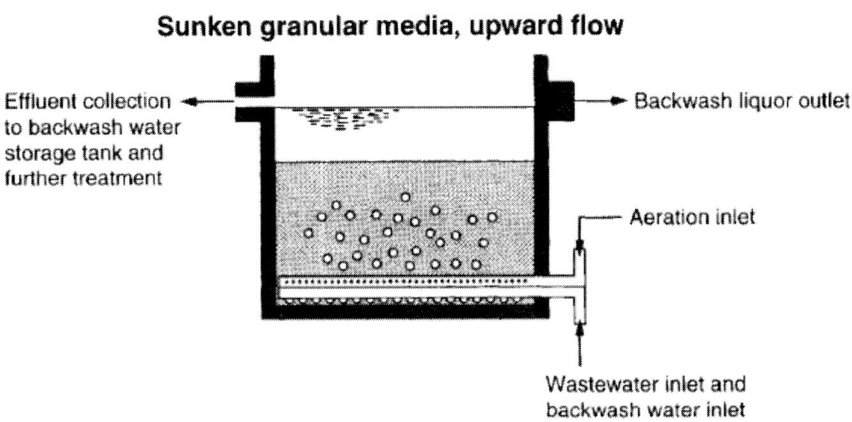

(a)

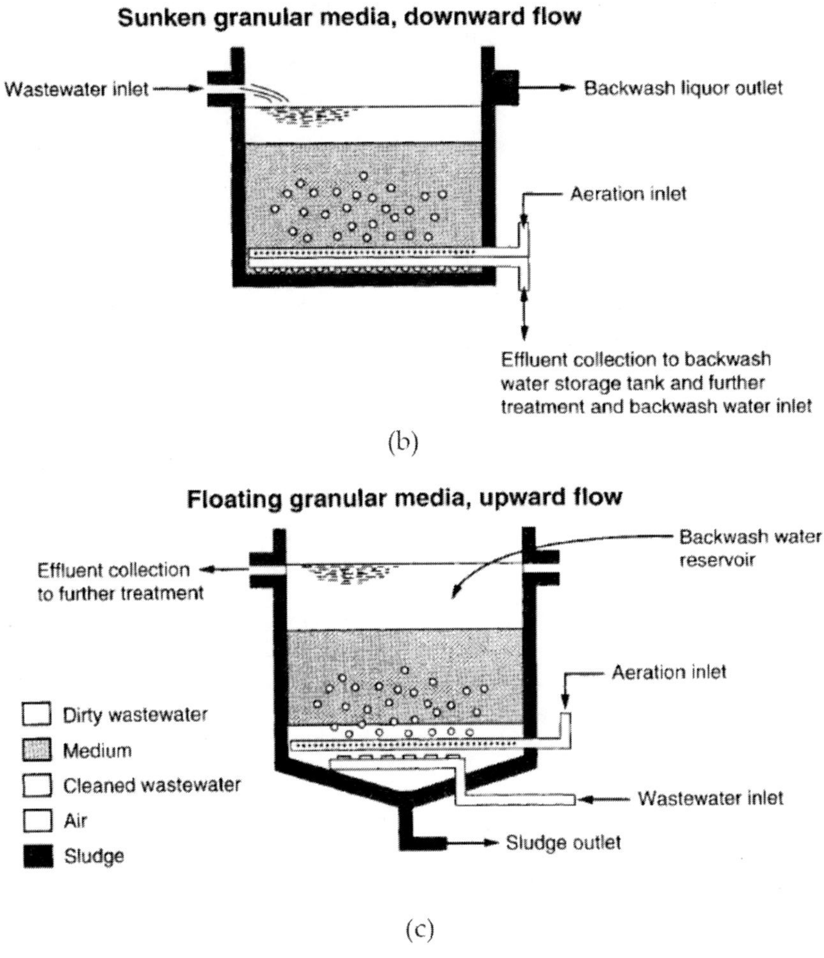

Figure 30. Biological aerated filters [1].

4.4.6. Rotating biological contactors

The rotating biological contactors (RBC) system (Figure 33) can be implemented to amend and improve the available treatment processes as the secondary or tertiary treatment processes. The RBC is successfully implemented in all three steps of the biological treatment, which are BOD_5 removal, nitrification, and denitrification. The process is a fixed-biofilm of either aerobic or anaerobic biological treatment system for removal of nitrogenous and carbonaceous compounds from wastewater (Figure 34). The RBC installations (Figure 35) were designed for removal of BOD_5 or ammonia nitrogen (NH_3-N), or both, from wastewater [1, 2].

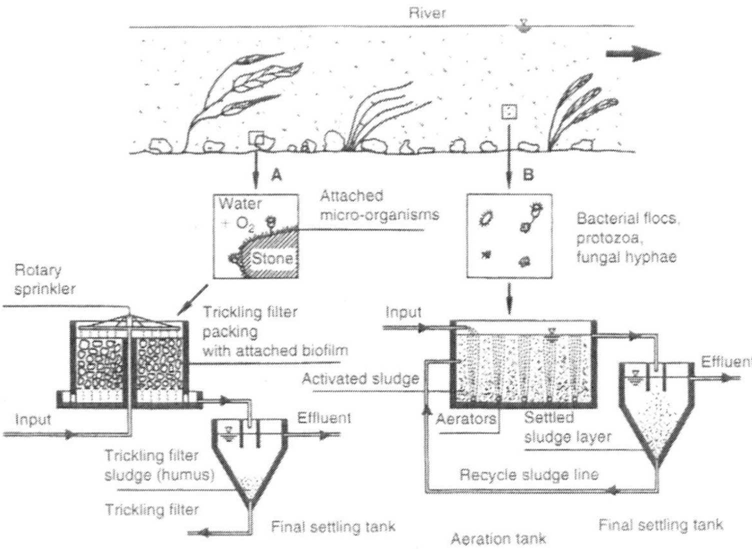

Figure 31. Relationship between the natural bacterial populations in rivers and the development of (A) trickling (percolating) filter and (B) activated sludge system [1].

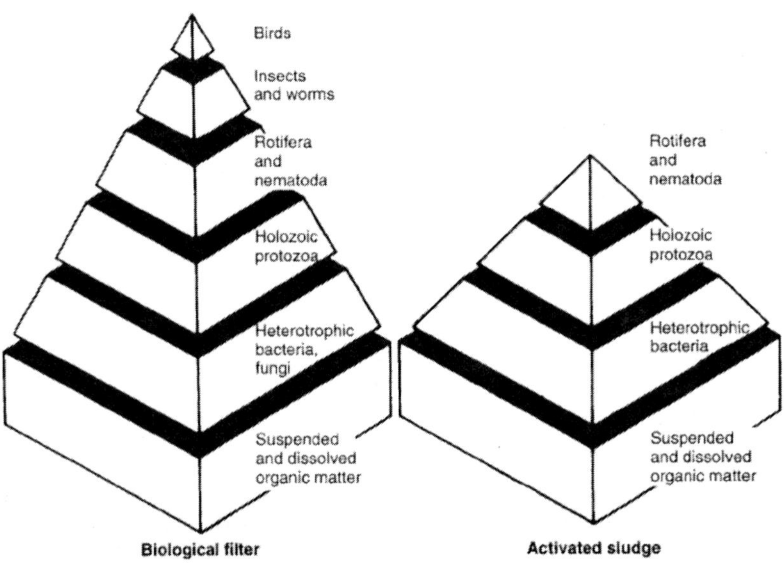

Figure 32. Comparison of the food chain pyramids for biological filters and activated sludge systems [1].

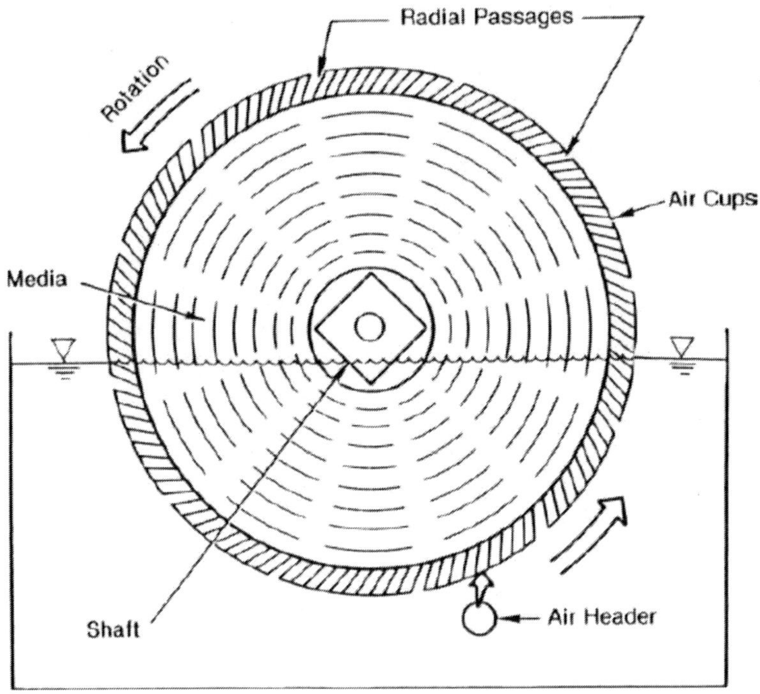

Figure 33. Schematic diagram of air-drive RBC [2].

The RBC consists of media, shaft, drive, bearings, and cover (Figure 34). The RBC hardware consists of a large diameter and closely spaced circular plastic media that is mounted on a horizontal shaft supported by bearings and is slowly rotated by an electric motor. The plastic media are made of corrugated polystyrene or polyethylene material with different designs, dimensions, and densities. The model designs are based on increasing surface area and firmness, allowing a winding wastewater flow path and stimulating air turbulence [1, 2].

4.4.7. Biological removal of nutrients

4.4.7.1. Biological phosphorous removal

It is widely agreed that microorganisms utilize acetate and fatty acids to accumulate polyphosphates as poly-β-hydroxybutyrate, which is an acid polymer. The precise mechanism is based on the production and regeneration of adenosine diphosphate (ADP) within the bacteria, and it involves the adenosine triphosphate (ATP). Phosphate removal requires true anaerobic conditions, which occur only when there is no other oxygen donor [3]. Figure 36 shows a phosphate removal process. This process needs long narrow tanks for maintenance of plug flow.

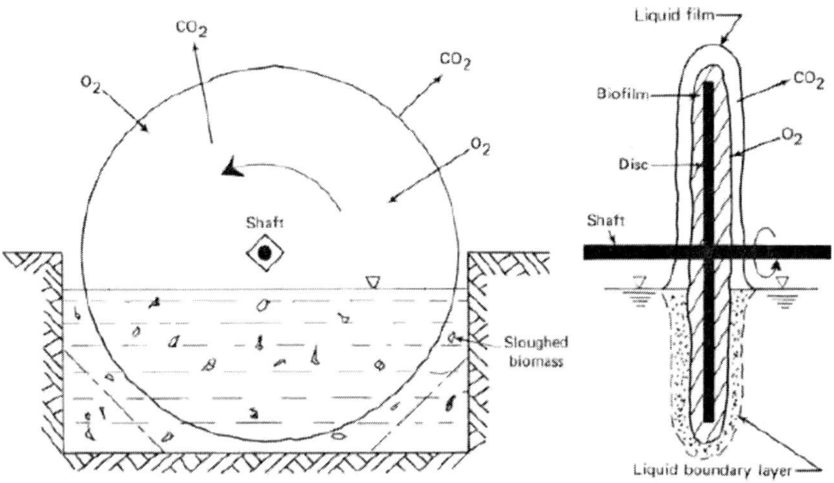

Figure 34. Mechanism of attached growth media in an RBC system [2].

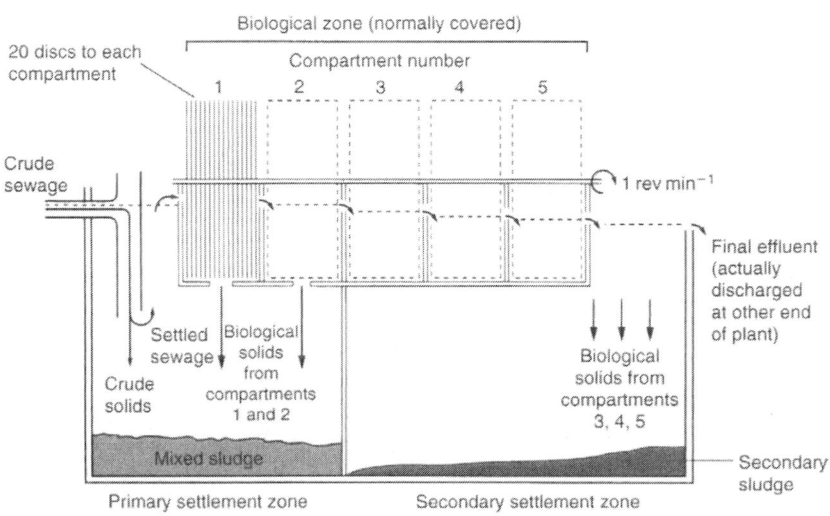

Figure 35. RBC system [1].

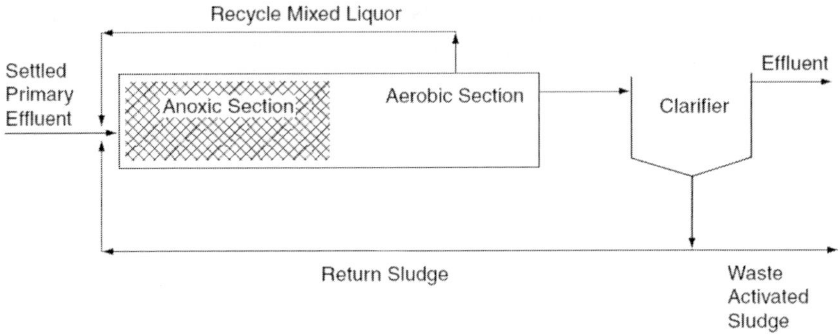

Figure 36. Phosphate removal process [3].

4.4.7.2. Biological removal of nitrogen

The nitrification and denitrification processes are responsible for N_2O production (Figure 37). Figure 38 shows a nitrification/denitrification system for biological removal of nitrogen.

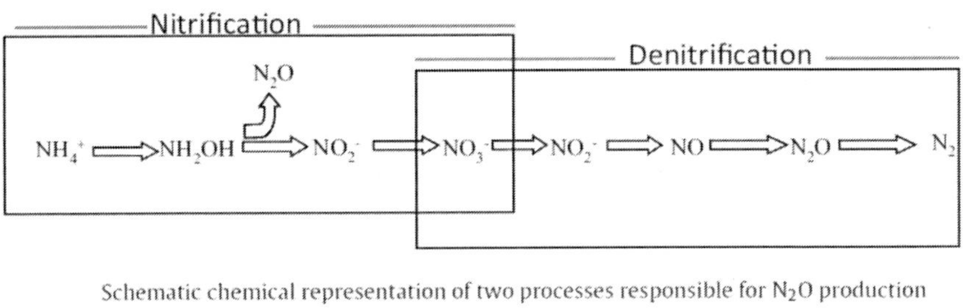

Figure 37. Schematic illustration of nitrification and denitrification processes that are responsible for N_2O release [16].

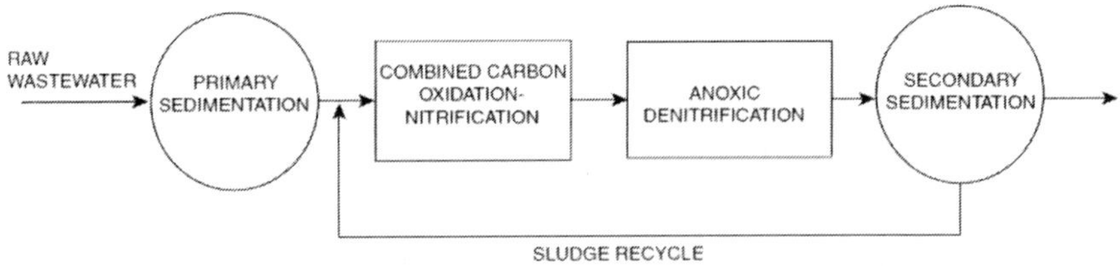

Figure 38. Nitrification/denitrification system for biological removal of nitrogen [3].

4.4.8. Phytoremediation

Phytoremediation is a treatment process that solves environmental problems by implementing plants that abate environmental pollution without excavating the pollutants and disposing them elsewhere. Phytoremediation is the abatement of pollutant concentrations in contaminated soils or water using plants that are able to accumulate, degrade, or eliminate heavy metals, pesticides, solvents, explosives, crude oils and its derivatives, and a multitude of other contaminants and pollutants from water and soils. Figures 39 through 44 show the designs of constructed wetlands where the phytoremediation takes place.

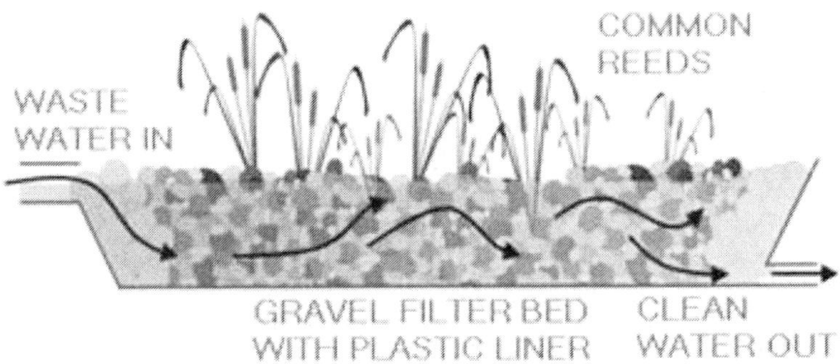

Figure 39. Cross-sectional view of a typical subsurface flow constructed wetland [17].

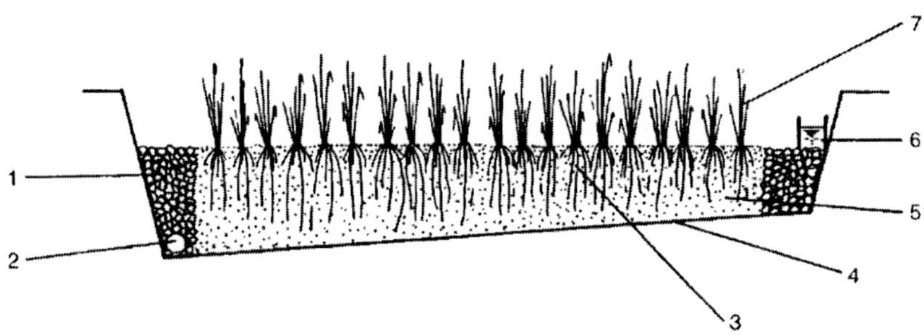

Figure 40. Components of a horizontal flow reed bed: (1) drainage zone consisting of large rocks, (2) drainage tube of treated effluent, (3) root zone, (4) impermeable liner, (5) soil or gravel, (6) wastewater distribution system, and (7) reeds [1].

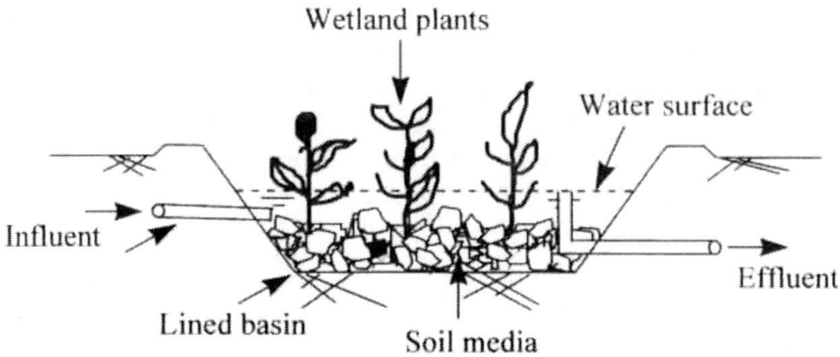

Figure 41. Free water surface system [18].

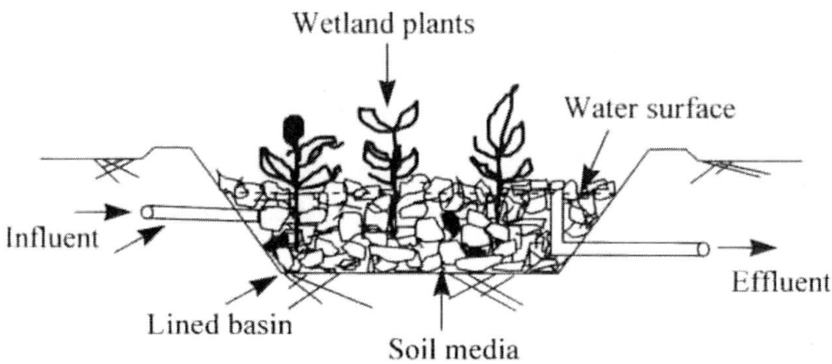

Figure 42. Sub-surface flow system [18].

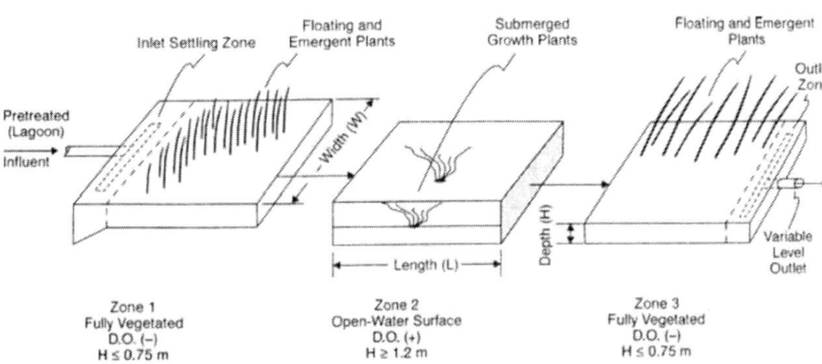

Figure 43. Components of a free water surface constructed wetland [2].

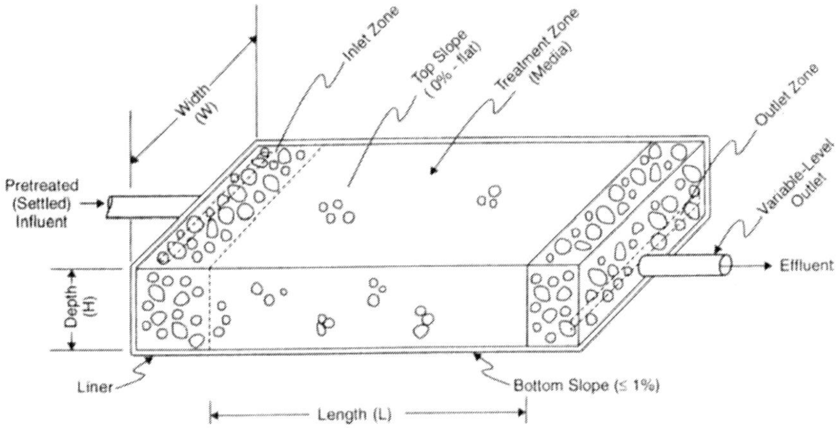

Figure 44. Components of a vegetated submerged bed system [2].

The incorporation of heavy metals, such as mercury, into the food chain may be a deteriorating matter. Phytoremediation is useful in these situations, where natural plants or transgenic plants are able to phytodegrade and phytoaccumulate these toxic contaminants in their above-ground parts, which will be then harvested for extraction. The heavy metals in the harvested biomass can be further concentrated by incineration and recycled for industrial implementation. Rhizofiltration is a sort of phytoremediation that involves filtering wastewater through a mass of roots to remove toxic substances or excess nutrients. Phytoaccumulation or phytoextraction implements plants or algae to remove pollutants and contaminants from wastewater into plant biomass that can be harvested. Organisms that accumulate over than usual amounts of pollutants from soils are termed hyperaccumulators, where a multitude of tables that show the different hyperaccumulators are available and should be referred to. In the case of organic pollutants, such as pesticides, explosives, solvents, industrial chemicals, and other xenobiotic substances, certain plants render these substances non-toxic by their metabolism and this process is called phytotransformation. In other cases, microorganisms that live in symbiosis with plant roots are able to metabolize these pollutants in wastewater. Figure 45 shows the tissues where the rhizofiltration, phytodegradation, and phytoaccumulation take place.

4.4.9. Vermifiltration

Vermiculture, or worm farming, is the implementation of some species of earthworm, such as *Eisenia fetida* (known as red wiggler, brandling, or manure worm) and *Lumbricus rubellus*, to make vermicompost, also known as worm compost, vermicast, worm castings, worm humus, or worm manure, which is the end-product of the breakdown of organic matter and considered to be a nutrient-rich biofertilizer and soil conditioner. Vermiculture can be implemented to transform livestock manure, food leftovers, and organic matters into a nutrient-rich biofertilizer.

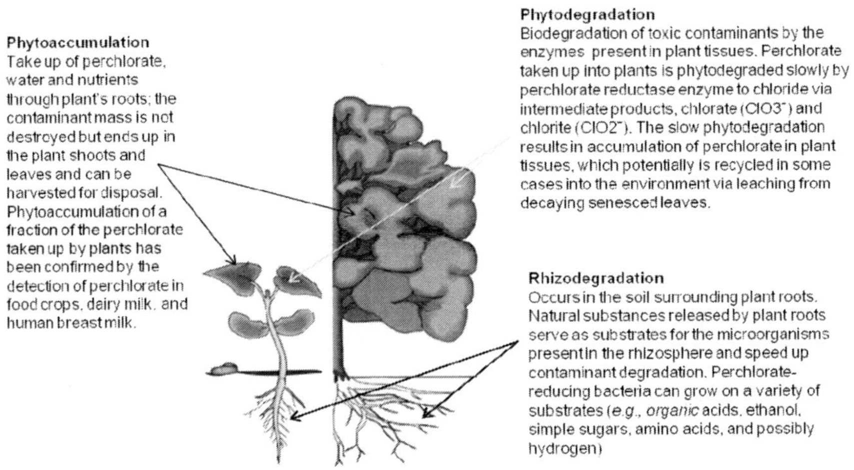

Figure 45. Rhizofiltration, phytodegradation, and phytoaccumulation [19].

The potential use of earthworms to break down and manage sewage sludge began in the late 1970s [20] and was termed vermicomposting. The introduction of earthworms to the filtration systems, termed vermifiltration systems, was advocated by José Toha in 1992 [21]. Vermifilter is widely used to treat wastewater, and appeared to have high treatment efficiency, including synchronous stabilization of wastewater and sludge [22, 23, 24]. Vermifiltration is a feasible treatment method to reduce and stabilize liquid-state sewage sludge under optimal conditions [24, 25, 26]. Vermicomposting involves the joint action of earthworms and microorganisms [24, 27, 28], and significantly enhances the breakdown of sludge. Earthworms operate as mechanical blenders and by comminuting the organic matter they modify its physical and chemical composition, steadily decreasing the C:N ratio, increasing the surface area exposed to microorganisms, and making it much more suitable for bacterial activity and further breakdown. Throughout the passageway is the earthworm gut, they move fragments and bacteria-rich excrements, consequently homogenizing the organic matter [29]. An intensified bacterial diversity was found in vermifilter, compared with conventional biofilter without earthworms [25]. The principle of using earthworms to treat sewage sludge is based on the perception that there is a net loss of biomass and energy when the food chain is extended [25]. Compared to other technologies of liquid-state sludge stabilization, such as anaerobic digestion and aerobic digestion [30], vermifiltration is a low-cost and an ecologically sound technique, and more suitable for sewage sludge treatment of small or developing-countries' WWTPs [23, 24, 25, 26, 31]. Figure 46 illustrates schematic diagram of a vermifilter, where the earthworms are in the filter bed.

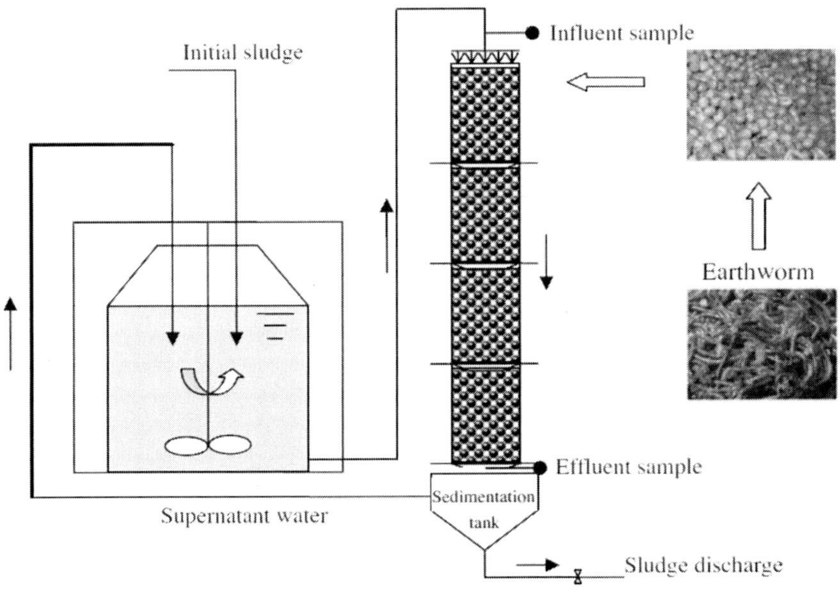

Figure 46. Schematic diagram of a vermifilter [24].

An important application is in livestock manure treatment as shown in Figure 47, where manure is flushed out from the livestock building to a raw effluent tank then the raw effluent is screened to separate the solid waste from manure. The screened effluent is then introduced to the vermifilter to produce the vermicompost. The vermifiltered effluent is then stored in a sedimentation tank. Afterwards, the vermifiltered effluent is introduced to constructed wetlands where the phytoremediation process takes place. The purified water can be then used to flush the water from the livestock building.

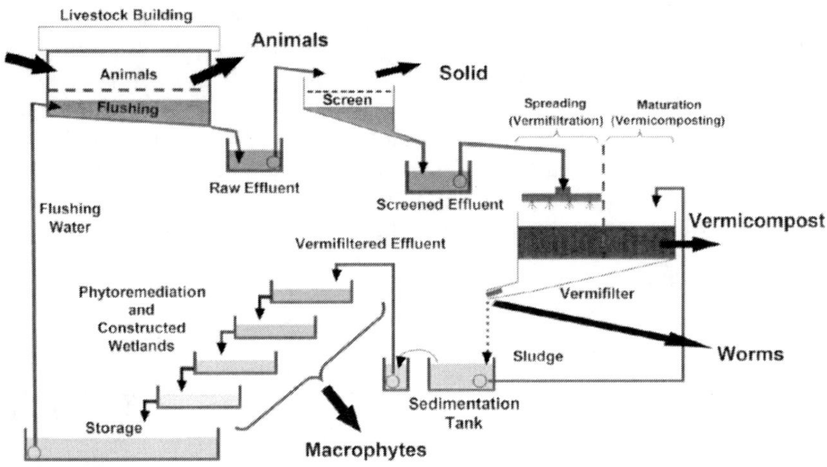

Figure 47. Schematic diagram of a manure treatment system containing vermifiltration and phytoremediation processes (Amended and redrawn from Morand et al. [32]).

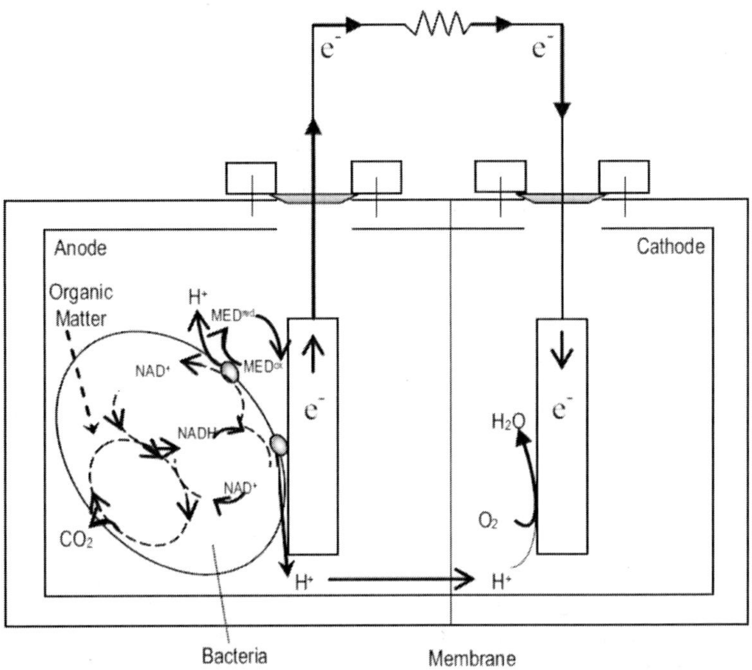

Figure 48. Schematic diagram of the essential components of an MFC [33].

4.4.10. Microbial fuel cells

The microbial fuel cells (MFCs) allow bacteria to grow on the anode by oxidizing the organic matter that result in releasing electrons. The cathode is sparked with air to provide dissolved oxygen for the reaction of electrons, protons, and oxygen on the cathode, which result in completing the electrical circuit and producing electrical energy (Figure 48).

5. CHEMICAL TREATMENT OF WASTEWATER

5.1. Chemical precipitation

The dissolved inorganic components can be removed by adding an acid or alkali, by changing the temperature, or by precipitation as a solid. The precipitate can be removed by sedimentation, flotation, or other solid removal processes [1]. Although chemical precipitation (coagulation, flocculation) is still implemented, it is highly recommended to substitute the chemical precipitation process by phytoremediation (see previous section), where the trend is to ramp up the implementation of bioremediation and phytoremediation to reduce the use of chemicals, which is in line with the "Green Development".

5.2. Neutralization

Neutralization is controlling the pH of the wastewater whether it is acidic or alkaline to keep the pH around 7. The lack of sufficient alkalinity will require the addition of a base (Table 3) to adjust the pH to the acceptable range. Lime (CaO), calcium hydroxide ($Ca(OH)_2$), sodium hydroxide (NaOH), and sodium carbonate (Na_2CO_3), also known as soda ash, are the most common chemicals used to adjust the pH [34]. The lack of sufficient acidity will require the addition of an acid to adjust the pH to the acceptable range. Sulfuric acid (H_2SO_4) and carbonic acid (H_2CO_3) are the most common chemicals used to adjust the pH.

5.3. Adsorption

Adsorption is a physical process where soluble molecules (adsorbate) are removed by attachment to the surface of a solid substrate (adsorbent). Adsorbents should have an extremely high specific surface area. Examples of adsorbents include activated alumina, clay colloids, hydroxides, resins, and activated carbon. The surface of the adsorbent should be free of adsorbate. Therefore, the adsorbent should be activated before use. A wide range of organic materials can be removed by adsorption, including detergents and toxic compounds. The most widely used adsorbent is activated carbon, which can be produced by pyrolytic carbonization of biomass [1]. Figure 49 illustrates the difference between absorption and adsorption. Activated carbon is the most implemented adsorbent and is a sort of carbon processed to be riddled with small, low-volume pores that enlarge the surface area available for adsorption. Owing to its high level of microporosity, 1 g of activated carbon has a surface area larger than 500 m^2, which was determined by gas adsorption. Figure 50 shows a bed carbon adsorption unit. Note that the carbon can be regenerated by thermal oxidation or steam oxidation and reused. The adsorption capacity, one of the most important characteristics of an adsorbent, can be calculated as follows:

$$\underset{mg/g}{\text{Adsorption Capacity}} = \underset{mg}{\text{Adsorbate}} / \underset{g}{\text{Adsorbent}}$$

Table 3. Neutralization: Case of acidic wastewater [34].

Neutralization reactions	
To neutralize sulfuric acid with	
Lime:	$H_2SO_4 + CaO \rightleftharpoons CaSO_4 + H_2O$
Calcium hydroxide:	$H_2SO_4 + Ca(OH)_2 \rightleftharpoons CaSO_4 + H_2O$
Sodium hydroxide:	$H_2SO_4 + NaOH \rightleftharpoons NaSO_4 + 2H_2O$
Soda ash:	$H_2SO_4 + Na_2CO_3 \rightleftharpoons Na_2SO_4 + H_2O + CO_2$
To neutralize hydrochloric acid with	
Lime:	$2HCl + CaO \rightleftharpoons CaCl_2 + H_2O$
Calcium hydroxide:	$HCl + Ca(OH)_2 \rightleftharpoons CaCl_2 + H_2O$
Sodium hydroxide:	$HCl + NaOH \rightleftharpoons NaCl + H_2O$
Soda ash:	$HCl + Na_2CO_3 \rightleftharpoons NaCl + H_2O + CO_2$

Note: a stoichiometric reaction will yield a pH of 7.0.

[i] - Note: a stoichiometric reaction will yield a pH of 7.0

The factors that affect adsorption are [3]:

1) Particle diameter: the adsorption is inversely proportional to the particle size of the adsorbent, and directly proportional to surface area.

2) Adsorbate concentration: the adsorption is directly proportional to adsorbate concentration.
3) Temperature: the adsorption is directly proportional to temperature.
4) Molecular weight: generally, the adsorption is inversely proportional to molecular weight depending upon the compound weight and configuration of pores diffusion control.
5) pH: the adsorption is inversely proportional to pH due to surface charge.
6) Individual properties of adsorbate and adsorbent are difficult to compare.
7) Iodine number: is the mass of iodine (g) that is consumed by 100 g of a substance.

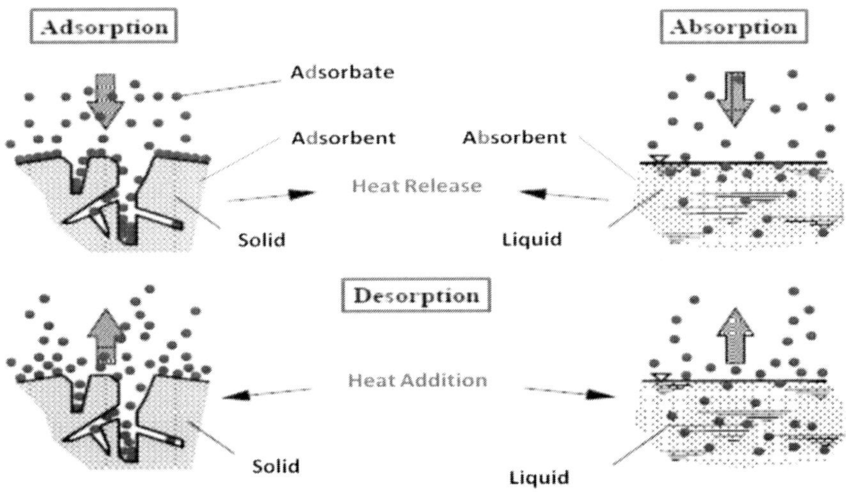

Figure 49. A comparison between absorption and adsorption.

5.4. Disinfection

The disinfection of wastewater is the last treatment step of the tertiary treatment process. Disinfection is a chemical treatment process conducted by treating the effluent with the selected disinfectant to exterminate or at least inactivate the pathogens. The rationales behind effluent disinfection are to protect public health by exterminating or inactivating the pathogens such as microbes, viruses, and protozoan, and to meet the wastewater discharge standards. The purpose of disinfection is the protection of the microbial wastewater quality. The ideal disinfectant should have bacterial toxicity, is inexpensive, not dangerous to handle, and should have reliable means of detecting the presence of a residual. The chemical disinfection agents include chlorine, ozone, ultraviolet radiation, chlorine dioxide, and bromine [3].

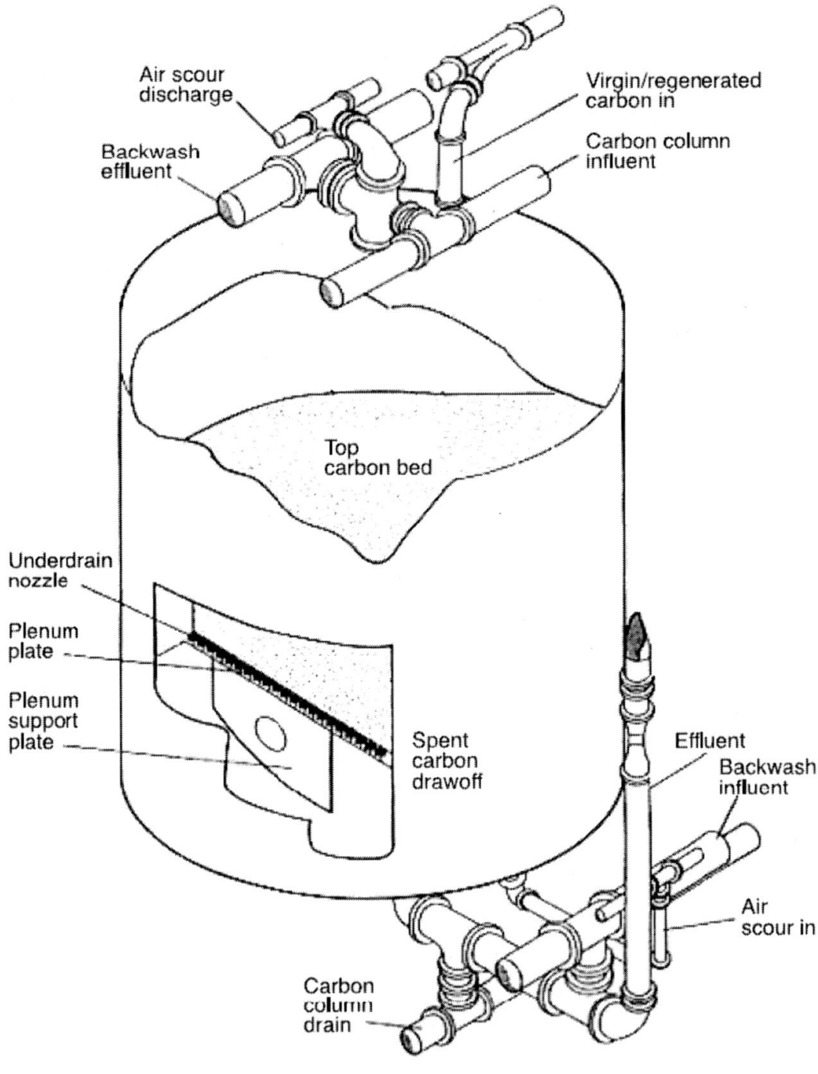

Figure 50. A bed carbon adsorption unit [35].

5.4.1. Chlorine

Chlorine is one of the oldest disinfection agents used, which is one of the safest and most reliable. It has extremely good properties, which conform to the aspects of the ideal disinfectant. Effective chlorine disinfection depends upon its chemical form in wastewater. The influencing factors are pH, temperature, and organic content in the wastewater [3]. When chlorine gas is dissolved in wastewater, it rapidly hydrolyzes to hydrochloric acid (HCl) and hypochlorous acid (HOCl) as shown in the following chemical equation:

$$Cl_2+H{=}O \leftrightarrow H^++Cl^-+HOCl$$

Free ammonia combines with the HOCl form of chlorine to form chloramines in a three-step reaction, as follows:

$$NH^3+HOCl \rightarrow NH_2Cl+H_2O$$

$$NH_2Cl+HOCl \rightarrow NHCl_2+H_2O$$

$$NHCL_2+HOCl \rightarrow NCl_3+H_2O$$

Figure 51 illustrates the chlorination curve, where the formation of chloramines occurs at the breakpoint. The free chlorine residual first rises then falls until the reaction with ammonia has been completed. As additional chlorine is applied and ammonia is consumed, the chlorine residual rises again.

Dechlorination is a very important process, where activated carbon, sulfur compounds, hydrogen sulfide, and ammonia can be implemented to minimize the residual chlorine in a disinfected effluent prior to discharge. Activated carbon and sulfur compounds are the most widely used [3]. The commonly used sulfur compounds are sulfur dioxide (SO_2), sodium metabisulfite (NaS_2O_5), sodium bisulfate ($NaHSO_3$), and sodium sulfite (Na_2SO_3). The dechlorination reactions with the abovementioned compounds are described in the following equations:

$$SO_2+2H_2O+Cl_2 \rightarrow H_2SO_4+2HCl$$

$$SO_2+H_2O+HOCl \rightarrow 3H^++Cl^-+SO^{2-}{}_4$$

$$Na_2S_2O_5+2Cl_2+3H_2O \rightarrow 2NaHSO_4+4HCl$$

$$NaHSO_3+H_2O+Cl_2 \rightarrow NaHSO_4+2HCl$$

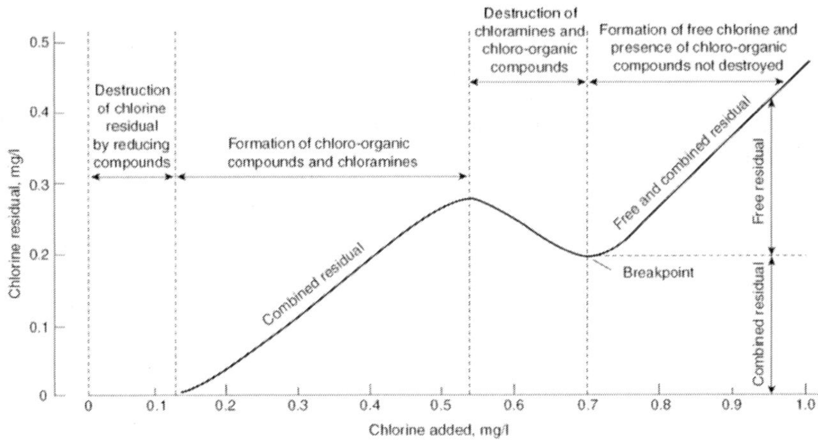

Break point chlorination by the formation of chloramines. The free chlorine residuals first rises then falls until the reaction with ammonia have been completed. As additional chlorine is applied and ammonia is consumed, the chlorine residual again rises.

Figure 51. Chlorination curve [3].

5.4.2. Ozone

Ozone (O_3) is a very strong oxidant typically used in wastewater treatment. Ozone is able to oxidize a multitude of organic and inorganic compounds in wastewater. These reactions cause an ozone demand in the treated wastewater, which should be fulfilled throughout wastewater ozonation prior to developing an assessable residual. Ozone should be generated at the point of application for use in wastewater treatment as ozone is an unstable molecule [3]. Figure 52 illustrates the corona discharge method for making ozone. Ozone is generally formed by combining an oxygen atom with an oxygen molecule (O_2) as follows:

5.4.3. Ultraviolet light

Ultraviolet (UV) radiation is a microbial disinfectant that leaves no residual. It requires clear, un-turbid, and non-colored water for its implementation. The commercial UV disinfection systems use low- to medium-powered UV lamps with a wavelength of 354 nm [3]. The UV dosage can be calculated as follows:

$$D = I \cdot t$$

where, D is the UV dose (mW. s/cm^2); I is the intensity (mW/cm^2); and t is the exposure time (s).

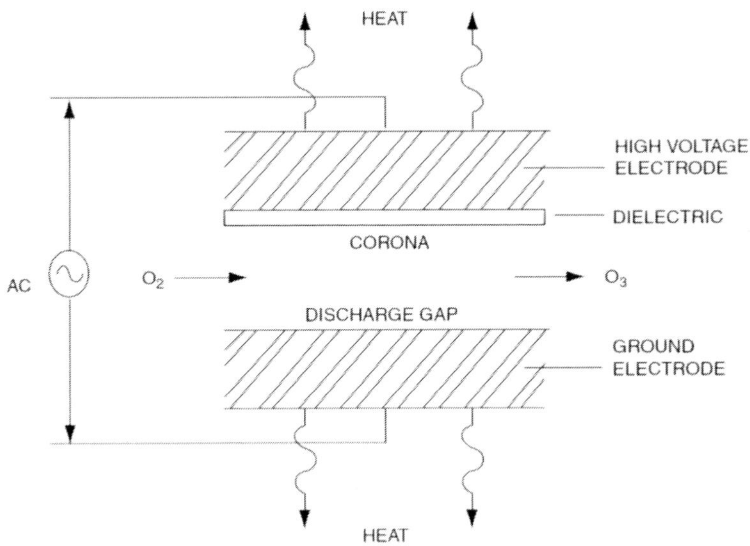

Figure 52. Schematic drawing of corona discharge method for making ozone [3].

The advantages of UV radiation are: (1) directly effective against the DNA of many microorganisms, (2) not reactive with other forms of carbonaceous demand, and (3) provides superior bactericidal kill values while not leaving any residues. The advantage is often the disadvantage, because power fluctuations, variations in hydraulic flow rates, and color or turbidity can cause the treatment to be ineffective [3]. Additionally, cell recovery and re-growth of the damaged organisms because of the inactivation of their predators and competitors has come to light.

5.5. Ion exchange

Ion exchange (IX) is a reversible reaction in which a charged ion in a solution is exchanged with a similarly charged ion which is electrostatically attached to an immobile solid particle. The most common implementation of ion exchange method in wastewater treatment is for softening, where polyvalent cations (e.g., calcium and magnesium) are exchanged with sodium [36]. Practically, wastewater is introduced into a bed of resin. The resin is manufactured by converting a polymerization of organic compounds into a porous matrix. Typically, sodium is exchanged with cations in the solution [34]. The bed is shut down when it becomes saturated with the exchanged ions, where it should be regenerated by passing a concentrated solution of sodium back through the bed. Figure 53 shows the schematic illustration of organic cation-exchange bead. Figure 54 shows a typical ion exchange resin column. Table 4 shows the ion preference and affinity for some selected compounds.

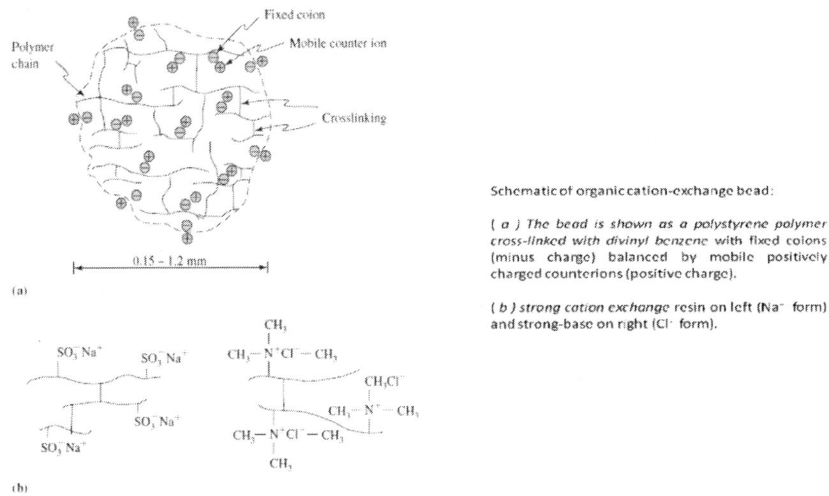

Figure 53. Schematic illustration of organic cation-exchange bead [34].

Table 4. Ion preference and affinity for some selected compounds [3].

Strong Acid Cation Exchanger	Strong Base Anion Exchanger	Weak Acid Cation Exchanger	Weak Base Anion Exchanger	Weak Acid Chelate Exchanger
Barium (2+)	Iodide (1−)	Hydrogen (1+)	Hydroxide (1−)	Copper (2+)
Lead (2+)	Nitrate (1−)	Copper (2+)	Sulfate (2−)	Iron (2+)
Mercury (2+)	Bisulfite (1−)	Cobalt (2+)	Chromate (2−)	Nickel (2+)
Copper (1+)	Chloride (1−)	Nickel (2+)	Phosphate (2−)	Lead (2+)
Calcium (2+)	Cyanide (1−)	Calcium (2+)	Chloride (1−)	Manganese (2+)
Nickel (2+)	Bicarbonate (1−)	Magnesium (2+)		Calcium (2+)
Cadmium (2+)	Hydroxide (1−)	Sodium (1+)		Magnesium (2+)
Copper (2+)	Fluoride (1−)			Sodium (1+)
Cobalt (2+)	Sulfate (2−)			
Zinc (2+)				
Cesium (1+)				
Iron (2+)				
Magnesium (2+)				
Potassium (1+)				
Manganese (2+)				
Ammonia (1+)				
Sodium (1+)				
Hydrogen (1+)				
Lithium (1+)				

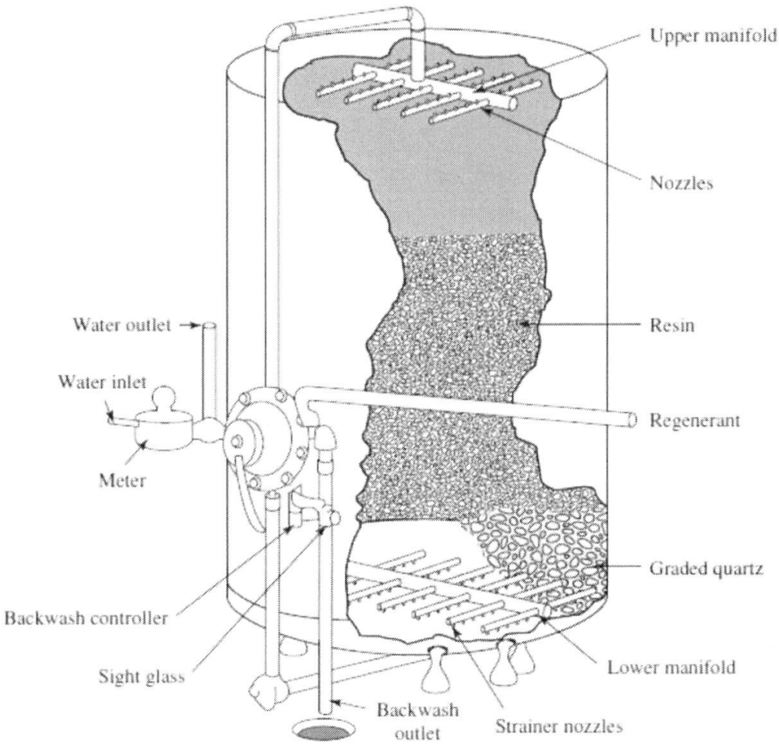

Typical ion exchange resin column.

(*Source:* U.S. EPA, 1981.)

Figure 54. Typical ion exchange resin column [37].

5.6. Physicochemical treatment processes

The principal advanced physicochemical wastewater treatment processes are elucidated in Table 5.

6. WASTEWATER TREATMENT PLANTS

This section shows some examples of WWTPs as shown in Figure 55 (a, b, and c) and Figure 56. On the other hand, there are some computer programs for planning and designing WWTPs (Figures 57, 58, and 59).

Table 5. Principal advanced physicochemical wastewater treatment processes [1].

Principal advanced physico-chemical wastewater treatment processes

Process	*Removal function*
Filtration	Suspended solids
Air stripping	Ammonia
Breakpoint chlorination	Ammonia
Ion exchange	Nitrate, dissolved inorganic solids
Chemical precipitation	Phosphorus, dissolved inorganic solids
Carbon adsorption	Toxic compounds, refractory organics
Chemical oxidation	Toxic compounds, refractory organics
Ultrafiltration	Dissolved inorganic solids
Reverse osmosis	Dissolved inorganic solids
Electrodialysis	Dissolved inorganic solids
Volatilization and gas stripping	Volatile organic compounds

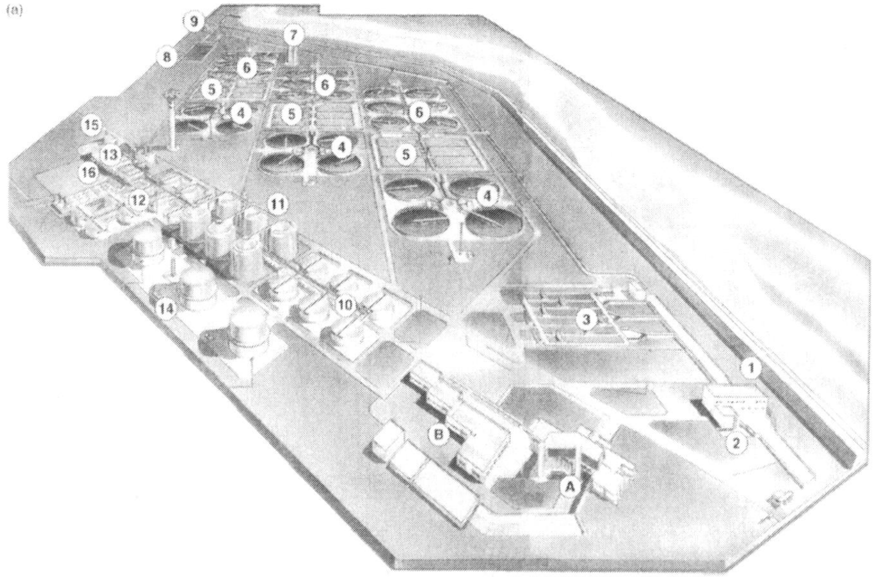

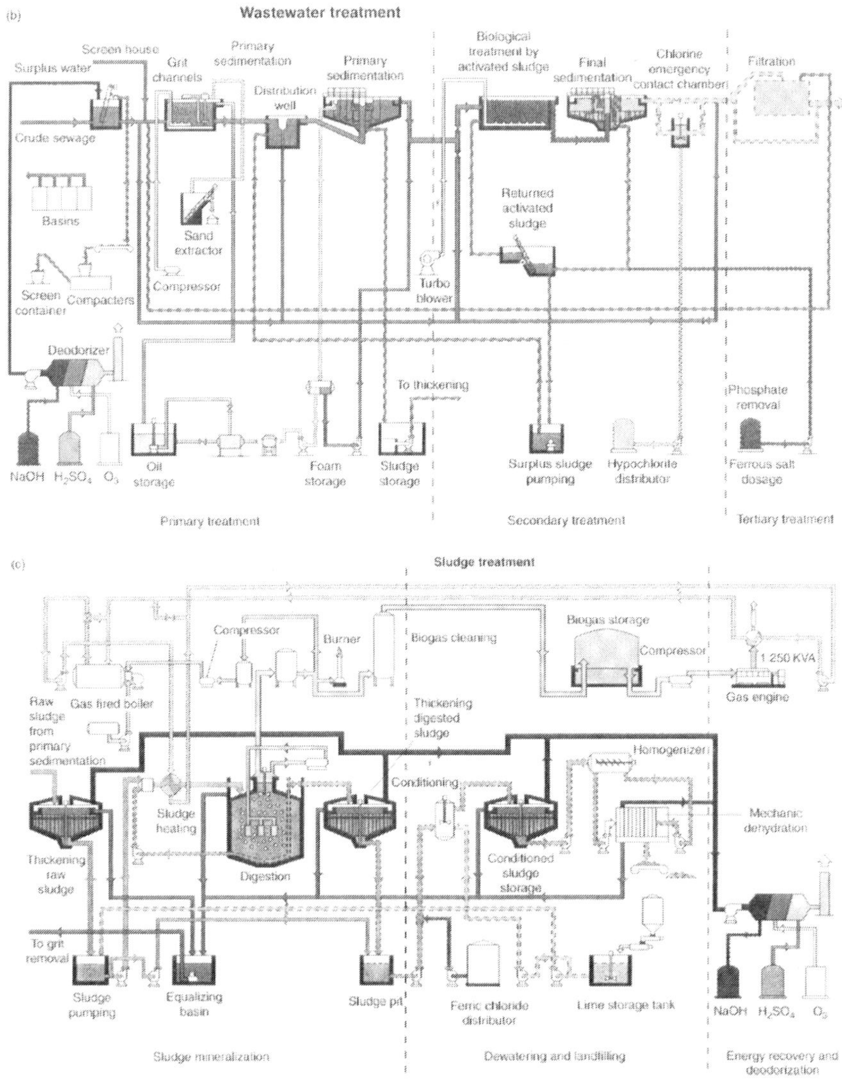

Figure 55. WWTP showing: (a) layout of the plant, (b) wastewater process flow diagrams, and (c) sludge process flow diagram. **Wastewater treatment**: 1. Storm water overflow; 2. screening; 3. grit removal; 4. primary sedimentation; 5. aeration tanks; 6. Secondary sedimentation; 7. emergency chlorination; 8. filtration; 9. effluent outfall. **Sludge treatment**: 10. raw sludge thickeners; 11. digestion tanks; 12. digested sludge thickeners; 13. power house; 14. biogas storage; 15. filter press house; 16. transformer station. A and B are administrative areas [1].

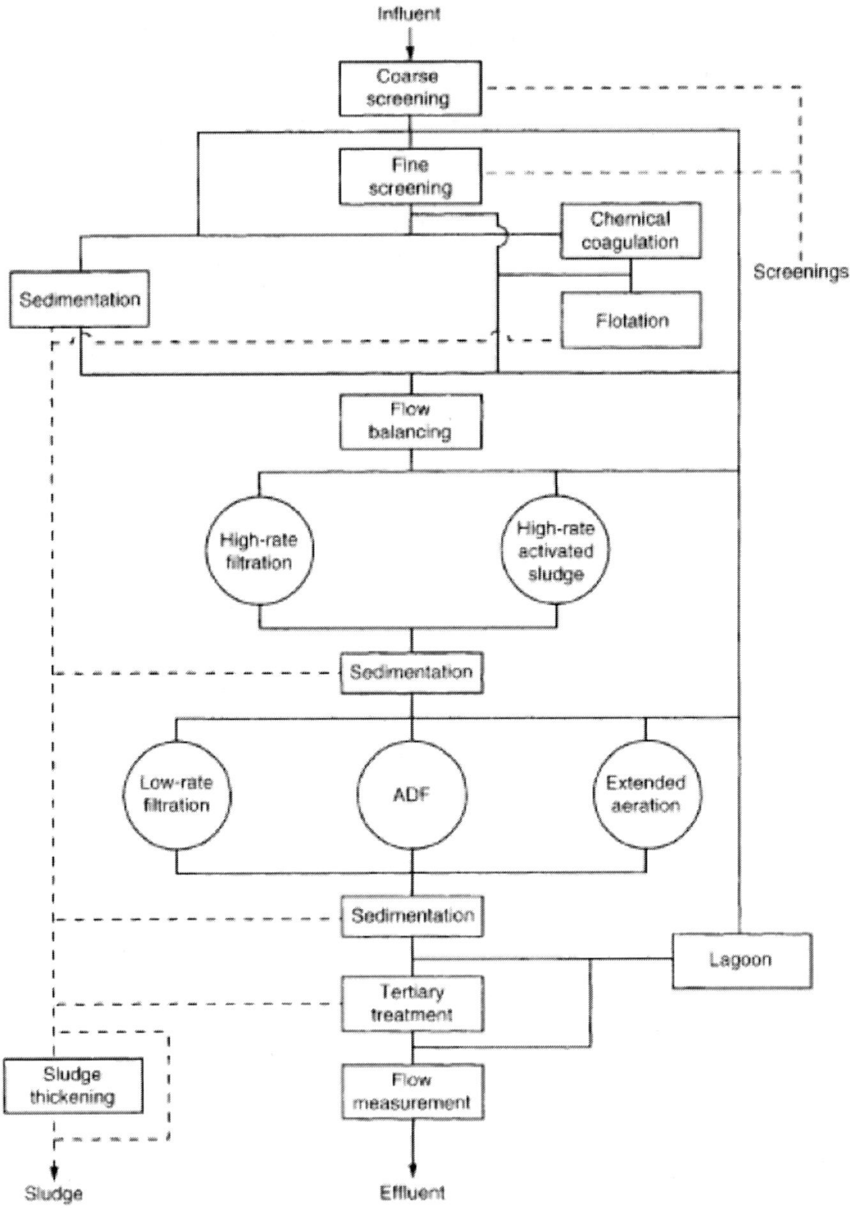

Figure 56. Summary of the main process options commonly employed at both domestic and industrial WWTPs. Not all of these unit processes may be selected, but the order of their use remains the same [1].

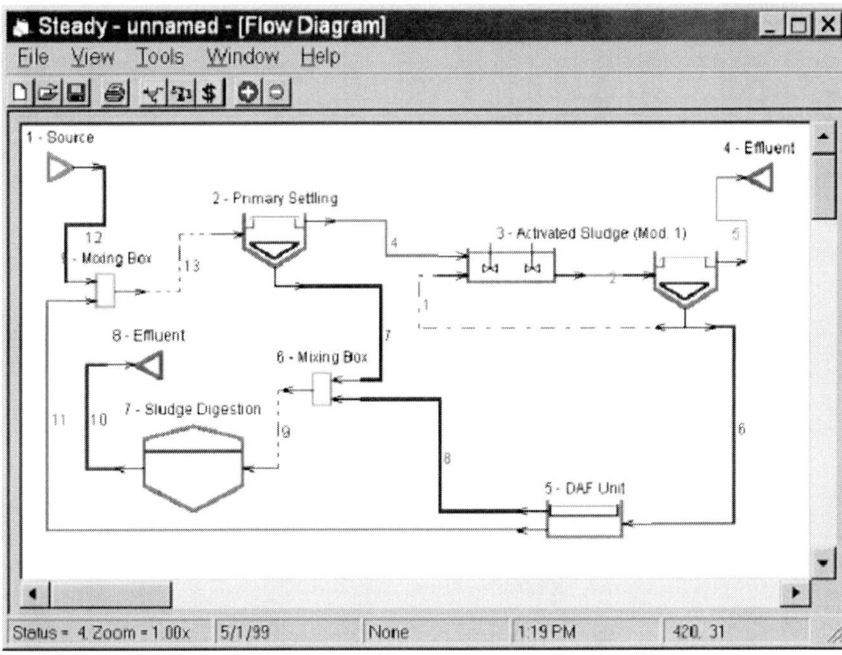

Figure 57. Screenshot of the STEADY program [3].

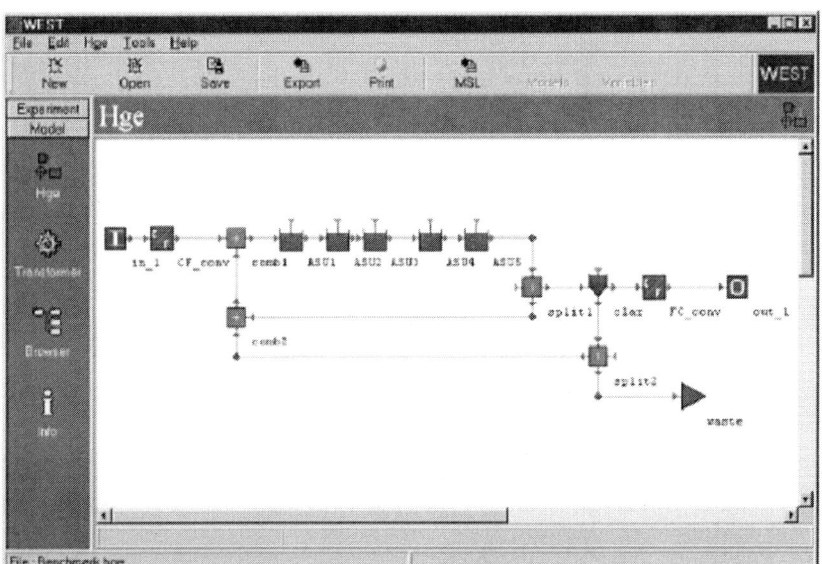

Figure 58. WEST software typical plant configuration [3].

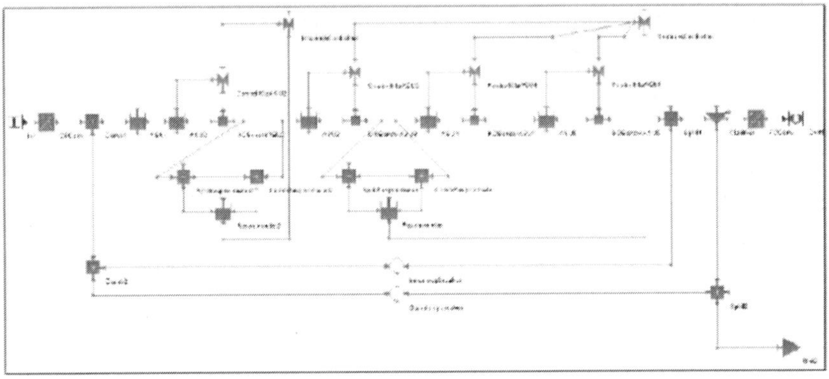

Figure 59. WEST configuration for multitank system [3].

7. CONCLUSIONS

According to this study, it can be conclude that:

1) The trend is to ramp up the implementation of bioremediation, phytoremediation, and mycoremediation to reduce the use of chemicals, which is in line with the "Green Development".
2) The recent developments elucidate that subsequent to the physical treatment processes (the primary treatment) the biological treatment processes come in turn as secondary treatment and precede the chemical treatment processes, which constitute the tertiary treatment.
3) Microbial fuel cells, phytoremediation, and mycoremediation are the focus of the future development in this field.

REFERENCES

1. Gray N. F. (2005). Water Technology: An Introduction for Environmental Scientists and Engineers (2nd Edition), Elsevier Science & Technology Books, ISBN 0750666331, Amsterdam, The Netherlands.
2. Lin, S. D. (2007). Water and Wastewater Calculations Manual (2nd Edition), McGraw-Hill Companies, Inc., ISBN 0-07-154266-3, New York, USA.
3. Russell D. L. (2006). Practical Wastewater Treatment, John Wiley & Sons, Inc., ISBN-13: 978-0-471-78044-1, Hoboken, New Jersey, USA.
4. Samer M. (2015). GHG Emission from Livestock Manure and its Mitigation Strategies, In: Climate Change Impact on Livestock: Adaptation and Mitigation, V. Sejian, J. Gaughan, L. Baumgard & C. Prasad (Eds.), In Press, Springer International, ISBN 978-81-322-2264-4, Germany.

5. Samer M., Mostafa E. & Hassan A. M. (2014). Slurry Treatment with Food Industry Wastes for Reducing Methane, Nitrous Oxide and Ammonia Emissions. Misr Journal of Agricultural Engineering, 31 (4): 1523–1548.
6. Samer M. (2011). How to Construct Manure Storages and Handling Systems? IST Transactions of Biosystems and Agricultural Engineering, Vol. 1, No. 1, pp. 1–7, ISSN 1913-8741.
7. Samer M., Grimm H., Hatem M., Doluschitz R. & Jungbluth T. (2008). Mathematical Modeling and Spark Mapping for Construction of Aerobic Treatment Systems and their Manure Handling System, Proceedings of International Conference on Agricultural Engineering, Book of Abstracts p. 28, EurAgEng, Crete, Greece, June 23-25, 2008.
8. Ersahin M. E., Ozgun H., Dereli R. K. & Ozturk I. (2011). Anaerobic Treatment of Industrial Effluents: An Overview of Applications, In: Waste Water - Treatment and Reutilization, F. Einschlag (Ed.), pp. 3–28, InTech, ISBN 978-978-953-307-249-4, Rijeka, Croatia.
9. Samer M. (2012). Biogas Plant Constructions, In: Biogas, S. Kumar (Ed.), pp. 343–368, InTech, ISBN 978-953-51-0204-5, Rijeka, Croatia.
10. Samer M. (2010). A Software Program for Planning and Designing Biogas Plants. Transactions of the ASABE, Vol. 53, No. 4, pp. 1277–1285, ISSN 2151-0032.
11. Pfost D. L. & Fulhage C. D. (2000). Anaerobic Lagoons for Storage/Treatment of Livestock Manure, EQ 387, MU Extension University of Missouri-Columbia, USA.
12. Westerman P. W., Shaffer K. A. & Rice J. M. (2008). Sludge Survey Methods for Anaerobic Lagoons, AG-639W, North Carolina Cooperative Extension Service, College of Agriculture & Life Sciences, North Carolina State University, USA.
13. Pavko A. (2011). Fungal Decolourization and Degradation of Synthetic Dyes Some Chemical Engineering Aspects, In: Waste Water - Treatment and Reutilization, F. Einschlag (Ed.), pp. 65–88, InTech, ISBN 978-978-953-307-249-4, Rijeka, Croatia.
14. Diano N. & Mita D.G. (2011). Removal of Endocrine Disruptors in Waste Waters by Means of Bioreactors, In: Waste Water - Treatment and Reutilization, F. Einschlag (Ed.), pp. 29-48, InTech, ISBN 978-978-953-307-249-4, Rijeka, Croatia.
15. Kabir M., Suzuki M. & Yoshimura N. (2011). Excess Sludge Reduction in Waste Water Treatment Plants, In: Waste Water - Treatment and Reutilization, F. Einschlag (Ed.), pp. 133–150, InTech, ISBN 978-978-953-307-249-4, Rijeka, Croatia.
16. Chadwick D., Sommer S., Thorman R., Fangueiro D., Cardenas L., Amon B. & Misselbrook T. (2011). Manure Management: Implications for Greenhouse Gas Emissions. Animal Feed Science and Technology, 166–167(2011): 514– 531, ISSN 0377-8401.
17. Morris R. H. & Knowles P. (2011). Measurement Techniques for Wastewater Filtration Systems, In: Waste Water - Treatment and Reutilization, F. Einschlag (Ed.), pp. 109–132, InTech, ISBN 978-978-953-307-249-4, Rijeka, Croatia
18. ESCWA (2003). Economic and Social Commission for Western Asia.
19. Kucharzyk K. H., Soule T., Paszczynski A. J. & Hess T. F. (2011). Perchlorate: Status and Overview of New Remedial Technologies, In: Waste Water - Treatment and Reutilization, F. Einschlag (Ed.), pp. 171–194, InTech, ISBN 978-978-953-307-249-4, Rijeka, Croatia.
20. Li, X., Xing M., Yang J. & Huang Z. (2011). Compositional and functional features of humic acid-like fractions from vermicomposting of sewage sludge and cow dung, J. Hazard. Mater. 185: 740–748.
21. Li Y., Robin P., Cluzeau D., Bouché M., Qiu J., Laplanche A., Hassouna M., Morand P., Dappelo C. & Callarec J. (2008). Vermifiltration as a stage in reuse of swine wastewater: monitoring methodology on an experimental farm, Ecol. Eng. 32: 301–309.
22. Li Y., Xiao Y., Qiu J., Dai Y. & Robin P. (2009). Continuous village sewage treatment by vermifiltration and activated sludge process, Water Sci. Technol. 60: 3001–3010.
23. Xing M., Yang J., Wang Y., Liu J. & Yu F. (2011). A comparative study of synchronous treatment of sewage and sludge by two vermifiltration using epigeic earthworm Eisenia fetida, J. Hazard. Mater. 185 (2–3): 881–888.
24. Xing M., Li X., Yang J., Lv B. & Lu Y. (2012). Performance and mechanism of vermifiltration system for liquid-state sewage sludge treatment using molecular and stable isotopic techniques. Chemical Engineering Journal 197: 143–150.

25. Zhao L., Wang Y., Yang J., Xing M., Li X., Yi D. & Deng D. (2010). Earthworm-microorganism interactions: a strategy to stabilize domestic wastewater sludge, Water Res. 44: 2572–2582.
26. Xing M., Zhao L., Yang J., Huang Z. & Xu Z. (2011). Distribution and transformation of organic matter during liquid-state vermiconversion of activated sludge using elemental analysis and spectroscopic evaluation, Environ. Eng. Sci. 28 (9): 619–626.
27. Suthar S. (2008). Development of a novel epigeic–anecic-based polyculture vermireactor for efficient treatment of municipal sewage water sludge, Int. J. Environ. Waster Manage. 2: 84–101.
28. Xing M., Li X., Yang J., Huang Z. & Lu Y. (2011). Changes in the chemical characteristics of water-extracted organic matter from vermicomposting of sewage sludge and cow dung, J. Hazard. Mater. http://dx.doi.org/10.1016/j.jhazmat.2011.11.070.
29. Domínguez J. & Edwards C.A. (2004). Vermicomposting organic wastes, in: S.H. Shakir Hanna, W.Z.A. Mikhaïl (Eds.), Soil Zoology for Sustainable Development in the 21st Century, Cairo, pp. 369–395.
30. Fytili D. & Zabaniotou A. (2008). Utilization of sewage sludge in EU application of old and new methods – a review, Renew. Sust. Energ. Rev. 12: 116–140.
31. Sinha R.K., Bharambe G. & Chaudhari U. (2008). Sewage treatment by vermifiltration with synchronous treatment of sludge by earthworms: A low-cost sustainable technology over conventional systems with potential for decentralization, Environmentalist. 28: 409–420.
32. Morand P., Robin P., Qiu J.P., Li Y., Cluzeau D., Hamon G., Amblard C., Fievet S., Oudart D., Pain le Quere C., Pourcher A.-M., Escande A., Picot B. & Landrain B. (2009). Biomass production and water purification from fresh liquid manure by vermiculture, macrophytes ponds and constructed wetlands to recover nutrients and recycle water for flushing in pig housing. Proceedings of the International Congress "Ecological engineering: From concepts to applications Foreword (EECA)", 2-4 December 2009, Paris, France.
33. Logan B. (2008). Microbial Fuel Cells, Wiley. Hoboken, NJ, EEUU. 200 pp.
34. Davis M. L. (2010). Water and Wastewater Engineering: Design Principles and Practice, McGraw-Hill Companies, Inc., ISBN 978-0-07-171385-6, New York, USA.
35. Sincero, A. P. & G. A. Sincero (2003). Physical-Chemical Treatment of Water and Wastewater, CRC Press LLC, ISBN 1-84339-028-0, Boca Raton, Florida, USA.
36. Clifford D. A. (1999). Ion Exchange and Inorganic Adsorption, In: Water Quality and Treatment (5th Edition), R. D. Letterman (Ed.), pp. 9.1–9.91., American Water Works Association, McGraw-Hill, New York, USA.
37. U.S. EPA (1981) Development Document for Effluent Limitations: Guideline and Standards for the Metal Finishing Point Source Category, U.S. Environmental Protection Agency, EPA/440/1-83-091, Washington, D.C., USA.
38. U.S. EPA (2002). Wastewater Technology Fact Sheet: Anaerobic Lagoons, U.S. Environmental Protection Agency, EPA 832-F-02-009, Washington, D.C., USA.

CHAPTER 11

Morphology of Modified Biochar and Its Potential for Phenol Removal from Aqueous Solutions

Kostas A.Komnitsas* and DimitraZaharaki

School of Mineral Resources Engineering, Technical University of Crete, Chania, Greece

Keywords: biochar, phenols, chemical modification, adsorption, kinetics, isotherms

1. INTRODUCTION

Biochar is a carbon rich and porous material which is produced by thermal conversion of biomass in low oxygen atmosphere at temperature usually not exceeding 600°C. Slow pyrolysis of biomass has the advantage of retaining almost half of its carbon in stable biochar (Manyà, 2012; Kim et al., 2013). Biochar can be produced in every country using various raw materials including energy crops (e.g., corn, cereals, wood pellets, and oilseed rape), agricultural wastes (e.g., wheat straw, rice husk, waste wood, pistachio, peanut, hazelnut shells, and manure) and other wastes including sewage sludge (Liu et al., 2012; Moussavi and Khosravi, 2012; Agrafioti et al., 2013, 2014; Inyang et al., 2014; Frišták et al., 2015).

Biochar is an environmentally friendly material and may exhibit similar properties with activated carbon. Due to its chemical and biological stability it may be utilized as carbon sink in soils for thousands of years. Also, biochar's ability to retain nutrients is considered very important in terms of improving soil fertility and crop productivity (Steiner et al., 2008; Sánchez et al., 2009; Cayuela et al., 2013; Kloss et al., 2014; Sánchez-García et al., 2014).

Due to its strong adsorption capacity, which is related to the large surface area and the presence of negatively charged organic functional groups, biochar can be used as adsorbent for the removal of various contaminants from wastewaters. Contaminants may include metals such as Pb, Cu, Zn, As, Ni, and Cd, as well as organic compounds, such as pesticides and phenols. Adsorption on biochar may be considered as a feasible approach due to its high efficiency, low cost, ease of application and minor by-product generation compared to traditional alternatives (Beesley et al., 2011; Yu et al., 2015).

Phenolic compounds which are present in various concentrations in several waste streams, e.g., 0.5–24 $g \cdot L^{-1}$ in olive mill wastewater (OMW), cause toxic effects on aquatic organisms, bacteria and yeast and prevent seed germination. They are reported as human carcinogens, exhibit poor biodegradability and may also cause long-term ecological damage (Bayramoglu et al., 2009; Di Bene et al., 2013; Inyang and Dickenson, 2015).

The removal of phenolic compounds with the use of biochar was reported so far in a limited number of studies. According to Mubarik et al. (2016), highly porous biochar produced from sugarcane bagasse can be used as efficient and low-cost adsorbent to remove 2,4,6-trichlorophenol from stock solutions.

The potential of biochar produced from switchgrass, hardwood and softwood on removing phenols was mainly related to its microporous surface (Han et al., 2013). Rice straw biochar was used to decrease the concentration of pentachlorophenol in leachates produced from contaminated sediments which were then used to increase the germination rate of wheat seeds (Lou et al., 2011). Biochar produced after pyrolysis of municipal wastewater biosolids removed effectively halogenated phenols, while its sorption capacity was improved at lower solution pH (Oh and Seo, 2016). Sewage sludge biochar showed excellent adsorption capacity for sulfonated methyl phenol resins which are used as additive in drilling fluids (Liu et al., 2015).

Also, biochar prepared from paper mill sludge was used for the removal of pentachlorophenol by simultaneous adsorption and dechlorination (Devi and Saroha, 2015).

According to the Federal Register of Environmental Protection Agency (Environmental Protection Agency, 1995), the benchmark established for the content of total phenols in stormwater discharge is 1 $mg \cdot L^{-1}$. In Greece, no limits for phenol disposal in water receivers have been so far established. Also, no health-based thresholds for the concentration of phenols in drinking water have been proposed; an indicative acceptable value of 0.5 $\mu g \cdot L^{-1}$ has been proposed in accordance with the Council Directive 98/83/EC on the quality of water intended for human consumption (Greek Government Gazette, 2007)[1]. According to the Canadian Soil Quality Guidelines the recommended concentration of phenols in soils to avoid adverse effects is 3.8 $mg \cdot kg^{-1}$ (Canadian Council of Ministers of the Environment, 2007).

Chemical modification of biochar with the use of acids, bases or polymers usually enhances its adsorption efficiency. The so called modified/activated/engineered biochars may be produced by treating the feedstock either prior to or after pyrolysis. The final product is characterized by increased surface area, modified chemical functionality and presence of high-affinity adsorption sites which enhance chemical bonding with contaminants (Mayer et al., 2014; Wang et al., 2015).

According to Qian et al. (2013), Al-modified crop straw biochar resulted in adsorption of larger amounts of As(V) from aqueous solutions at $pH < 7$, in comparison with the respective non-modified biochar. High As adsorption potential was also shown for modified corn straw biochar which was prepared by pyrolysis of the raw material that was initially soaked in aqueous $KMnO_4$ solution (0.079 $mol \cdot L^{-1}$) (Yu et al., 2015). Sun et al. (2015) have shown that eucalyptus saw dust biochar modified with citric, tartaric, and acetic acids can efficiently remove methylene blue from aqueous solution.

Also, more than 90% NO_3 removal was attained using biochar produced from conocarpus green waste which was chemically modified with magnesium and iron oxides prior to pyrolysis (Usman et al., 2016). Finally, hydrogel-biochar composites modified with NaCl were found to improve sorption of phenols (Karakoyun et al., 2011).

Various models have been widely applied to describe adsorption of contaminants from solutions and elucidate the mechanisms involved. These models are based either on solution concentration, e.g., first-order, second-order reversible or irreversible and pseudo-first, pseudo-second order or on adsorption efficiency, e.g., Lagergren's first-order equation, Zeldowitsch's model and Elovich's equation (Ho and McKay, 1999; Ho, 2006a,b; Dari et al., 2015; Zahir et al., 2015).

The present paper aims to characterize the morphology of biochar modified with the use of either 1M KOH or 1M $FeCl_3$ and also assess, through kinetic and equilibrium experiments, the effect of modification on the removal of phenols from synthetic solutions.

2. METHODOLOGY

2.1 Raw Materials

The raw materials used for the production of biochar were (i) pistachio *(Pistacia vera L.)* shells (PIr) from trees cultivated in Aegina island, Greece (ii) pecan *(Carya illinoinensis)* shells (PEr) from trees cultivated in the region of Chania, island of Crete, Greece and (iii) pine wood sawdust (SDr) obtained from a carpentry workshop located in Akrotiri, Chania, Crete.

2.2 Biochar Production and Characterization

The raw materials were oven dried (ON–02, MEDLINE) for 24 h prior to use in order to remove moisture. Then, slow pyrolysis was carried out in a modified laboratory furnace N–8L SELECTA at 400°C for 60 min using porcelain capsules. The heating rate was 10°C·min^{-1}. Nitrogen was purged in the oven for 60 min at a rate of 100 cm^3·min^{-1} to remove air. The obtained biochars from pistachio shells, pecan shells and sawdust are hereinafter mentioned as PI, PE, and SD, respectively.

All biochars were pulverized using a FRITSCH pulverizer to obtain grain size < 100 μm, as evaluated using a Mastersizer S (Malvern Instruments) particle size analyzer. The following parameters were determined in feedstocks and biochars. Pyrolysis yield (y_P) was determined from the % weight loss after heating. pH was measured using a solid:liquid ratio of 1:10 with a Hanna 211 pH/Eh meter. Porosity (%) was measured by mercury intrusion porosimetry using a Micromeritics AutoPore 9400 porosimeter. Brunauer–Emmett–Teller (BET) surface area was measured using a NOVA Surface Area Analyzer (Quantachrome instruments). The elemental C, H, S, and N analysis was carried out in a Flash 2000 Elemental Analyzer Thermo Scientific calibrated using BBOT standards (2,5-Bis(5-tert-butyl-2-benzo-oxazol-2-yl)thiophene) containing carbon; the oxygen content was subsequently calculated as the difference. Biochars were subjected to thermogravimetric (TG) analysis using a differential thermogravimetric analyzer TGA-6/DTG of Perkin Elmer as described in detail in Komnitsas et al. (2015, 2016); volatile matter (VM) and ash content were determined, while char fraction (100%–%VM) and fixed carbon (FC) content (% char–% ash content) were calculated.

The following analyses were also carried out for feedstock and biochar characterization. X–ray diffraction (XRD) analysis was performed using a Bruker AXS (D8 Advance type) diffractometer with Cu tube, scanning range from 4° to 70° 2θ, step 0.02° and measuring time 0.2 s/step. The qualitative analysis was assessed with the use of the DIFFRAC^{plus} EVA v. 2006 software and the Powder Diffraction File (PDF-2) database. SEM analysis was performed with a JEOL 6380LV scanning electron microscope equipped with an EDS INCA microanalysis system with low vacuum, pressure 30 Pa, voltage 20 kV and 10–12 mm working distance from the detector. FTIR analysis was carried out using pellets produced after mixing a pulverized sample of feedstock or biochar with KBr at a ratio of 1:100 w/w, using a Perkin-Elmer Spectrum 1000 spectrometer (USA).

2.3 Modification of Biochar

Pulverized biochars PI, PE, and SD were subjected to chemical modification. In brief, 10 g of each biochar were added in 250 mL of 1M KOH (pH 13.5) or 1M $FeCl_3$ (pH 0.4) and the solution stirred with magnetic stirrers (Agimatic-ED, Spain) at 350 rpm for 24 h at room temperature. The supernatant solution was discarded and the solid material was collected and washed repeatedly with distilled water to remove excess of KOH or $FeCl_3$ and then was oven-dried at 40°C for 24 h. The biochars modified with KOH or $FeCl_3$ were named as PI-K, PE-K, SD-K or PI-F, PE-F, SD-F, respectively, and characterized as previously described.

2.4 Kinetic and Equilibrium Studies

Kinetic and equilibrium experiments were carried out using biochars PI, PE, and SD and modified biochars PI-K, PE-K, SD-K and PI-F, PE-F, SD-F. A synthetic phenol solution 100 mg·L^{-1} was prepared and used in all experiments by dissolving the required quantity of phenol C_6H_5OH (Sigma Aldrich, Germany) in distilled water.

For the kinetic experiments the adsorbent-biochar concentration used was 5 g·L^{-1}. The selection of this specific concentration was based on previous studies carried out by the authors (Komnitsas et al., 2015, 2016). Agitation took place in 200 mL glass beakers at room temperature at 350 rpm, using a Vibromatic (Spain) rocking mixer. Reference tests were also carried out using 5 g·L^{-1} of activated carbon (AC) (Donau Chemie, Austria), which is an established commercial adsorbent. At various time intervals (0.5, 1, 2, 5, 18, 24, 48, and 72 h) 10 mL of liquid samples were withdrawn and filtered through Whatman filters (0.45 μm) for the determination of phenol concentration in solution using the Folin–Ciocalteu method (Box, 1983) and a SMART3 Lamotte, USA, colorimeter. All tests and measurements were carried out in triplicate.

Equilibrium experiments were carried out using four adsorbent-biochar concentrations, namely 0.5, 1, 2, and 5 g·L^{-1}, while the adsorbed phenol concentration was calculated as the difference between the initial and the concentration at equilibrium in liquid phase. The other experimental conditions were similar to those described previously for the kinetic experiments.

3. RESULTS AND DISCUSSION

3.1 Characterization of Biochars

The characterization of the three raw materials, namely pistachio shells (PIr), pecan shells (PEr), and sawdust (SDr) as well as of the biochars produced after pyrolysis at 400°C for 60 min (PI, PE, SD) and the modified biochars using 1M KOH (PI-K, PE-K, SD-K) or 1M $FeCl_3$ (PI-F, PE-F, SD-F) is presented in Table 1. It is shown that the pyrolysis yield varies between 17.7 and 34.7% depending on pyrolysis temperature, heating rate and residence time (Komnitsas et al., 2015; Tripathi et al., 2016).

The paste pH (biochar:water ratio 1:10 w/w) of biochars PI, PE and SD is 6.4, 6.1, and 4.8, respectively, which is higher compared to the paste pH of the respective raw materials. Slightly lower values are obtained for the KOH-modified biochars, while a further slight decrease of pH is shown for the $FeCl_3$-modified biochars due to the acidity of the ferric chloride solution.

The volatile matter (VM) content of all biochars is substantially decreased compared to the raw materials, while the VM content of biochar PI is higher compared to PE and SD biochars. The char and fixed carbon (FC) contents increased in all biochars due to their higher content of bioavailable carbon. The ash content (difference between char and FC content), which expresses the inorganic matter content, is low and varies in all raw materials and biochars between 1.6 and 1.8%. The C content in all biochars increases substantially compared to the raw materials, while the hydrogen, nitrogen, and oxygen contents decrease accordingly.

Decreased H/C and O/C ratios were obtained compared to those present in the raw materials, while chemical modification resulted in a further decrease of the O/C ratios, as also shown in other studies (Dehkhoda et al., 2016). Finally, it is mentioned that no sulfur was identified in any of the raw materials used.

Increased porosity of all biochars was also shown (29.3, 25.3, and 22.5%, for PI, PE, and SD, respectively) compared to the raw materials due to the transformation of aliphatic C structures to aromatic C structures (Brewer et al., 2014; Gray et al., 2014). The porosity was substantially increased for all and especially the KOH-modified biochars.

Table 1. Characterization of pistachio shells (PIr), pecan shells (PEr), sawdust (SDr), and produced biochars.

	PIr	PI	PI-K	PI-F	PEr	PE	PE-K	PE-F	SDr	SD	SD-K	SD-F
y_P, %	–	31.4	–	–	–	34.7	–	–	–	17.7	–	–
pH	4.2	6.4	6.1	5.7	4.8	6.1	5.8	5.6	3.7	4.8	4.6	4.4
VM, %	86	46.3	–	–	71.3	38.9	–	–	89.6	39.9	–	–
Char, %	14	53.7	–	–	28.7	61.1	–	–	10.4	60.1	–	–
FC, %	12.4	52	–	–	27	59.3	–	–	8.8	58.4	–	–
Ash, %	1.6	1.7	–	–	1.7	1.8	–	–	1.6	1.7	–	–
% C	45.93	73.42	78.24	75.4	47.4	71.6	75.3	72.8	46.7	65.2	72.9	67.3
% H	6.04	2.93	3.07	2.95	5.39	2.65	2.75	2.63	5.81	2.08	2.19	2.11
% N	0.42	0.21	0.74	0.71	0.63	0.6	0.76	0.72	0.51	0.32	0.39	0.36
% O	47.61	23.44	17.95	20.94	46.58	25.15	21.09	23.85	46.98	32.4	24.52	30.23
H/C ratio	0.132	0.040	0.039	0.039	0.114	0.037	0.037	0.036	0.124	0.032	0.030	0.031
O/C ratio	1.037	0.319	0.229	0.278	0.983	0.351	0.281	0.328	1.006	0.497	0.336	0.449
Porosity, %	15.2	29.3	49.8	41.4	12.8	25.3	34.7	29.8	10.4	22.5	27.5	26.5
Surface area, $m^2 \cdot g^{-1}$	–	196.4	572.4	421.5	–	142.4	397.3	351.6	–	48.7	124.6	110.8

The surface area of the non-modified biochars followed the sequence PI>PE>SD, and was also significantly increased after modification following the same trend, namely PI-K (572.4 $m^2 \cdot g^{-1}$) > PE-K (397.3 $m^2 \cdot g^{-1}$) > SD-K (124.6 $m^2 \cdot g^{-1}$) and PI-F (421.5 $m^2 \cdot g^{-1}$) > PE-F (351.6 $m^2 \cdot g^{-1}$) > SD-F (110.8 $m^2 \cdot g^{-1}$). Porosity and surface area are parameters of major importance in terms of adsorption efficiency, as discussed in the following sections.

The XRD analysis of the starting materials and the non-modified biochars has been presented in detail in previous studies (Komnitsas et al., 2015, 2016). In brief, in all raw materials (PIr, PEr, and SDr) the characteristic amorphous peaks of cellulose, which is one of the structural components of the primary cell wall of green plants, are detected. However, after pyrolysis the intensity of these peaks was reduced. The peaks of residual inorganic phases, such as calcite, quartz, whewellite, halite, and thermonatrite are only visible in non-modified biochars, after partial decomposition of organic matter.

Figure 1 shows the XRD patterns of biochars modified with KOH or $FeCl_3$ (PI-K, PE-K, SD-K or PI-F, PE-F, SD-F, respectively), which exhibit a more or less elevated background between 15 and 30° 2-theta, due to the presence of organic matter (Cao and Harris, 2010). The cellulose peak is broader and of lower intensity in all biochars, compared to the starting materials, indicating a less ordered structure. Residual inorganic phases, such as quartz and calcite are more easily detected in modified biochars due to the partial decomposition of the organic matter during modification.

Sodium carbonate is present in PE-F and PE-K biochars, as a result of atmospheric carbonation; this phase is soluble and disappears when samples are washed with water.

SEM and FTIR analyses were carried out for biochars produced from pistachio shells, which show the highest phenol adsorption capacity compared to biochars produced from pecan shells and wood sawdust, as deduced from the experimental results discussed in the following sections. Indicative SEM images of raw pistachio shells (PIr), biochar PI, KOH-modified biochar PI-K and $FeCl_3$-modified biochar PI-F, are shown in Figure 2. The matrix of raw pistachio shells PIr is heterogeneous, while after pyrolysis for the production of PI biochar agglomeration takes place and a porous structure is formed due to the release of volatiles. SEM images of modified biochars (PI-K and PI-F) reveal their microporous structure as a result of their high porosity (49.8% and 41.4%, respectively, Table 1). Moreover, very fine particles with an average size of 15 μm are present in modified biochars resulting in a substantial increase of the surface area (572.4 and 421.5 $m^2 \cdot g^{-1}$ for PI-K and PI-F, respectively).

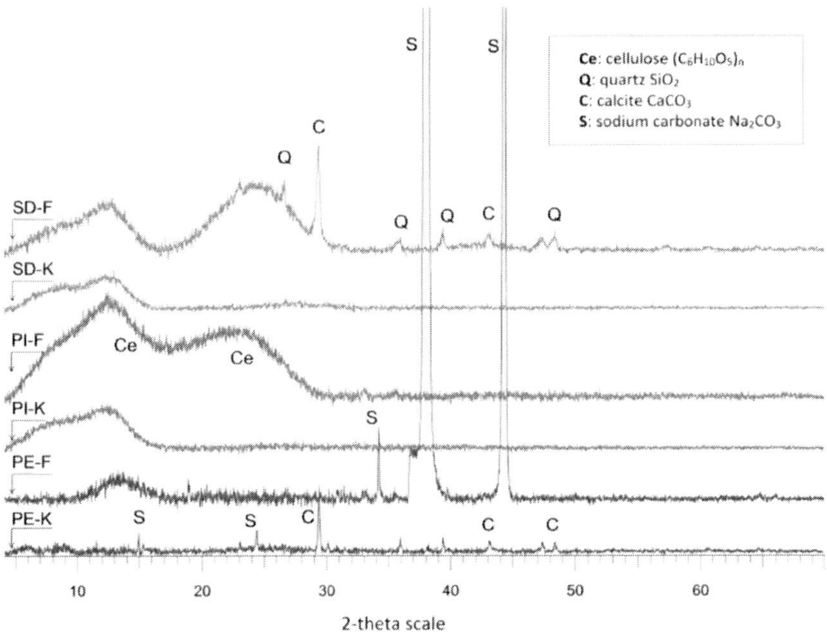

Figure 1. XRD patterns of biochars modified with 1M KOH (PI-K, PE-K, SD-K) or 1M FeCl3 (PI-F, PE-F, SD-F).

The FTIR spectra of raw pistachio shells (PIr), biochar PI, KOH-modified biochar PI-K and $FeCl_3$-modified biochar PI-F, are presented in Figure 3. The respective FTIR spectra band assignments are listed in detail in Table 2. The broad peaks shown in all samples at around 3420 cm^{-1}indicate the presence of hydroxyl group (−OH) stretching and strong hydrogen bonding, while those at 2920 and 2850 cm^{-1} are due to aliphatic C−H deforming vibration. The band at 1735 cm^{-1} for PIr is slightly shifted for all biochars to 1700 cm^{-1} and is assigned to ν(C = O) vibration in carbonyl group or the presence of carboxylic bonds. The band of PIr at 1630 cm^{-1}, which is due to C = O or C = C stretching in aromatic groups, is slightly shifted to 1612 cm^{-1} in modified biochars indicating that chemical interactions occurred on the modified biochar surface. The shifting of the peaks around 1600–1700 cm^{-1} is also related to the conjunction of the carbonyl groups with the aromatic ring. The band at 1385 cm^{-1}, shown in all patterns, is attributed to d(C = H) vibration in alkanes and alkyl groups and is more intense for the modified biochars PI-K and PI-F. In PIr, the band at 1260 cm^{-1} is attributed to C = C stretching while the sharp peak at 1045 cm^{-1} is mainly attributed to alcohol C−O or Si−O−Si functional groups. Both bands almost disappear in all biochars, either due to decomposition of cellulose after pyrolysis at 400°C or chemical modification. The bands at 900–600 cm^{-1} shown mainly in PIr and non-modified biochar PI, are assigned to C–H wagging vibrations due to the presence of aromatic and heteroaromatic compounds.

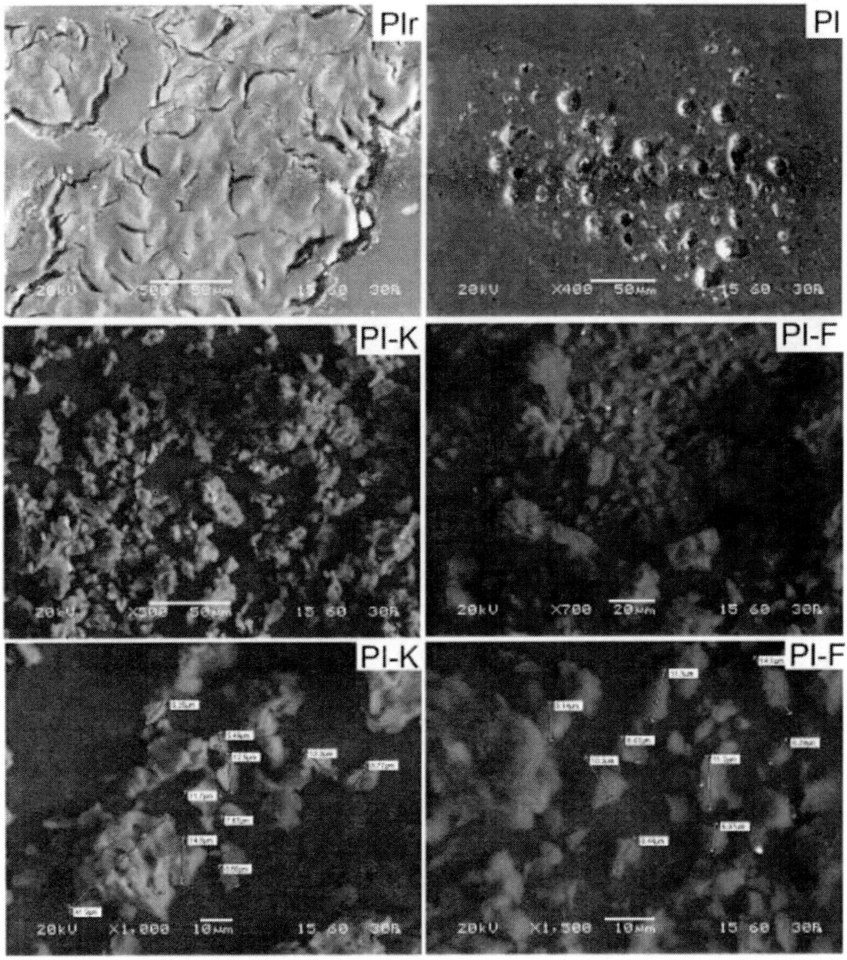

Figure 2. SEM images of raw pistachio shells (PIr) and biochars PI, PI-K (x500, x1000), and PI-F (x700, x1500).

3.2 Kinetic Studies

The kinetics of phenol adsorption on biochars PI, PE, SD and modified biochars PI-K, PE-K, SD-K and PI-F, PE-F, SD-F was studied using the pseudo-first and pseudo-second order Lagergren kinetic models which are described by the linear Equations (1) and (2), respectively:

$$\log(q_e - q_t) = \log q_e - \frac{k_1 t}{2.303} \tag{1}$$

$$\frac{t}{q_t} = \frac{1}{k_2 q_e^2} + \frac{t}{q_e} \tag{2}$$

where q_t and q_e ($mg \cdot g^{-1}$) is the uptake of phenols per unit weight of adsorbent at time t and at equilibrium, respectively, k_1 (h^{-1}) and k_2 (h^{-1}) are the rate constants for the pseudo-first and pseudo-second order kinetics, respectively.

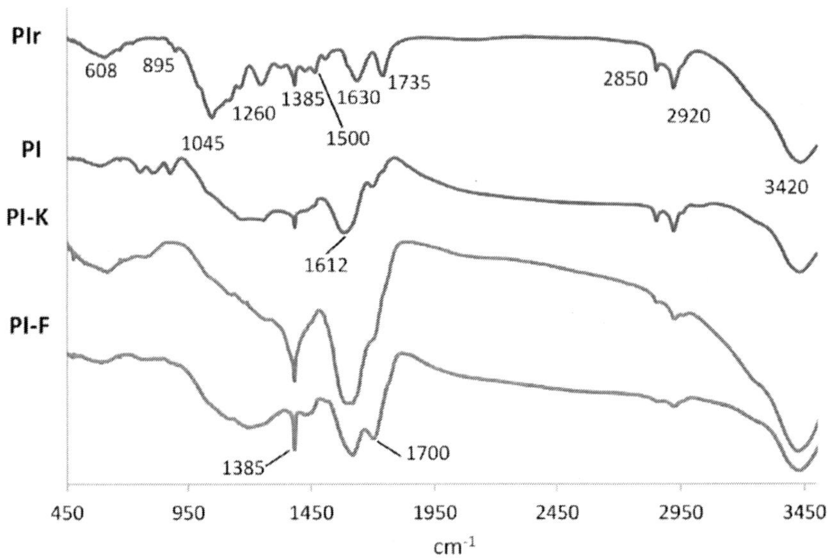

Figure 3. FTIR spectra of raw pistachio shells (PIr) and biochars PI, PI-K, and PI-F.

The linear plot of $\log(q_e - q_t)$ vs. t, according to Equation (1), provides k_1 and q_e values for the pseudo-first order model. The plot of t/q_t vs. t, according to Equation (2), provides k_2 and q_e for the pseudo-second order model. Parameters for the pseudo-first and pseudo-second order of Lagergren kinetic models, as derived from the experimental tests, are presented in Table 3. Equilibrium in all tests was achieved in 24 h.

Table 2. FTIR spectra band assignments presented in Figure 3.

Band, cm^{-1}	Assignment	References
3420	Hydroxyl group (–OH) stretching	Liu et al., 2013
2850, 2920	Aliphatic C–H deforming vibration	Angin and Şensöz, 2014
2350	Asymmetrical stretching of CO_2 (denotes atmospheric carbonation)	Socrates, 2001
1735, 1700	ν(C=O) vibration in carbonyl group or presence of carboxylic bonds	Sharma et al., 2004; Cao and Harris, 2010
1630, 1612, 1500	Aromatic C=O ring stretching (likely –COOH) or C=C stretching of aromatic groups present in lignin	Mimmo et al., 2014; Yang and Jiang, 2014
1385	δ(C=H) vibration in alkanes and alkyl groups	Samsuri et al., 2013
1260	C=C stretching	Cao and Harris, 2010
1045	Alcohol C–O or aromatic stretching peak, O–H deformation vibrations, b-glycosidic bond present in cellulose and hemicellulose or Si–O–Si functional groups	Angin et al., 2013; Mimmo et al., 2014
900-600	C–H wagging vibrations	Ghani et al., 2013

As shown in Table 3, for all biochars, modified or not, the pseudo-second order model fits the data much better (R^2 values are higher than 0.998) than the pseudo-first order model. These results indicate that the reaction rate is proportional to the number of the active sites present on the adsorbent surface. It is known that adsorption in batch systems is affected by both surface or pore diffusion (Mohan et al., 2007; Li et al., 2010). Additional modeling is required to define which diffusion type is the rate limiting step (Hui et al., 2003).

3.3 Adsorption of Phenols on Biochars

Table 4 shows the % phenol adsorption from synthetic solution (initial concentration 100 $mg \cdot L^{-1}$) on non-modified biochars PI, PE, SD as well as on biochars modified with 1M KOH (PI-K, PE-K, SD-K) or 1M $FeCl_3$ (PI-F, PE-F, SD-F). It is shown that for all adsorbent concentrations considered (0.5, 1,

2, and 5 g L^{-1}), biochars modified with 1M KOH exhibit the highest phenol adsorption efficiency.

Table 3. Parameters for the pseudo-first and pseudo-second order of Lagergren kinetic models for phenols adsorption on biochars PI, PE, SD and modified biochars PI-K, PE-K, SD-K and PI-F, PE-F, SD-F.

	Pseudo-first order			Pseudo-second order		
	$k_1(min^{-1})$	$q_e(mg \cdot g^{-1})$	R^2	$k_2(g \cdot mg^{-1} \cdot min^{-1})$	$q_e(mg \cdot g^{-1})$	R^2
PI	0.222	6.789	0.806	0.070	10.799	0.999
PE	0.242	5.198	0.811	0.097	8.251	0.999
SD	0.318	4.120	0.862	0.151	6.614	1.000
PI-K	0.242	7.716	0.741	0.084	15.773	0.999
PE-K	0.350	7.876	0.962	0.097	12.821	1.000
SD-K	0.188	5.158	0.802	0.102	9.597	0.998
PI-F	0.213	7.697	0.702	0.069	14.025	0.999
PE-F	0.419	8.792	0.994	0.084	11.561	1.000
SD-F	0.242	5.177	0.869	0.099	8.104	0.999

Table 4. Percentage phenol adsorption on biochars PI, PE, SD and modified biochars PI-K, PE-K, SD-K or PI-F, PE-F, SD-F for various adsorbent concentrations (0.5, 1, 2, and 5 g·L−1).

Adsorbent concentration, $g \cdot L^{-1}$	Percentage phenol adsorption									
	PI	PE	SD	PI-K	PE-K	SD-K	PI-F	PE-F	SD-F	AC*
0.5	14.6	11.1	9.9	32.8	24.4	18.6	23.4	17.6	11.6	15.2
1	29.5	23.2	20.2	53.9	37.6	28.5	43.9	28.4	21.8	28.6
2	36.9	27.4	25.8	61.3	51.8	33.7	50.3	47.1	28.5	36.1
5	51.3	39.3	31.8	76.6	61.7	46.2	67.4	55.4	38.6	52.7

*Activated carbon.

Experimental data shows that phenols are adsorbed more efficiently on KOH-modified biochars following the sequence PI-K>PE-K>SD-K, and the maximum adsorption reaches 76.6, 61.7, and 46.2%, respectively, when the adsorbent concentration is 5 g·L^{-1}. Lower adsorbent concentrations result in reduced adsorption efficiency due to the presence of less surface adsorption sites (Wang et al., 2006).

The same trend is also noticed for the $FeCl_3$-modified biochars and non-modified biochars, namely PI-F>PE-F>SD-F and PI>PE>SD, respectively. Activated carbon, which was used as reference adsorbent, shows quite similar with PI biochar adsorption efficiency, for all adsorbent concentrations used. An important finding of this study is that modification of PI biochar with KOH or $FeCl_3$ results in better phenol adsorption efficiency than activated carbon. It is mentioned that biochar is a cheap, environment friendly and readily available product that can be easily chemically activated. On the other hand, commercial activated carbon, which is usually prepared from coal, lignite, coconut shells or wood and is widely used for the cleanup of contaminated water and wastewater, is usually an expensive material. Its production cost is mainly related to the reactivation process and is increased when high quality activated carbon is needed (Mohan and Chander, 2006; Quintelas et al., 2008; Al-Lagtah et al., 2016).

Figures 4–6 show the adsorption rates of phenols, in $mg{\cdot}g^{-1}{\cdot}h^{-1}$, for non-modified, KOH-modified and $FeCl_3$-modified biochars, respectively. It is seen from these figures that:

- For all adsorbents used the adsorption rate of phenols increases with decreasing adsorbent concentration.
- The highest adsorption rate of phenols, 2.74 $mg{\cdot}g^{-1}{\cdot}h^{-1}$ at equilibrium, is shown for 0.5 $mg{\cdot}L^{-1}$ PI-K biochar concentration (Figure 5).
- The KOH-modified biochars exhibited the highest phenol removal capacity compared to $FeCl_3$-modified and non-modified biochars. More specifically, biochar PI-K exhibited the highest adsorption rate compared to biochars PE-K and SD-K, for all adsorbent concentrations tested (Figure 5). A similar trend was also noticed for $FeCl_3$-modified biochars (Figure 6) and non-modified biochars (Figure 4).
- Activated carbon shows almost identical adsorption rates with PI biochar, namely 1.28, 1.20, 0.78, and 0.45 $mg{\cdot}g^{-1}{\cdot}h^{-1}$, for 0.5, 1, 2, and 5 $g{\cdot}L^{-1}$, respectively (Figure 4).

Biochar PI-K shows the highest phenol adsorption capacity due to its highest surface area and porosity i.e., 572.4 $m^2{\cdot}g^{-1}$ and 49.8%, respectively. For biochars exhibiting lower surface area and porosity such as PE-K (397.3 $m^2{\cdot}g^{-1}$ and 34.7%, respectively) the adsorption capacity decreases accordingly. Similar behavior has been reported by Dai et al. (2013) for biochars produced from wetland plant residues. It is known that the potential of biochar as adsorbent in batch systems depends on its structure, its specific surface area and the available active sorption sites, properties which are affected by modification (Qu et al., 2013).

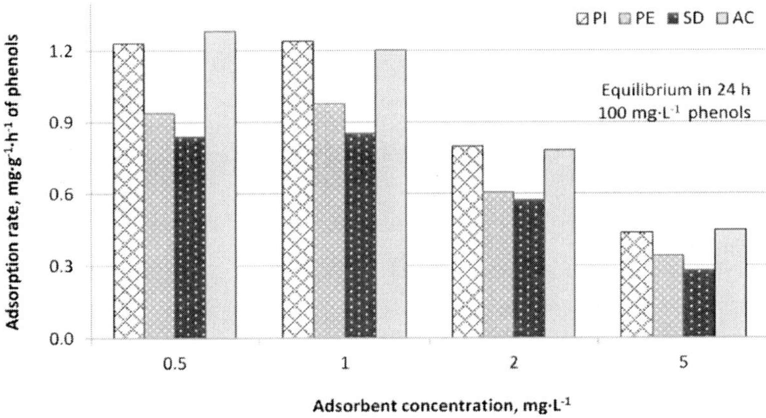

Figure 4. Adsorption rates, mg·g−1·h−1, of phenols on biochars PI, PE, and SD for adsorbent concentration 0.5, 1, 2, and 5 g·L−1 (AC: activated carbon, used as reference adsorbent).

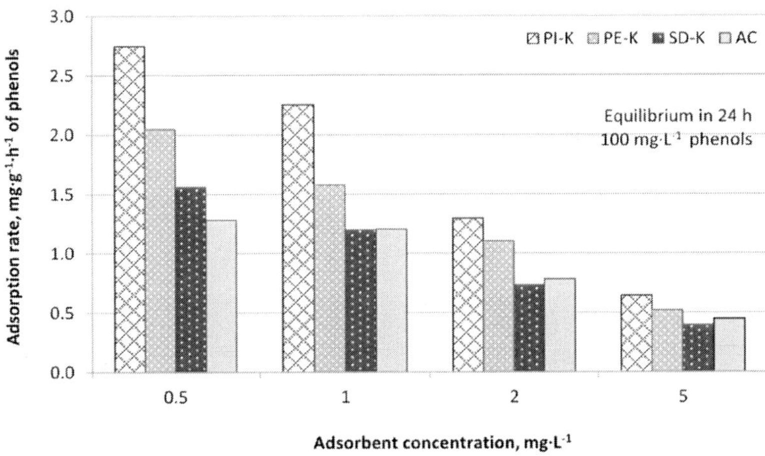

Figure 5. Adsorption rates, mg·g−1·h−1, of phenols on biochars PI-K, PE-K, and SD-K for adsorbent concentration 0.5, 1, 2, and 5 g·L−1 (AC: activated carbon, used as reference adsorbent).

The main mechanism affecting adsorption of phenols on biochars is most probably the hydrophobic sorption on biochar surface which is also related to their increased carbon content and the number of oxygenated functional groups (Yu et al., 2015; Oh and Seo, 2016). Also aromatic hydrocarbons present in phenols can be covalently bonded to surfaces of biochar (Cornelissen et al., 2005).

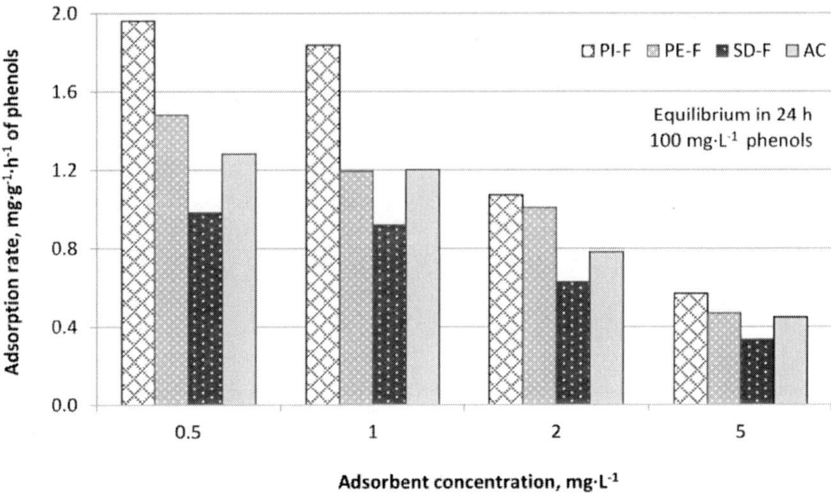

Figure 6. Adsorption rates, mg·g−1·h−1, of phenols on biochars PI-F, PE-F, and SD-F for adsorbent concentration 0.5, 1, 2, and 5 g·L−1 (AC: activated carbon, used as reference adsorbent).

3.4 Phenol Sorption Isotherm Models

In order to describe adsorption of phenols on biochars, the Freundlich and Langmuir models have been used. The Freundlich model, which assumes a multilayer sorption on heterogeneous adsorbent surface, is described by Equation (3):

$$\log q_e = \log K_f + \frac{1}{n} \log C_e \tag{3}$$

where q_e ($mg \cdot g^{-1}$) is the uptake of phenol per unit weight of adsorbent in equilibrium, C_e ($mg \cdot L^{-1}$) is the equilibrium concentration of phenol in solution, K_f ($L \cdot g^{-1}$) is the constant related to the adsorption capacity of the adsorbent and 1/n is the constant related to the adsorption intensity.

The Langmuir model, which assumes a monolayer adsorption on a homogenous surface when the adsorbent has a constant number of adsorption sites, is described by Equation

$$\frac{C_e}{q_e} = \frac{1}{bq_{max}} + \frac{1}{q_{max}} C_e \tag{4}$$

where C_e ($mg \cdot L^{-1}$) is the equilibrium concentration of phenol in solution, q_e ($mg \cdot g^{-1}$) is the uptake of phenol per unit weight of adsorbent in equilibrium, q_{max} is the maximum adsorption capacity of the adsorbent ($mg \cdot g^{-1}$) and b ($L \cdot mg^{-1}$) is the Langmuir constant related to the energy of the adsorption.

The Freundlich and Langmuir isotherms for phenol adsorption on biochars PI, PE, SD and modified biochars PI-K, PE-K, SD-K or PI-F, PE-F, SD-F, are presented in Figures 7, 8, respectively. The respective equations and the correlation coefficients R^2 are also shown in these figures which were prepared using the equilibrium concentration of phenols in solution for four different adsorbent concentrations (Table 4). It is shown that for all investigated biochars the Freundlich model gives a better fit compared to Langmuir model, while the highest R^2 values correspond to the PI-K biochar, which shows the highest phenol adsorption efficiency.

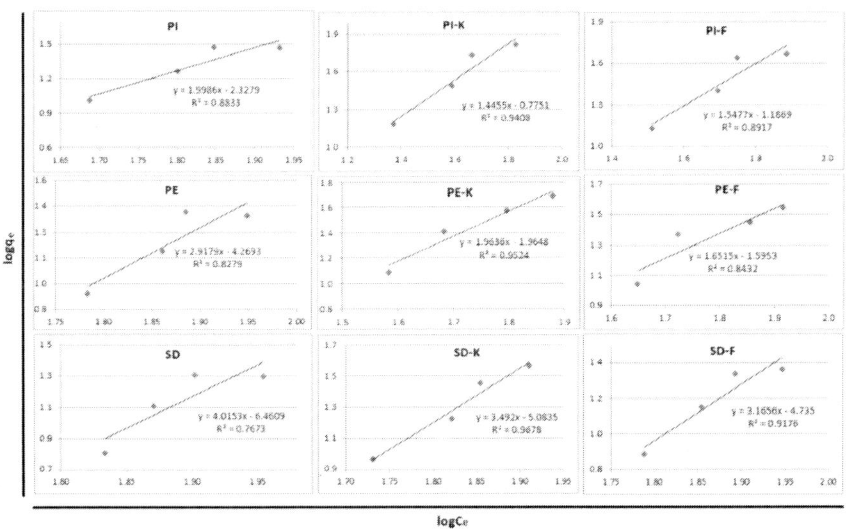

Figure 7. Freundlich isotherms for phenols adsorption on biochars PI, PE, SD, and modified biochars PI-K, PE-K, SD-K or PI-F, PE-F, SD-F.

The Freundlich model describes very well sorption probably due to surface heterogeneity of the produced biochars, including pore size distribution and presence of functional groups (Girods et al., 2009). The Langmuir model is not suitable to describe adsorption of phenols on biochars as also discussed in other relevant studies (Huang and Chen, 2010; Han et al., 2013; Agrafioti et al., 2014).

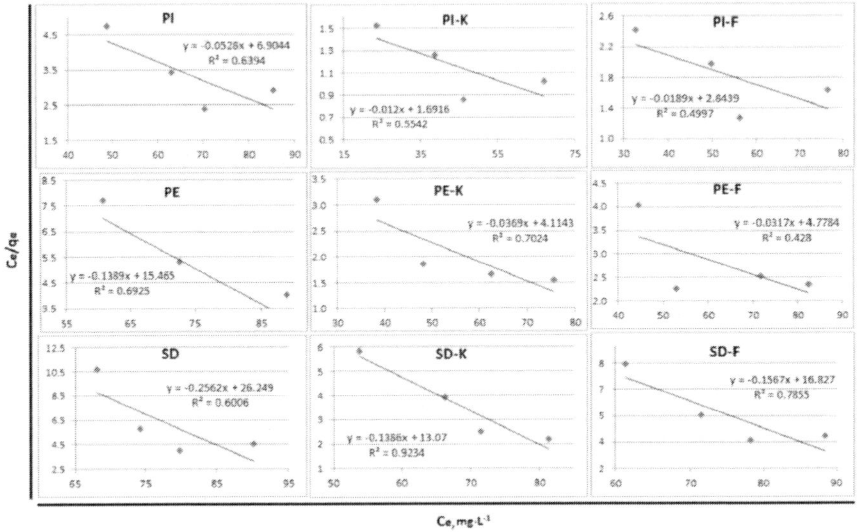

Figure 8. Langmuir isotherms for phenols adsorption on biochars PI, PE, SD, and modified biochars PI-K, PE-K, SD-K or PI-F, PE-F, SD-F.

Thus, the Freundlich Equation (3) for PI-K biochar ($\log K_f = -0.78$, $1/n = 1.45$) becomes

$$\log q_e = 1.45 \log C_e - 0.78 \Leftrightarrow$$
$$q_e = 0.17\, C_e^{1.45} \quad (5)$$

4. CONCLUSIONS

Three different agricultural wastes, namely pistachio shells, pecan shells and pine wood sawdust were pyrolyzed for the production of biochars which were also chemically modified with the use of 1M KOH or 1M $FeCl_3$ and tested for the removal of phenols from synthetic solutions.

The results of this study show that KOH is a more efficient biochar modifier compared to $FeCl_3$, since it results in increased carbon content, porosity and surface area. Phenols are adsorbed on KOH-modified biochars following the sequence PI-K>PE-K>SD-K.

Modification of PI-biochar with KOH or $FeCl_3$ results in better phenol adsorption efficiency compared to commercial activated carbon.

Porosity and surface area are considered the main parameters that affect biochar properties and thus define their potential in environmental applications, including adsorption of contaminants from solutions. The use of analytical techniques, XRD, SEM, and FTIR, offers significant insights regarding the morphology of the produced biochars in terms of content of organic carbon, degree of homogeneity, and existence of porous structure, factors which determine the number of available active sites and assess their adsorption efficiency.

Finally, kinetic and equilibrium studies show that the pseudo-second order model fits best adsorption kinetics data, while the Freundlich model describes very well adsorption of phenols on biochars.

FOOTNOTES

1. Council Directive 98/83/EC of 3 November 1998. On the quality of water intended for human consumption.

REFERENCES

1. Agrafioti, E., Bouras, G., Kalderis, D., and Diamadopoulos, E. (2013). Biochar production by sewage sludge pyrolysis. *J. Anal. Appl. Pyrol.* 101, 72–78. doi: 10.1016/j.jaap.2013.02.010
2. Agrafioti, E., Kalderis, D., and Diamadopoulos, E. (2014). Arsenic and chromium removal from water using biochars derived from rice husk, organic solid wastes and sewage sludge. *J. Environ. Manage.* 133, 309–314. doi: 10.1016/j.jenvman.2013.12.007
3. Al-Lagtah, N. M. A., Al-Muhtaseb, A. H., Ahmad, M. N. M., and Salameh, Y. (2016). Chemical and physical characteristics of optimal synthesised activated carbons from grass-derived sulfonated lignin versus commercial activated carbons. *Microporous Mesoporous Mater.* 225, 504–514. doi: 10.1016/j.micromeso.2016.01.043
4. Angin, D., Köse, T. E., and Selengil, U. (2013). Production and characterization of activated carbon prepared from safflower seed cake biochar and its ability to absorb reactive dyestuff. *Appl. Surf. Sci.* 280, 705–710. doi: 10.1016/j.apsusc.2013.05.046
5. Angin, D., and Şensöz, S. (2014). Effect of pyrolysis temperature on chemical and surface properties of biochar of rapeseed (*Brassica napus L.*). *Int. J. Phytoremediat.* 16, 684–693. doi: 10.1080/15226514.2013.856842
6. Bayramoglu, G., Gursel, I., Tunali, Y., and Arica, M. Y. (2009). Biosorption of phenol and 2-chlorophenol by Funalia trogii pellets. *Bioresour. Technol.* 100, 2685–2691. doi: 10.1016/j.biortech.2008.12.042
7. Beesley, L., Moreno-Jimenez, E., Gomez-Eyles, J. L., Harris, E., Robinson, B., and Sizmur, T. (2011). A review of biochars' potential role in the remediation, revegetation and restoration of contaminated soils. *Environ. Pollut.* 159, 3269–3282. doi: 10.1016/j.envpol.2011.07.023
8. Box, J. D. (1983). Investigation of the Folin–Ciocalteau phenol reagent for the determination of the polyphenolic substances in natural waters. *Water Res.* 17, 511–525. doi: 10.1016/0043-1354(83)90111-2

9. Brewer, C. E., Chuang, V. J., Masiello, C. A., Gonnermann, H., Gao, X., Dugan, B., et al. (2014). New approaches to measuring biochar density and porosity. *Biomass Bioenerg.* 66, 176–185. doi: 10.1016/j.biombioe.2014.03.059
10. Canadian Council of Ministers of the Environment (2007). *Canadian Soil Quality Guidelines for the Protection of Environmental and Human Health.* Winnipeg, MB: Excerpt from publication No 1299.
11. Cao, X., and Harris, W. (2010). Properties of dairy-manure-derived biochar pertinent to its potential use in remediation. *Bioresour. Technol.* 101, 5222–5228. doi: 10.1016/j.biortech.2010.02.052
12. Cayuela, M. L., Sánchez-Monedero, M. A., Roig, A., Hanley, K., Enders, A., and Lehmann, J. (2013). Biochar and denitrification in soils: when, how much and why does biochar reduce N_2O emissions? *Sci. Reports* 3:1732. doi: 10.1038/srep01732
13. Cornelissen, G., Gustafsson, Ö., Bucheli, T. D., Jonker, M. T. O., Koelmans, A. A., and van Noort, P. C. M. (2005). Extensive sorption of organic compounds to black carbon, coal, and kerogen in sediments and soils: mechanisms and consequences for distribution, bioaccumulation, and biodegradation. *Environ. Sci. Technol.* 39, 6881–6895. doi: 10.1021/es050191b
14. Dai, Z., Meng, J., Muhammad, N., Liu, X., Wang, H., He, Y., et al. (2013). The potential feasibility for soil improvement, based on the properties of biochars pyrolyzed from different feedstocks. *J. Soils Sediments* 13, 989–1000. doi: 10.1007/s11368-013-0698-y
15. Dari, B., Nair, V. D., Colee, J., Harris, W. G., and Mylavarapu, R. (2015). Estimation of phosphorus isotherm parameters: a simple and cost-effective procedure. *Front. Environ. Sci* 3:70. doi: 10.3389/fenvs.2015.00070
16. Dehkhoda, A. M., Ellis, N., and Gyenge, E. (2016). Effect of activated biochar porous structure on the capacitive deionization of NaCl and $ZnCl_2$ solutions. *Micropor. Mesopor. Mat.* 224, 217–228. doi: 10.1016/j.micromeso.2015.11.041
17. Devi, P., and Saroha, A. K. (2015). Simultaneous adsorption and dechlorination of pentachlorophenol from effluent by Ni–ZVI magnetic biochar composites synthesized from paper mill sludge. *Chem. Eng. J.* 271, 195–203. doi: 10.1016/j.cej.2015.02.087
18. Di Bene, C., Pellegrino, E., Debolini, M., Silvestri, N., and Bonari, E. (2013). Short- and long-term effects of olive mill wastewater land spreading on soil chemical and biological properties. *Soil Biol. Biochem.* 56, 21–30. doi: 10.1016/j.soilbio.2012.02.019
19. Environmental Protection Agency (1995). *Final National Pollutant Discharge Elimination System Storm Water Multi-Sector General Permit for Industrial Activities, Federal Register, Vol. 60.* Washington, DC: Environmental Protection Agency.
20. Frišták, V., Pipíška, M., Lesný, J., Soja, G., Friesl-Hanl, W., and Packová, A. (2015). Utilization of biochar sorbents for Cd^{2+}, Zn^{2+}, and Cu^{2+} ions separation from aqueous solutions: comparative study. *Environ. Monit. Assess.* 187, 4093. doi: 10.1007/s10661-014-4093-y
21. Ghani, W. A. W. A. K., Mohd, A., Da Silva, G., Bachmann, R. T., Taufiq-Yap, Y. H., Rashid, U., et al. (2013). Biochar production from waste rubber-wood-sawdust and its potential use in C sequestration: chemical and physical characterization. *Ind. Crop. Prod.* 44, 18–24. doi: 10.1016/j.indcrop.2012.10.017
22. Girods, P., Dufour, A., Fierro, V., Rogaume, Y., Rogaume, C., Zoulalian, A., et al. (2009). Activated carbons prepared from wood particleboard wastes: characterisation and phenol adsorption capacities. *J. Hazard. Mater.* 166, 491–501. doi: 10.1016/j. jhazmat. 2008.11.047
23. Gray, M., Johnson, M. G., Dragila, M. I., and Kleber, M. (2014). Water uptake in biochars: the roles of porosity and hydrophobicity. *Biomass Bioenerg.* 61, 196–205. doi: 10.1016/j.biombioe.2013.12.010
24. Greek Government Gazette (2007). *Modification of CMD (Common Ministerial Decision) on Quality of Drinking Water, in Accordance with 98/83/EC Directive.* 630 (issue B′), 26-4-2007.
25. Han, Y., Boateng, A. A., Qi, P. X., Lima, I. M., and Chang, J. (2013). Heavy metal and phenol adsorptive properties of biochars from pyrolyzed switchgrass and woody biomass in

correlation with surface properties. *J. Environ. Manage.* 118, 196–204. doi: 10.1016/j.jenvman.2013.01.001
26. Ho, Y.-S. (2006a). Review of second-order models for adsorption systems. *J. Hazard. Mater. B* 136, 681–689. doi: 10.1016/j.jhazmat.2005.12.043
27. Ho, Y.-S. (2006b). Second-order kinetic model for the sorption of cadmium onto tree fern: a comparison of linear and non-linear methods. *Water Res.* 40, 119–125. doi: 10.1016/j.watres.2005.10.040
28. Ho, Y.-S., and McKay, G. (1999). Pseudo-second order model for sorption processes. *Process Biochem.* 34, 451–465. doi: 10.1016/S0032-9592(98)00112-5
29. Huang, W., and Chen, B. (2010). Interaction mechanisms of organic contaminants with burned straw ash charcoal. *J. Environ. Sci.* 22, 1586–1594. doi: 10.1016/S1001-0742(09)60293-X
30. Hui, C.-W., Chen, B., and McKay, G. (2003). Pore-Surface Diffusion Model for Batch Adsorption Processes. *Langmuir* 19, 4188–4196. doi: 10.1021/la026624v
31. Inyang, M., and Dickenson, E. (2015). The potential role of biochar in the removal of organic and microbial contaminants from potable and reuse water: a review. *Chemosphere* 134, 232–240. doi: 10.1016/j.chemosphere.2015.03.072
32. Inyang, M., Gao, B., Zimmerman, A., Zhang, M., and Chen, H. (2014). Synthesis, characterization, and dye sorption ability of carbon nanotube- biochar nanocomposites. *Chem. Eng. J.* 236, 39–46. doi: 10.1016/j.cej.2013.09.074
33. Karakoyun, N., Kubilay, S., Aktas, N., Turhan, O., Kasimoglu, M., Yilmaz, S., et al. (2011). Hydrogel–Biochar composites for effective organic contaminant removal from aqueous media. *Desalination* 280, 319–325. doi: 10.1016/j.desal.2011.07.014
34. Kim, P., Johnson, A. M., Essington, M. E., Radosevich, M., Kwon, W.-T., Lee, S.-H., et al. (2013). Effect of pH on surface characteristics of switchgrass-derived biochars produced by fast pyrolysis. *Chemosphere* 90, 2623–2630. doi: 10.1016/j.chemosphere.2012.11.021
35. Kloss, S., Zehetner, F., Oburger, E., Buecker, J., Kitzler, B., Wenzel, W. W., et al. (2014). Trace element concentration in leachates and mustard plant tissue (*Sinapis alba L.*) after biochar application to temperate soil. *Sci. Total Environ.* 481, 498–508. doi: 10.1016/j.scitotenv.2014.02.093
36. Komnitsas, K., Zaharaki, D., Bartzas, G., Kaliakatsou, G., and Kritikaki, A. (2016). Efficiency of pecan shells and sawdust biochar on Pb and Cu adsorption. *Desalin. Water Treat.* 57, 3237–3246. doi: 10.1080/19443994.2014.981227
37. Komnitsas, K., Zaharaki, D., Pyliotis, I., Vamvuka, D., and Bartzas, G. (2015). Assessment of pistachio shell biochar quality and its potential for adsorption of heavy metals. *Waste Biomass Valor.* 6, 805–816. doi: 10.1007/s12649-015-9364-5
38. Li, Y., Du, Q., Wang, X., and Xia, Y. (2010). Removal of lead from aqueous solution by activated carbon prepared from Entermorpha prolifera by zinc chloride activation. *J. Hazard. Mater.* 183, 583–589. doi: 10.1016/j.jhazmat.2010.07.063
39. Liu, P., Liu, W.-J., Jiang, H., Chen, J.-J., Li, W.-W., and Yu, H.-Q. (2012). Modification of bio-char derived from fast pyrolysis of biomass and its application in removal of tetracycline from aqueous solution. *Bioresource Technol.* 121, 235–240. doi: 10.1016/j. biortech.2012.06.085
40. Liu, Y., Chen, J., Chen, M., Zhang, B., Wu, D., and Cheng, Q. (2015). Adsorption characteristics and mechanism of sewage sludge-derived adsorbent for removing sulfonated methyl phenol resin in wastewater. *RSC Adv.* 5, 76160–76169. doi: 10.1039/C5RA17125C
41. Liu, Z., Quek, A., Hoekman, S. K., and Balasubramanian, R. (2013). Production of solid biochar fuel from waste biomass by hydrothermal carbonization. *Fuel* 103, 943–949. doi: 10.1016/j.fuel.2012.07.069
42. Lou, L., Wu, B., Wang, L., Luo, L., Xu, X., Hou, J., et al. (2011). Sorption and ecotoxicity of pentachlorophenol polluted sediment amended with rice-straw derived biochar. *Bioresource Technol.* 102, 4036–4041. doi: 10.1016/j.biortech.2010.12.010
43. Manyà, J. J. (2012). Pyrolysis for biochar production. A review to establish current knowledge gaps and research needs. *Environ. Sci. Technol.* 46, 7939–7954. doi: 10.1021/es301029g

44. Mayer, Z. A., Eltom, Y., Stennett, D., Schroder, E., Apfelbacher, A., and Hornung, A. (2014). Characterization of engineered biochar for soil management. *Environ. Prog. Sustain. Energy* 33, 490–496. doi: 10.1002/ep.11788
45. Mimmo, T., Panzacchi, P., Baratieria, M., Davies, C. A., and Tonon, G. (2014). Effect of pyrolysis temperature on miscanthus (*Miscanthus X giganteus*) biochar physical, chemical and functional properties. *Biomass Bioenerg.* 62, 149–157. doi: 10.1016/j. biombioe. 2014.01.004
46. Mohan, D., and Chander, S. (2006). Removal and recovery of metal ions from acid mine drainage using lignite-A low cost sorbent. *J. Hazard. Mater.* 137, 1545–1553. doi: 10.1016/j.jhazmat.2006.04.053
47. Mohan, D., Pittman, C. U., Bricka, M., Smith, F., Yancey, B., Mohammad, J., et al. (2007). Sorption of arsenic, cadmium, and lead by biochars produced from fast pyrolysis of wood and bark during bio-oil production. *J. Colloid Interface Sci.* 310, 57–73. doi: 10.1016/j. jcis.2007.01.020
48. Moussavi, G., and Khosravi, R. (2012). Preparation and characterization of a biochar from pistachio hull biomass and its catalytic potential for ozonation of water recalcitrant contaminants. *Bioresour. Technol.* 119, 66–71. doi: 10.1016/j.biortech.2012.05.101
49. Mubarik, S., Saeed, A., Athar, M. M., and Iqbal, M. (2016). Characterization and mechanism of the adsorptive removal of 2,4,6-trichlorophenol by biochar prepared from sugarcane baggase. *J. Ind. Eng. Chem.* 33, 115–121. doi: 10.1016/j.jiec.2015.09.029
50. Oh, S.-Y., and Seo, Y.-D. (2016). Sorption of halogenated phenols and pharmaceuticals to biochar: affecting factors and mechanisms. *Environ. Sci. Pollut. Res.* 23, 951–961. doi: 10.1007/s11356-015-4201-8
51. Qian, W., Zhao, A.-Z., and Xu, R.-K. (2013). Sorption of As(V) by Aluminum-Modified crop straw-derived biochars. *Water Air Soil Pollut.* 224:1610. doi: 10.1007/s11270-013-1610-5
52. Qu, X., Alvarez, P. J. J., and Li, Q. (2013). Applications of nanotechnology in water and wastewater treatment. *Water Res.* 47, 3931–3946. doi: 10.1016/j.watres.2012.09.058
53. Quintelas, C., Fernandes, B., Castro, J., Figueiredo, H., and Tavares, T. (2008). Biosorption of Cr(VI) by three different bacterial species supported on granular activated carbon-A comparative study. *J. Hazard. Mater.* 153, 799–809. doi: 10.1016/j.jhazmat.2007.09.027
54. Samsuri, A. W., Sadegh-Zadeh, F., and Seh-Bardan, B. J. (2013). Adsorption of As(III) and As(V) by Fe coated biochars and biochars produced from empty fruit bunch and rice husk. *J.E.C.E.* 1, 981–988. doi: 10.1016/j.jece.2013.08.009
55. Sánchez, M. E., Lindao, E., Margaleff, D., Martínez, O., and Morán, A. (2009). Pyrolysis of agricultural residues from rape and sunflowers: production and characterization of bio-fuels and biochar soil management. *J. Anal. Appl. Pyrolysis* 85, 142–144. doi: 10.1016/j.jaap.2008.11.001
56. Sánchez-García, M., Roig, A., Sánchez-Monedero, M. A., and Cayuela, M. L. (2014). Biochar increases soil N_2O emissions produced by nitrification-mediated pathways. *Front. Environ. Sci.* 2:25. doi: 10.3389/fenvs.2014.00025
57. Sharma, R. K., Wooten, J. B., Baliga, V. L., Lin, X., Geoffrey Chan, W., and Hajaligol, M. R. (2004). Characterization of chars from pyrolysis of lignin. *Fuel* 83, 1469–1482. doi: 10.1016/j.fuel.2003.11.015
58. Socrates, G. (2001). *Infrared and Raman Characteristic Group Frequencies, 3rd Edn.* Chichester: John Wiley & Sons Ltd.
59. Steiner, C., Glaser, B., Teixeira, W. G., Lehmann, J., Blum, W. E. H., and Zech, W. (2008). Nitrogen retention and plant uptake on a highly weathered central Amazonian Ferralsol amended with compost and charcoal. *J. Plant Nutr. Soil Sci.* 171, 893–899. doi: 10.1002/jpln.200625199
60. Sun, L., Chen, D., Wanb, S., and Yu, Z. (2015). Performance, kinetics, and equilibrium of methylene blue adsorption on biochar derived from eucalyptus saw dust modified with citric, tartaric, and acetic acids. *Bioresource Technol.* 198, 300–308. doi: 10.1016/j.biortech.2015.09.026

61. Tripathi, M., Sahu, J. N., and Ganesan, P. (2016). Effect of process parameters on production of biochar from biomass waste through pyrolysis: a review. *Renew. Sust. Energ. Rev.* 55, 467–481. doi: 10.1016/j.rser.2015.10.122
62. Usman, A. R. A., Ahmad, M., El-Mahrouky, M., Al-Omran, A., Ok, Y. S., Sallam, A. S., et al. (2016). Chemically modified biochar produced from conocarpus waste increases NO_3 removal from aqueous solutions. *Environ. Geochem. Health* 38, 511–521. doi: 10.1007/s10653-015-9736-6
63. Wang, S., Gao, B., Li, Y., Mosa, A., Zimmerman, A. R., Ma, L. Q., et al. (2015). Manganese oxide-modified biochars: preparation, characterization, and sorption of arsenate and lead. *Bioresource Technol.* 181, 13–17. doi: 10.1016/j.biortech.2015.01.044
64. Wang, X., Sato, T., and Xing, B. (2006). Competitive sorption of pyrene on wood chars. *Environ. Sci. Technol.* 40, 3267–3272. doi: 10.1021/es0521977
65. Yang, G.-X., and Jiang, H. (2014). Amino modification of biochar for enhanced adsorption of copper ions from synthetic wastewater. *Water Res.* 48, 396–405. doi: 10.1016/j.watres.2013.09.050
66. Yu, Z., Zhou, L., Huang, Y., Song, Z., and Qiu, W. (2015). Effects of a manganese oxide-modified biochar composite on adsorption of arsenic in red soil. *J. Environ. Manage.* 163, 155–162. doi: 10.1016/j.jenvman.2015.08.020
67. Zahir, H., Naidoo, M., Kostadinova, R.-M., Ortiz, K. A., Sun-Kou, R., and Navarro, A. E. (2015). Decolorization of hair dye by lignocellulosic waste materials from contaminated waters. *Front. Environ. Sci.* 2:28. doi: 10.3389/fenvs.2014.00028

CHAPTER 12

Removal of BTX Contaminants with O_3 and O_3/UV Processes

Ján Derco1*, Katarína Šimovičová2, Jozef Dudáš1 and Mária Valičková1

1 Institute of Chemical and Environmental Engineering, Faculty of Chemical and Food Technology, Slovak University of Technology, Bratislava, Slovak Republic

2 Water Research Institute, Bratislava, Slovak Republic

ABSTRACT

The legal basis for the monitoring of priority and priority hazardous substances in water, sediment, and biota follows from Directive 2008/105/EC which defines the good chemical status to be achieved by all Member States together with the Water Framework Directive 2000/60/EC. The BTX compounds are considered to be the most toxic components of gasoline. Thus, organic petroleum components can induce a serious problem to public health and the aquatic environment. The effect of ozone and ozone/UV on degradation of the BTX in a model water was studied. The results indicate that the highest BTX removal rates were observed during the first 5 min of the process for all investigated pollutants. The treatment efficiencies above 90% were observed in all investigated pollutants after 40 min of ozonation. The results show a significant proportion of stripping in the removal of BTX components. Higher overall efficiency was observed by O_3/UV process after abstracting share of stripping process. Application of investigated processes appears to be a promising procedure for removal of petrol aromatic hydrocarbons from aquatic environment. However, for practical application, an improvement of process removal efficiency and investigation of impact of ozonation intermediates and products on aquatic microorganisms are required.

Keywords: BTX, jet-loop reactor, ozone, O_3/UV, wastewater treatment

1. INTRODUCTION

The adoption of the Framework Directive on water [1] provides a policy tool that enables sustainable protection of water resources. The Decision No. 2455/2001/EC of the European Parliament and the Council of November 2001 [2] established the list of 33 priority substances or group of substances, including the priority hazardous substances, presenting a significant risk to water pollution or *via* the aquatic environment including such risks to water used for the abstraction of drinking water.

Hazardous substances are defined as substances or groups of substances that are toxic, persistent, and liable to bioaccumulation, and other substances or groups of substances which give rise to an equivalent level of concern. The EC member countries have extended this list with relevant pollutants for individual countries. Thus, in the supplement of the Water Act [3] there have been identified altogether 59 relevant substances for SR.

The BTEX contaminants consist of benzene, ethyl benzene, toluene, and three isomers of xylene. These compounds are the volatile organic compounds (VOCs) found in petroleum derivatives such as petrol (gasoline). They represent one of the main groups of soluble organic compounds that is present in wastewater from refinery. They are the most toxic components of gasoline. These substances can lead to serious health problems ranging from irritation of eyes, skin, and mucous membranes and ending with weakened nervous system, decreased bone marrow function, and cancer. Benzene in particular is highly toxic. The World Health Organization classifies the substance as carcinogenic. It is also on the list of the priority substances [4].

Many oil substances have acute toxic effect on aquatic microorganisms with possible chronic consequences [5].

Commonly used wastewater treatment processes usually apply physical and physiochemical processes. Thus, the discharge of organic pollutants may create some environmental problems, particularly at microlevel. The aromatic oil fraction consists mainly of polyaromatic hydrocarbons (PAHs) and is more toxic and persistent than the aliphatic hydrocarbons [6]. Leakages including release of petroleum products, e.g., gasoline, diesel fuel, lubricating, and heating oil, from leaking oil tanks are the most frequent sources of soil and groundwater contaminations with BTEX substances. They are polar and readily soluble, and thus they are able to penetrate into soil and groundwater and cause serious environmental problems. These compounds dispose of acute and long-term toxic effects [4].

BTEX are among the most frequently detected contaminants in US public drinking-water systems that rely on groundwater sources [7]. These organic compounds make up a significant percentage of petroleum. The most contaminated locality of hazardous BTEX substances in the Slovak Republic is the airport at Sliač, Sliač-Vlkanová territory, contaminated by the Soviet Army, and the gas station in Rajecké Teplice where BTEX contaminants were identified as dominant in the groundwater. BTEX were also found in groundwater in Bratislava due to poor technical conditions of technological equipment (old stocks of aviation fuel) and the subsequent uncontrolled release of oil into the rock mass at the Airport of M. R. Štefanik [8].

Ozone is a very powerful oxidizing agent (E° = 2.07 V). Ozone may react with organic compounds in two ways: by direct reaction as molecular ozone or by indirect reaction through formation of secondary oxidants like free radical species [9–11]. In practice, both mechanisms may occur depending on the type of chemical wastewater pollution.

At low pH, the predominant reaction mechanisms are the direct electrophilic attack by molecular ozone [12], i.e., ozonolysis.

Under such conditions, ozone is a selective oxidant and reacts with multiple bonds (C=C, C=N, N=N, etc.) , but only at low rates with single bonds (C-C, C-O, O-H). At high pH, indirect reaction occurs, i.e., organics are degraded by secondary oxidants/chain reaction involving powerful radicals including OH, which are produced by ozone decomposition. These radicals are very strong and nonselective oxidants. Hydroxyl radicals can be formatted by increasing pH or by decomposition of O_3 with homogeneous and heterogeneous catalysts.

The main goal of our research was to study the feasibility of ozone and combination of O_3/UV processes for removal of selected benzene, toluene, and xylenes (BTX) from water/wastewater. Investigation of process kinetics and stripping of volatile substances were also performed.

2. EXPERIMENTAL METHODS

2.1. Experimental equipment

Ozonation trials were performed in a laboratory ozonation reactor. A schematic of the ozonation apparatus is illustrated in **Figure 1** [13]. Ozonation jet-loop reactor was operated in batch mode with regard to wastewater and in continuous mode with regard to gas. Active volume of the reactor was 3.5 L. The treated wastewater was transferred into ozonation reactor before starting operation of the reactor. A membrane pump was used to maintain external circulation of liquid reaction mixture. Pulsation of recirculated external flow was minimized with diaphragm pulsation damper (SERA 721.1, Seybert & Rahier, Immenhausen, Germany). Ozone was generated using a Lifetech generator with maximum production of 5 g h^{-1}. Ozone was generated at 50% of the maximum ozone generator's power and maintaining continuous oxygen flow of 60 L h^{-1}. A mixture of O_3 and O_2 was injected into a wastewater sample through a Venturi ejector. At the same time, the ejector sucked the mixture of O_3 and O_2 from the reactor headspace.

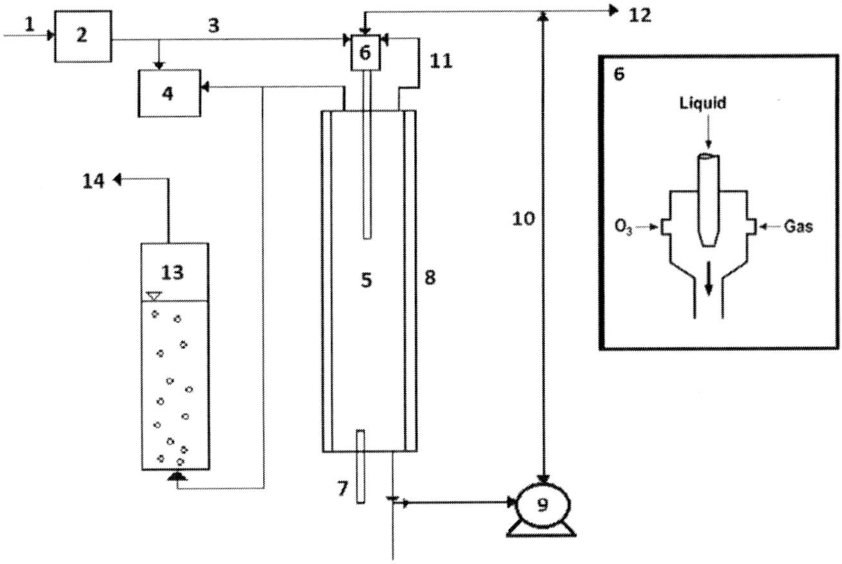

Figure 1. Schematic scheme of the ozonation equipment. 1—O_2, 2—O_3 generator, 3—O_2/O_3 gas flow, 4—UV detector of O_3, 5—O_3 reactor, 6—Venturi nozzle, 7—UV source, 8—tempering jacket, 9—recirculation pump, 10—external reaction mixture recirculation loop, 11—sucking of gas from head space, 12—sampling device, 13—outfall of gas, 14—destruction device for residual O_3.

This, together with external circulation, should improve the efficiency of ozone utilization in the ozonation reactor. Pen-Ray UV lamp with wavelength 254 nm was used to generate hydroxyl radicals in the reactor. The outfall of reaction-gas mixture was transported into a destruction glass column by a fine-bubble porous distributive device. The destructive reactive column contained a potassium iodide solution. The active volume of the destructive reactive column was 1.0 dm^3. An excess O_3 was destructed in this device [13].

2.2. Analytical methods

For determination BTEX compounds in model wastewater, gas chromatography was used with MS detector in connection with headspace autosampler.

Gas chromatographic method after static headspace extraction was used for quantification of the BTX in water. Headspace part was analyzed by gas chromatography with MS detector (Agilent Technologies 7890A GC Systems). All substances used for preparation of model wastewater and standard stock solutions were purchased from Dr. Ehrenstorfer.

2.3. Processing of experimental data

Desorption of volatile pollutants from water solution is proportional to its concentration in water solution

$$\frac{dS}{dt} = -k \,.\, S \tag{1}$$

where k/h^{-1} is desorption rate.

Experimental data of BTX degradation were fitted by the zero (Eq. (2))-, first (Eq. (3)-, and second (Eq. (4))-order reaction kinetic models. For a batch reaction system, under the assumption of a constant reaction volume, the following relationships were obtained.

$$S_t = S_0 - k_0 t \tag{2}$$

$$S_t = S_0 \exp(-k_1 t) \tag{3}$$

$$S_t = \frac{S_0}{(1 + S_0 k_2 t)} \tag{4}$$

where S_t ($g\ m^{-3}$) stands for the content of BTX substances in model wastewater in time t, S_0 ($g\ m^{-3}$) is the beginning content of BTX substances in model wastewater, k_0 ($g\ m^{-3}\ h^{-1}$), k_1 (h^{-1}), and k_2 ($g^{-1}\ m^3 h^{-1}$) are the rate constants for the kinetics of the zero, the first, and the second reaction order, respectively [13].

The grid search optimization method was applied to calculate values of parameters of the used mathematical models. The objective function was defined as the sum of squares between the measured and calculated values of BTX components divided with the number of measurements reduced by the number of estimated parameters [13, 14].

3. RESULTS OF THE WORK

The removals of studied compounds with the ozonation time are presented in **Figure 2**. The initial concentrations of benzene, toluene, o-xylene, and p-xylene were 800, 1600, 800, and 600 μg l^{-1}, respectively. From **Figure 2**, it is obvious that for all studied pollutants measured, the highest removal rates were observed within the first 5 min of the process. The highest affinity of ozone was measured toward p-xylene (59.6% removal efficiency). On the other hand, the lowest treatment efficiency was measured for benzene (20%) within the same ozonation time.

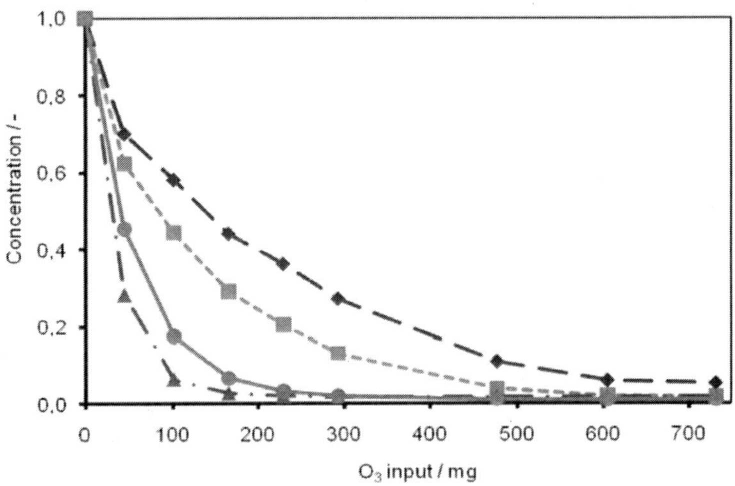

Figure 2. BTX (dimensionless values) removal during ozonation of model wastewater (◆ benzene, ■ toluene, ▲ p-xylene, ● o-xylene).

The treatment efficiencies of BTX components increased with the increase of ozonation time [15]. The highest treatment efficiency was observed for p-xylene (81.3% during the first 20 min of ozonation). The second highest treatment efficiency (72.2%) was measured for o-xylene. Final removal efficiencies of BTX constituents were observed in the range from 86.4 to 90%. The values of removal efficiency during O_3 process are summarized in **Table 1**.

Table 1. Removal efficiency values of BTX compounds with ozone.

t [min]	Benzene [%]	Toluene [%]	p-Xylene [%]	o-Xylene [%]
5	20.1	26.4	59.6	44.4
20	53.9	68.3	85.7	86.4
60	84.6	89.6	90.0	89.9

The best fit of experimental degradation data of all studied pollutants was obtained by the first-order kinetic model.

The removals of studied BTX by O_3/UV treatment time are presented in **Figure 3**. It is obvious that the removal rates of BTX by O_3/UV process are higher in comparison with the removal rates observed with ozone alone. This is confirmed also by the values of removal efficiency given in **Table 2**.

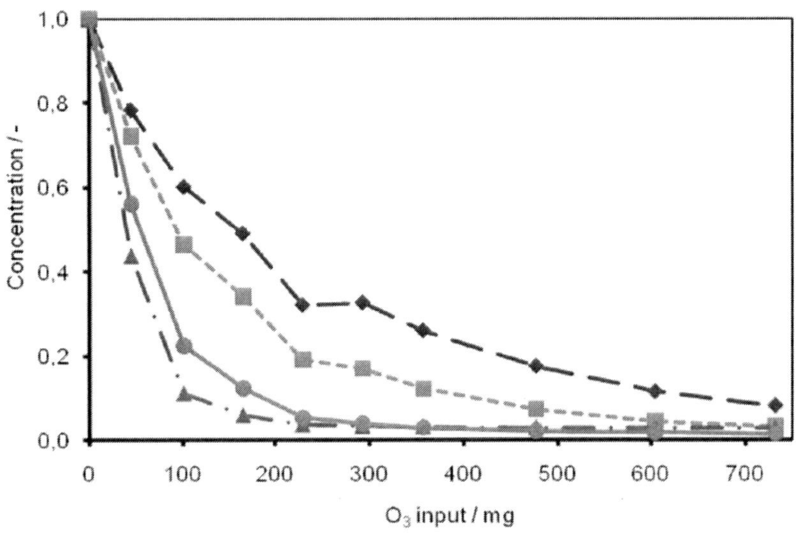

Figure 3. BTX (dimensionless values) removal by O_3/UV treatment of model wastewater (◆ benzene, ■ toluene, ▲ p-xylene, ● o-xylene).

Table 2. Removal efficiency values of BTX compounds by O_3/UV process.

t [min]	Benzene [%]	Toluene [%]	p-xylene [%]	o-xylene [%]
5	21.7	27.9	56.9	43.9
20	67.8	80.8	96.3	94.7
60	91.9	96.9	97.2	98.6

Comparisons of p-xylene and o-xylene removals using O_3 only for oxidation and O_3/UV treatment of model wastewater are presented in **Figures 4** and **5**, respectively. Slightly higher removal rates of these two pollutants were measured when treated with ozone alone.

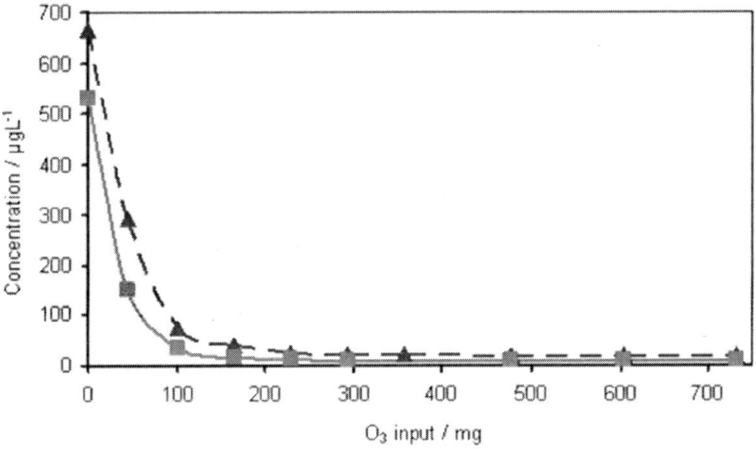

Figure 4. p-Xylene concentration profiles during O_3 and O_3/UV treatment of model wastewater (■ p-xylene-O_3, ▲ p-xylene-O_3/UV).

However, insignificant differences in the removal rates and treatment efficiencies follow from the treatment of the investigated BTX compounds using O_3 and O_3/UV. The data presented above represent overall removal of pollutants during O_3, or O_3/UV treatment, i.e., an effect of stripping of pollutants content is also included in the data.

The effect of gas striping of the investigated BTX compounds at the conditions of ozonation and O_3/UV trials was also studied. Volatility of substances depends on the size of molecules as well as on the vapor pressure [16]. With the increase of the molecular weight, the solubility of substance in water decreases. Important factor influencing solubility in water is hydrophobicity of substances. Solubility in water decreases with increase hydrophobicity of substance [17]. The information on hydrophobicity gives octanol-water partitioning coefficient value. Evaporation of substances correlates with vapor pressure [17] and is strongly influenced by temperature and the pressure of the system [18].

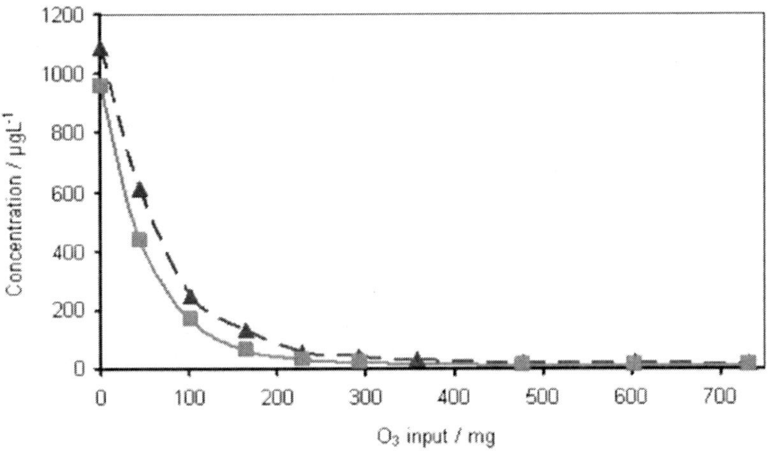

Figure 5. o-YLENE concentration profiles during O_3 and O_3/UV treatment of model wastewater (■ o-xylene-O_3, ▲ o-xylene-O_3/UV).

Volatility of substances can be quantified by the values Henry low. With the increase of the Henry's constant value, solubility of substance in water decreases. The values of basic physical-chemical properties of BTX components are given in **Table 3** [19].

Table 3. Basic characteristics for studied organic components of BTX.

	Benzene	Toluene	o-Xylene	p-Xylene
Molecular weight[g mol^{-1}]	78.11	92.13	106.16	106.16
Water solubility[mg L^{-1}]	1700	515	175	198
Vapor pressure(at 20°C) [kPa]	12.6923	3.7863	0.8799	0,8799
Boling point[°C]	80.1	110.6	144.4	138.4
Octanol-water partition coefficient (at 20°C)[log Kow]	2.13	2.69	2.77	3.15
Henry's law constant(at 25°C) [kPa m^3 mol^{-1}]	0.55	0.67	0.50	0.71
AA-EQS [µg L^{-1}]	10[a]	100[a]	10[b]	10 [b]

Notes: AA-EQS, Annual Average Value of Environmental Quality Standard (EQS).
[a] European Commission (2008) List of EQS [20].
[b] Regulation of the Slovak Government [21] List a EQS of total xylenes of 10 µg L^{-1}.

Comparison of o-xylene, p-xylene, benzene, and toluene concentration profiles measured during the stripping only and ozonation treatment of model water are presented in **Figures 6–9**, respectively. As it was already mentioned, 10 min of ozonation corresponds to input of 45 mg O_3 per liter of active volume of the jet-loop ozonation reactor.

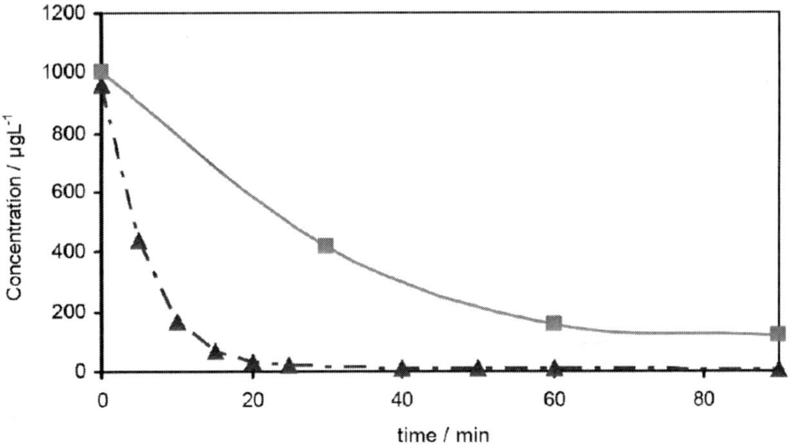

Figure 6. o-Xylene concentration profiles during stripping and ozonation of model wastewater (■ o-xylene-striping, ▲ o-xylene-ozonation).

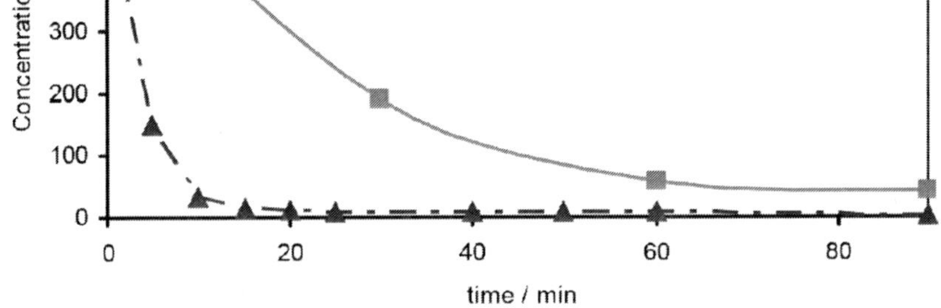

Figure 7. p-Xylene concentration profiles during stripping and ozonation of model wastewater (■ p-xylene-striping, ▲ p-xylene-ozonation).

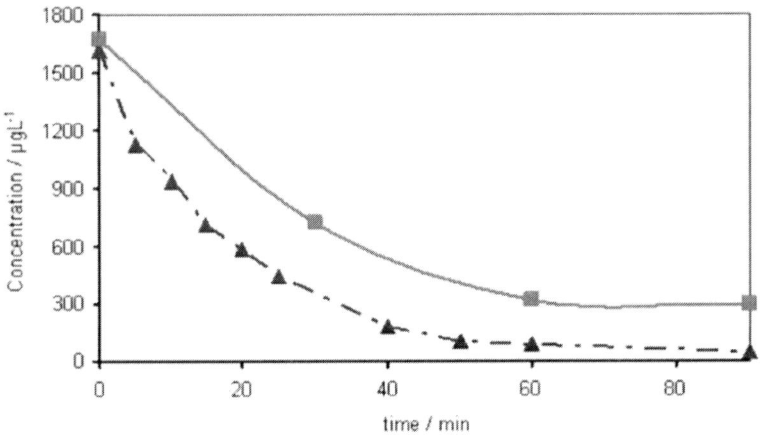

Figure 8. Benzene concentration profiles during stripping and ozonation of model wastewater (■ benzene-striping, ▲ benzene-ozonation).

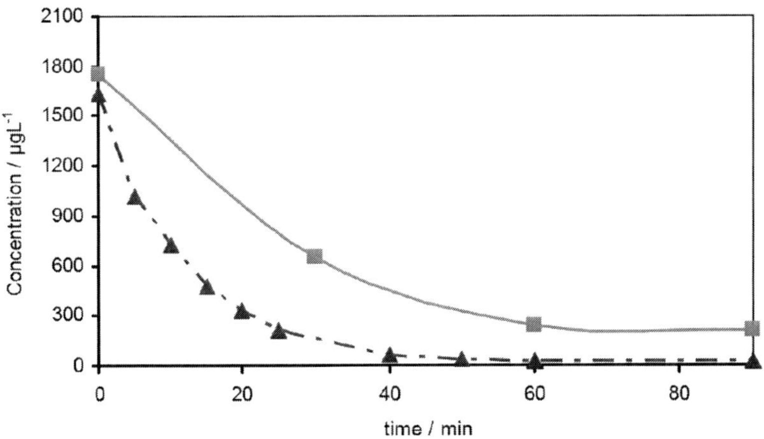

Figure 9. Toluene concentration profiles during stripping and ozonation of model wastewater (■ toluene-striping, ▲ toluene-ozonation).

It is obvious from the presented results that stripping can significantly contribute to the removal of investigated compounds during the ozonation and O_3/UV treatments. The higher contribution of stripping to the overall removal of the component during the ozonation is for benzene. This observation correlates very well with the physical properties of the different components (**Table 3**). Similar results were obtained for toluene (36.8 µg L^{-1} min^{-1}) in comparison to benzene (32.0 µg L^{-1} min^{-1}).

Concentration profiles of the studied BTX substances during ozonation, i.e., after excluding removal of the compounds by stripping process during ozonation of the model wastewater, are presented in **Figure 10**. In other words, the values plotted in **Figure 10** were obtained by subtraction of the concentrations of individual components due to stripping by oxygen flow from the total concentrations obtained during ozonation of the model wastewater.

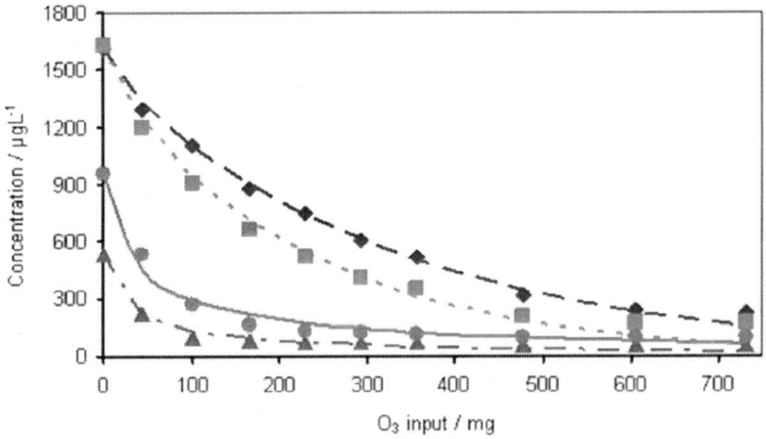

Figure 10. Experimental (◆ benzene, ■ toluene, ▲ p-xylene, ● o-xylene) and calculated (lines) BTX concentration profiles during ozonation of model wastewater.

Ozonation and stripping experimental trials were performed at the same operational conditions except presence of ozone in the system for stripping tests. The removal efficiencies for BTX components due to ozone oxidation of the model wastewater sample are given in **Table 4**.

Table 4. Removal efficiency values of BTX compounds by O_3 (stripping excluded).

t [min]	Benzene [%]	Toluene [%]	p-Xylene [%]	o-Xylene [%]
5	20.1	26.4	59.6	44.4
20	53.9	68.3	85.7	86.4
60	84.6	89.6	90.0	89.9

Experimental data were processed using kinetic models to evaluate an order of the reaction. The calculated concentration profiles obtained by kinetic models corresponded to the best fit of experimental data (**Table 4**) for ozone oxidation, i.e., after excluding contribution of stripping to overall BTX concentrations during ozonation.

The highest removal rates were observed during the first 5 minutes of ozonation for all investigated pollutants. However, there is approximately 10% difference in removal efficiency (**Table 1**) and it is caused by ozonation reaction only (**Table 4**).

The best fit of experimental degradation data of benzene and toluene was obtained by the first-order kinetic model. On the other hand, the second-order kinetic model was more appropriate for description degradation of xylenes. The rate constant values and the values of the correlation coefficient rXYrXY corresponding to treated kinetic data after subtracting volatilized portions due to oxygen stripping are summarized both for O_3 and O_3/UV treatments in **Table 6**.

Table 5. Removal efficiency values of BTX compounds by O_3/UV.

t [min]	Benzene [%]	Toluene [%]	p-Xylene [%]	o-Xylene [%]
5	21.7	27.9	56.9	43.9
20	67.8	80.8	96.3	94.7
60	91.9	96.9	97.2	98.6

Table 6. Kinetic parameters and statistical characteristics–striping excluded.

Pollutants		k_1 [h^{-1}]	k_2 [g m^{-3} h^{-1}]	
Benzene	O_3	3.92×10^{-2}	–	0.9964
	O_3/UV	5.42×10^{-2}	–	0.9770
Toluene	O_3	5.55×10^{-2}	–	0.9885
	O_3/UV	4.56×10^{-2}	–	0.9977
o-Xylene	O_3	–	2.38×10^{-4}	0.9793
	O_3/UV	–	1.78×10^{-4}	0.9636
p-Xylene	O_3	–	5.78×10^{-4}	0.9839
	O_3/UV	–	3.82×10^{-4}	0.9980

Similar concentration profiles for the studied BTX compounds during the O_3/UV treatment of the model wastewater are presented in **Figure 11**. The values presented in **Figure 11** were also obtained by subtraction of concentrations of individual components caused by gas stripping. The calculated concentration profiles were also obtained by kinetic models corresponding to the best fit of experimental values (**Table 6**) for O_3/UV treatment, i.e., after excluding contributions of stripping of individual compounds. The removal efficiencies values are given in **Table 5**.

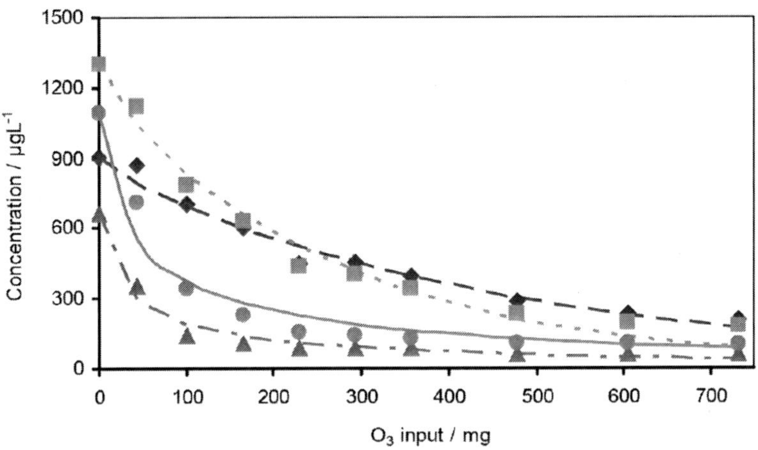

Figure 11. Experimental (◆ benzene, ■ toluene, ▲ p-xylene, ● o-xylene) and calculated (lines) BTX concentration profiles during O_3/UV treatment of model wastewater.

Removal efficiencies are very close to those given in **Table 2**. In the case of o-xylene and p-xylene, the best fit was obtained by the second-order kinetic model (**Table 6**). The first-order kinetic models for benzene and toluene may indicate significant influence of gas stripping on total removal of these compounds from solution and the process is probably determined by physical phenomena rather than chemical.

Figure 12 illustrates ozone concentration profiles in the jet-loop reactor during treatment of model wastewater with O_3 and O_3/UV.

Lower ozone concentration in water or higher ozone consumption in the system is obvious for O_3/UV treatment. Lower experimental ozone concentration for O_3/UV reaction system can be explained by decomposition and hydroxyl radicals formation. However, insignificant increase of removal rates was observed as a result of radical reaction mechanism. On the other hand, the values of kinetic constant for all compounds (**Table 6**) are slightly higher when ozone alone was applied in comparison to O_3/UV treatment. Thus, the higher removal rates for ozone treatment of BTX were observed in comparison to the O_3/UV treatment process.

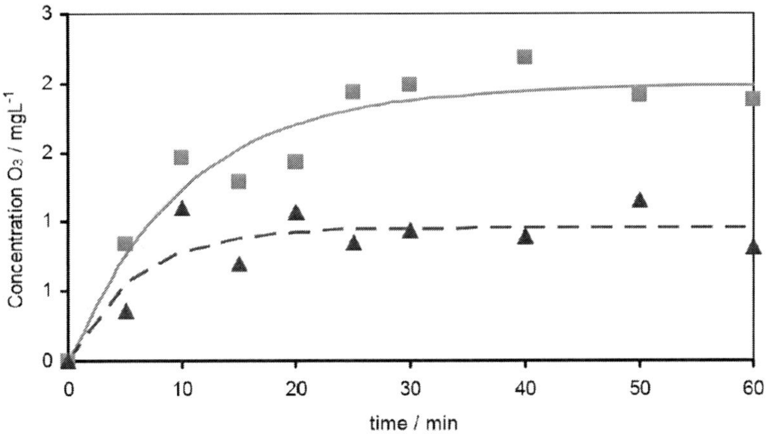

Figure 12. Ozone concentration profiles in the jet-loop reactor during treatment of model wastewater with O_3 (■) and O_3/UV (▲).

4. CONCLUSION

Effect of ozone and O_3/UV treatments on BTX components were investigated in this study. Investigation of stripping of volatile substances was also performed.

The highest removal rates for all investigated BTX components were observed during the first 5 min of processing for both ozonation and O_3/UV treatment processes.

Ozone showed the highest affinity to p-xylene. The lowest removal efficiency was measured for benzene. Treatment efficiencies above 90% were observed for all investigated pollutants after 40 min of ozonation. Longer ozonation time resulted in very low enhancements of removal efficiencies of both ozonation and O_3/UV treatment processes.

Application of O_3/UV treatment had no significant effect in comparison with ozonation only, particularly for benzene. In case of o-xylene and p-xylene removal efficiencies, over 90% were observed after 20 min of the process. Forty minutes of the process were needed for more than 90% removal efficiency of ethylbenzene.

Due to high volatility of BTX components their removal from liquid phase can be significantly influenced by stripping. According to physical characteristics, the highest stripping can be expected for benzene and toluene [15].

From the processing of experimental data by simple kinetic models [13, 14], removal of o-xylene and p-xylene was best achieved by the second-order kinetic model. On the other hand, best fit of experimental data for benzene and toluene was obtained using the first-order kinetic model.

From the result of the study one can conclude that ozonation is a prospective process and a promising procedure for the removal of BTX components from aquatic environment. However, further research should be performed to enhance process efficiency and to study the impact of reaction intermediates and products on aquatic ecosystem.

ACKNOWLEDGEMENTS

This work was supported by the Slovak Research and Development Agency under the contract No. APVV-0656-12. The authors would like to thank also for the support from the VEGA Grant 1/0859/14.

REFERENCES

1. European Commission (2000). Directive 2000/60/EC of the European Parliament and of the Council of 23 October 2000 establishing a framework for community action in the field of water policy. Off. J. Eur. Communities, L 327/1.
2. European Commission (2001). Decision No. 2455/2001/EC of the European Parliament and of the Council establishing the list of priority substances in the field of water policy and amending Directive 2000/60/EC, 2001. Off. J. Eur. Union, OJ L 44, pp. 1–5.
3. Water Act (2004). Act No. 364/2004 on water (Water Act) (in Slovak).
4. Mathur AK, Majumder CB, Chatterjee S. Combined removal of BTEX in air stream by using mixture of sugar cane bagasse, compost and GAC as biofilter media. J. Hazard. Mater. 2007; 148: 64–74. DOI: 10.1016/j.jhazmat.2007.02.030
5. Wiszniowski J, Ziembinska A, Ciesielski S. Effect of petroleum organic compounds in synthetic wastewater on treatment performance and microbial community structures in membrane biological reactor (MBR). Special Abstracts/J. Biotechnol. 2010;150 S: 277–278. DOI: 10.1016/j.jbiotec.2010.09.198.
6. Reddy CM, Quinn JG. GC-MS analysis of total petroleum hydrocarbons and polycyclic aromatic hydrocarbons in. Mar. Pollut. Bull. 1999; 38(2): 126–35. DOI: 10.1016/S0025-326X(98)00106-4.
7. USEPA, Occurrence Estimation Methodology and Occurrence Findings Report for the Six-Year Review of Existing National Primary Drinking Water Regulations. Washington, DC: EPA-815-R-03-006. Office of Water. 2003.
8. Frankovská J, Kordík J, Slaninka I, Jurkovič L, Greif V, Šottník P, Dananaj I, Mikita S, Dercová K, Jánová V. The Atlas of Remediation Methods for Contaminated Sites. Štátny geologický ústav Dionýza Štúra. Bratislava. 2010. ISBN 978-80-8934338-6 (in Slovak).

9. Hoigné J, Bader H. Rate constants of reactions of ozone with organic and inorganic compounds in water. I. Non-dissociating organic compounds. Water Res. 1983; 17: 173–183. DOI: 10.1016/0043-1354(83)90098-2.
10. Arslan-Alaton I. The effect of pre-ozonation on the biocompatibility of reactive dye hydrolysates. Chemosphere. 2003; 51(9): 825–833. DOI: 10.1016/S0045-6535(03)00231-5.
11. Sánchez-Polo M, Rivera-Utrilla J, Prados-Joya G, Ferro-García MA, Bautista-Toledo I. Removal of pharmaceutical compounds, nitroimidazoles, from waters by using the ozone/carbon system. Water Res. 2008; 42(7): 4163–4171. DOI: 10.1016/ j.watres. 2008.05.034.
12. Alvares ABC, Diaper C, Parsons SA. Partial oxidation by ozone to remove recalcitrance from wastewaters–a review. Environ. Technol. 2001; 22: 409–427. DOI: 10. 1080/09593332208618273.
13. Derco J, Melicher M, Kassai A, Dudáš J, Valičková M. Removal of benzothiazols by ozone pretreatment. Environ. Eng. Sci. 2011; 28(11): 781–785. DOI: 10.1089/ees.2010.0399.
14. Derco J, Valičková M, Šilhárová K, Dudáš J, Luptaková A. Removal of selected chlorinated micropollutants by ozonation. Chem Pap. 2013; 67: 1585–1593. DOI: 10.2478/ s11696-013-0324-x.
15. Šimovičová K, Derco J, Valičková M, Melicher M, Sumegová L. Reducing of component of gasoline in water by O3/H2O2 process. 46th Internatiol Conference on Petroleum Processing, Bratislava, 2013, ISBN 978-80-969792-4-0.
16. Bedient PB. Groundwater contamination, transport and remediation. Englewood Cliffs, New Jersey: Prentice Hall PTR. 1994.
17. Srijata M, Pranab R. BTEX : a serious ground-water contaminant. Res. J. Environ. Sci. 2011; 5: 394–398. DOI: 10.3923/rjes.2011.394.398.
18. Ullmann F. Ullmann's Encyclopedia of Industrial Chemistry (6th, completely rev. ed.) Electronic Release. Weinheim, Germany: Wiley – VCH. 2011. ISBN: 9783527303878.
19. Matěju V. ed. The Compendium of Clean-up Technologies. 2006; Vodní zdroje Ekomonitor spol. s.r.o., Chrudim. ISBN: 80-86832-15-5 (in Czech).
20. European Commission (2008). Directive 2008/105/EC of the European Parliament and of 10 the Council of 16 December 2008 on environmental quality standards in the field of water policy, amending and subsequently repealing Council Directives 82/176/EEC, 83/513/EEC, 84/156/EEC, 84/491/EEC, 86/280/EEC and amending Directive 2000/ 60/EC.
21. Regulation of the government of the Slovak Republic N° 269/2010: Quality objectives of surface waters and on limit values for pollution indicators of waste waters and special waters (in Slovak).

CHAPTER 13

Treatment of high salt oxidized modified starch waste water using micro-electrolysis, two-phase anaerobic aerobic and electrolysis for reuse

Xuenong Yi[1], Yulin Wang[2]

1 School of Environment and Architecture, University of Shanghai for Science and Technology, No. 516, Jungong Road, Yangpu District, Shanghai, China

2 Environment Protection Department, Shanghai Anyo Energy Efficiency Co., Ltd, No. 171, Garden Road, Hongkou District, Shanghai, China

ABSTRACT

A combined process of micro-electrolysis, two-phase anaerobic, aerobic and electrolysis was investigated for the treatment of oxidized modified starch wastewater (OMSW). Optimum ranges for important operating variables were experimentally determined and the treated water was tested for reuse in the production process of corn starch. The optimum hydraulic retention time (HRT) of micro-electrolysis, methanation reactor, aerobic process and electrolysis process were 5, 24, 12 and 3 h, respectively. The addition of iron-carbon fillers to the acidification reactor was 200 mg/L while the best current density of electrolysis was 300 A/m^2. The biodegradability was improved from 0.12 to 0.34 by micro-electrolysis. The whole treatment was found to be effective with removal of 96 % of the chemical oxygen demand (COD), 0.71 L/day of methane energy recovery. In addition, active chlorine production (15,720 mg/L) was obtained by electrolysis. The advantage of this hybrid process is that, through appropriate control of reaction conditions, effect from high concentration of salt on the treatment was avoided. Moreover, the process also produced the material needed in the production of oxidized starch while remaining emission-free and solved the problem of high process cost.

Keywords: High salt Oxidized starch, wastewater Combined process, Reuse, Emission-free

1. INTRODUCTION

Industrial demand for modified starch triggers the rise in number and expanding in scale of starch factories. The modified starch processes generate wastewater containing organic and inorganic matters that affect wastewater treatment, especially in chemical modification such as oxidation. A large amount of NaCl was produced during the oxidized modified starch production processes in the wastewater (Chavalparit and Limpaseni 1995). This is because of the addition of chemical substances to the process and the adjustment of pH with HCl. The presence of high salt concentration has been traditionally considered as inhibitory for biological wastewater treatment (Fusuwankaya et al. 2009; Kokabian et al. 2013) and caused paralysis of the original wastewater treatment system.

In recent days, studies on treatment of starch wastewater treatments mainly include flocculation and sedimentation, anaerobic biological treatment and aerobic biological process (Abeling and Seyfried 1993; Deng et al. 2003; Rajasimman and Karthikeyan 2007; Li et al. 2011). But there have been few cases on modified starch wastewater treatment, and those under investigation have considered neither the removal nor the reuse of salt generated. In one approach, an anaerobic hybrid reactor was used to treat the wastewater and its feasibility under high salt condition was proved but with no further discussion on salt removing (Fusuwankaya et al. 2009). In another approach, combined process of sedimentation, microfiltration and reverse osmosis was developed, but it only discussed the impact of high salt, and the cost was high, and utilization of resources was limited (Cancino-Madariaga and Aguirre 2001).

Considering the poor biodegradability, high salt concentration and complexity of oxidized modified starch wastewater, a combination process of micro electrolysis, two-phase anaerobic, aerobic and electrolysis was designed and a micro-electrolysis anaerobic two-phase reactor was developed. Effective control of the reaction conditions was necessary to avoid the inhibition of high salt to the processes. Hazardous materials and pollutants with poor biodegradability can be eliminated by the preset micro electrolysis (Huang et al. 2013) and the ratio of BOD/COD can be greatly improved so that microorganism activity is not restrained by hazardous materials in the wastewater (Li et al. 2010; Tao and Kang 2014). Other organic pollutants, ammonia nitrogen, and biodegradable pollutants can be removed by anaerobic and aerobic biological treatment (Abou-Elela et al. 2010; Shi et al. 2014). The remaining organics can be removed by electrolysis (Kim et al. 2013), and then active chlorine produced in the process can be reused in oxidized starch production process (Lin et al. 1998; Zaviska et al. 2012), as well as the treated water. This combination saves production process costs and maximizes resource utilization.

2. MATERIALS AND METHODS

The source and quality of wastewater

Table 1 is the original composition of oxidized starch wastewater obtained from a corn starch factory.

Table 1 Characteristics of OMSW process

Parameters	OMSW
COD (mg/L)	8050
BOD (mg/L)	966
NH+4−NNH4+−N (mg/L)	5–7
Salinity–NaCl (%)	22
pH	3.9

Figure 1 shows the flow chart of the combined process.

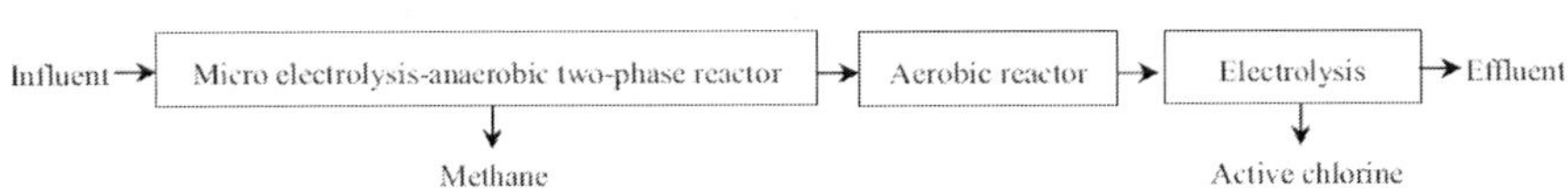

Figure. 1 Technological flow chart of the combination process

2.1 Reactor

A cylindrical-shaped reactor operated in a fed-batch mode as the micro-electrolysis anaerobic two-phase reactor is shown in Fig. 2. It had a working volume of 1.5 L and was divided into two laps. The volume ratio of inner to outer lap was 1:2. The inner lap was further divided into two along diameter. The left serves as a micro electrolysis reactor where iron-carbon fillers were placed, and the right is an acidification reactor with five compartments, where iron-carbon fillers, as the biocarriers to enrich the biofacies, were also placed. The outer lap, as a methane reactor, has 12 compartments. Inside both the methane reactor and the acidification reactor were installed an inclined tube. The iron-carbon fillers had a surface area of 120 m^2/g, a porosity of 65 %, Fe content of 75 %, C content of 20 % and the catalytic element of 5 %. Halophilic bacteria including acid-forming bacteria and methanogen were needed to be acclimated for anaerobic.

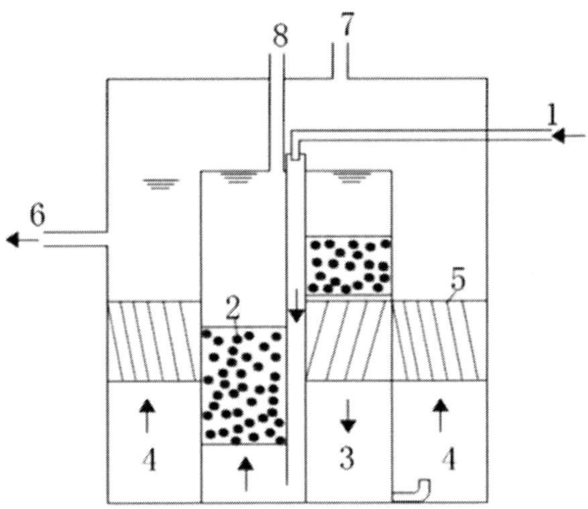

Figure. 2 Micro-electrolysis-anaerobic two-phase reactor. *1* inlet; *2* iron-carbon fillers; *3* acidification reactor; *4* methane reactor; *5* inclined tube; *6* outlet; *7* methane outlet; *8* sample connection for micro electrolysis effluent

2.2 Micro electrolysis, two-phase anaerobic and aerobic procedures

Bacteria extracted in starch wastewater outfall was used for the biochemical procedure after halotolerancy improvement in the wastewater with gradually increasing salt concentration. The fluid was removed from the reactor in the middle and into the micro electrolysis reactor from the bottom. The volume ratio of solution to fillers was 1:1. Because of the high salinity, the raw wastewater worked as an electrolyte and promoted micro electrolysis (Tao and Kang 2014). After a certain time of mixing and reaction, the pH value was regulated to 8 with NaOH. After sedimentation, the hydroxide precipitation was separated from the reaction system, and the supernatant overflowed into the up-flow and down-flow of the first interlayer in acidification reactor alternately at a HRT of 1–6 h. The methane reactor was between the two interlayers. And the second interlayer had a similar running mode with the first one at a HRT of 12–48 h. The inclined tube was in the up-flow area.

2.3 Electrolytic experiments

To avoid deposition on the electrode during the electrolysis, poly aluminium chloride (PAC) was added into the aerobically treated water for the purpose of removing microorganisms and suspended solids. The electrolysis experiments used titanium electrode coated with RuO_2 (Ti-RuO_2) as anode and cathode with an effective surface area of 24 cm^2 and the interval between electrode plates was 2 cm. The test was conducted under room temperature in an installation which had a total working volume of 300 mL. H_2 f process was directly discharged and Cl_2 was transferred into the electrolytes by air tube in favor of sodium hypochlorite generation.

2.4 Analytical methods

After the steady-state conditions were reached, the physic-chemical analysis investigated covers the following parameters. COD was determined using the potassium dichromate method based on the Standard of the People's Republic of China for Environmental Protection (Zhang et al. 2006). Five-day biological oxygen demand (BOD_5) was measured by the respirometric method (WTW Oxitop®IS6, Germany). The methane production and pH values were daily determined. The biogas production was first passed through a bottle filled with 3 % NaOH solution to adsorb the CO_2, H_2S and other trace amounts gas, and then the methane production was measured volumetrically using a wet gas flow meter (Sun et al. 2012). During the electrolysis process, the current density and temperature were monitored in real time. Chloride of the effluent was measured using argentometric method (APHA and AWWA 1998) and active chlorine was measured by iodometry (Lide 2012).

3. RESULTS AND DISCUSSION

3.1 Optimum HRT of micro electrolysis

The pH did not need to be adjusted, because the wastewater feed pH was 3.9 which was conducive to the micro electrolysis (Li et al. 2010). When those fillers were in contact with wastewater, numerous microscopic galvanic cells formed between the iron and carbon (Qin et al. 2012). The reactions can be represented as:

Iron anode (oxidation):

$$Fe \rightarrow Fe2^{+}+2e$$

$$Fe2^{+} \rightarrow Fe^{3+}+e$$

Carbon cathode (reduction) in the presence of oxygen:

$$2H^{+}+2e \rightarrow 2[H] \rightarrow H_2$$

$$O_2+2H_2O+4e \rightarrow 2OH^{-}$$

Fe^{2+} and [H] generated during micro-electrolysis have high chemical activity (Lai et al. 2012) and are believed to break down the carbon chains of organic contaminants. As shown in Fig. 3, the reaction times were 1, 2, 3, 4, 5 and 6 h, removing COD of 13.2, 24.7, 33.4, 42.7, 46.4 and 49.1 %, respectively. From the 4th hour, COD removal showed down as the increase in pH value retards the reaction rate of micro electrolysis. As HRT is too long, since the biopolymers in the reactor contained large amount of anion groups such as OH^{-}, CO_3^{2-} and other negatives groups, the iron ions could easily precipitate in term of $Fe(OH)_X$. They can cover the surface of iron-carbon fillers, and inhibit the process of micro electrolysis. To ensure good results on COD removing under neutral environment during the subsequent biological treatment and minimize the loss of fillers, the optimal reaction time was determined as 5 h.

3.2 Improvement of wastewater biodegradability by micro electrolysis

The ratio of BOD/COD was measured as a representation of biodegradability. The detection values of micro-electrolysis effluent were shown in Fig. 4.

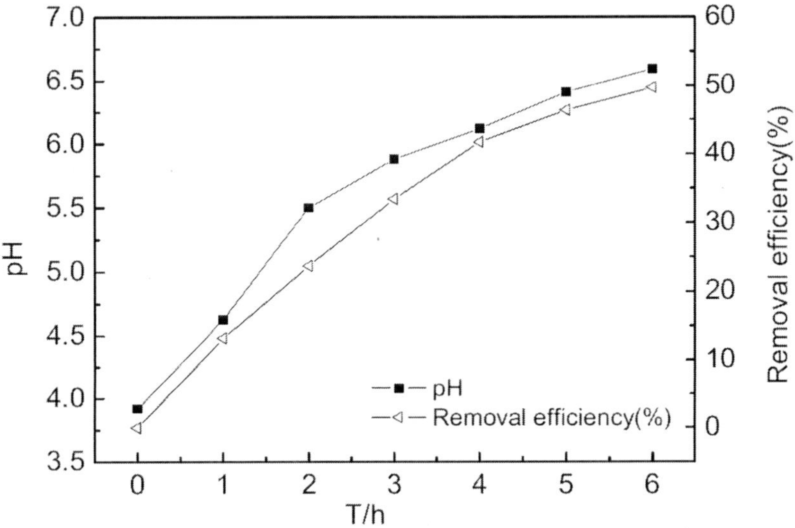

Figure. 3 Changes of pH and removal efficiency of COD by micro-electrolysis

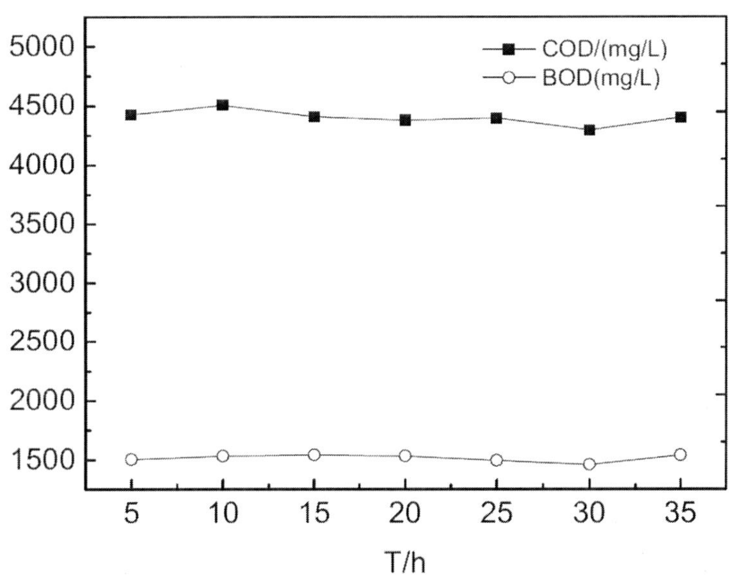

Figure. 4 The ratio of BOD/COD of micro-electrolysis effluent for seven batch tests

The water samples of inner reactor were replaced every 5 h for batch tests. The value of COD for seven batch tests dropped from 8050 mg/L to 4428.6, 4506.3, 4410.2, 4379.8, 4398.4, 4296.5 and 4403.8 mg/L, respectively. The BOD/COD ratio of wastewater changed from 0.12 to 0.34, indicating that those poorly biodegradable pollutants became easily biodegradable and toxic substances became less toxic after micro electrolysis treatment. This also provided a foundation for the follow-up bioprocess.

3.3 Influence of dosage of iron-carbon fillers in the acidification reactor

COD changes of the anaerobic effluent under different dosage at 0, 100, 200 and 300 mg/L were shown in Fig. 5. The COD removal efficiency were 28.2, 34.9, 43.5 and 42 %. When dosage of iron-carbon fillers was 100 mg/L, compared with no added fillers, the removal efficiency increased by about 1/4, indicating that the presence of iron-carbon fillers strengthens the anaerobic treatment. Iron could improve the degradability of anaerobic sludge, because iron could lower the solution oxidation–reduction potential beneficial to anaerobic organism and could be oxidized to Fe^{2+} and Fe^{3+} which were the necessary elements for anaerobic microorganism (Zhang et al. 2007), and some anaerobic organism could also accelerated the corrosion of iron (Duan et al. 2008).

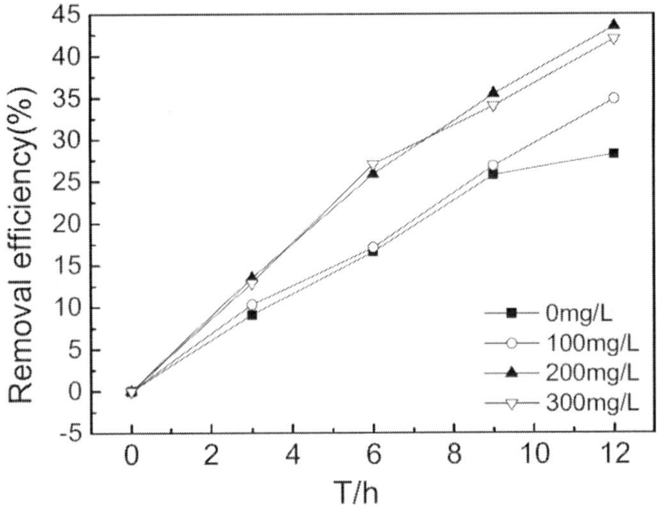

Figure. 5 COD removal efficiency of the anaerobic effluent at various dosage of iron-carbon fillers

That was a possible reason for the improved COD removal at 100 mg/L dosage. And when the dosage was 200 mg/L, the removal efficiency of COD reached the highest value.

3.4 Optimization of HRT in the methane reactor

In this study, a reactor was used as the micro-electrolysis-anaerobic two-phase reactor with the inner lap designed as micro electrolysis and acidification reactor and the outer as a two-phase anaerobic reactor. Two parts of acidogenic phase and methanogenic phase separated in the reactor, which could ensure the acid-forming bacteria and methanogenic bacteria functioning in their corresponding circumstances. The sludge sedimentation in the inclined tube did not only effectively reduce the loss of sludge, but also formed a layer of suspended sludge at the base so that sludge concentration improved. The water operated continually. The HRT at acidification reactor was similar with the micro-electrolysis reactor. At the optimal conditions of previous process, the effect of HRT (HRT in the methane reactor) on the removal efficiencies of COD was investigated. As shown in the Fig. 6, the COD removal efficiency were 40, 60.1, 60.2, 60.5 % respectively, at the HRT of 12, 24, 36 and 48 h, accordingly. Organic pollutants cannot be adequately degraded at a short HRT, but too long HRT caused self-degradation of microorganisms. It's clear that HRT at 24 h is recommended for the reactor.

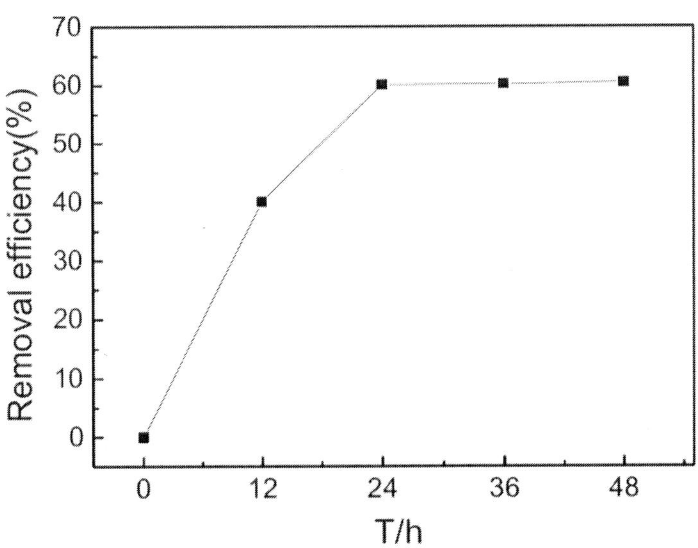

Figure. 6 COD removal efficiency at different HRT of the methane reactor

3.5 The methane gas yield during anaerobic process

To investigate the effect of Fe^{2+} and Fe^{3+} on the anaerobic performance, a micro-electrolysis anaerobic two-phase reactor (R1) and its control reactor (R2) were operated in parallel. There were no iron fillers in R2. Collection of methane gas from the micro-electrolysis anaerobic two-phase reactor started when the reactor had started up for 10 days. As shown in the Fig. 7, at a influent COD concentration of 1500 mg/L from 1st day to 7th day, the methane production in the two reactors were 0.71 ± 0.19 L/d (R1) and 0.32 ± 0.11 L/d (R2). According to the calculation based on the theoretical methane production of 0.35 mL/mg COD removed at standard state (Toprak 1995), the methane production efficiency (methane COD/removed COD) were 79.8 and 36.0 % in the R1 and R2. This result indicated that the addition of Fe^{x+} in R1 increased methane production by two times compared with R2, which was in agreement with the report of Coates et al. (2005) who demonstrated that stimulated microbial Fe^{3+}reduction can enhance methane production. For the idea of zero emissions and the maximization of utility in resource, the methane gas can be collected for recycling and generating economic effects.

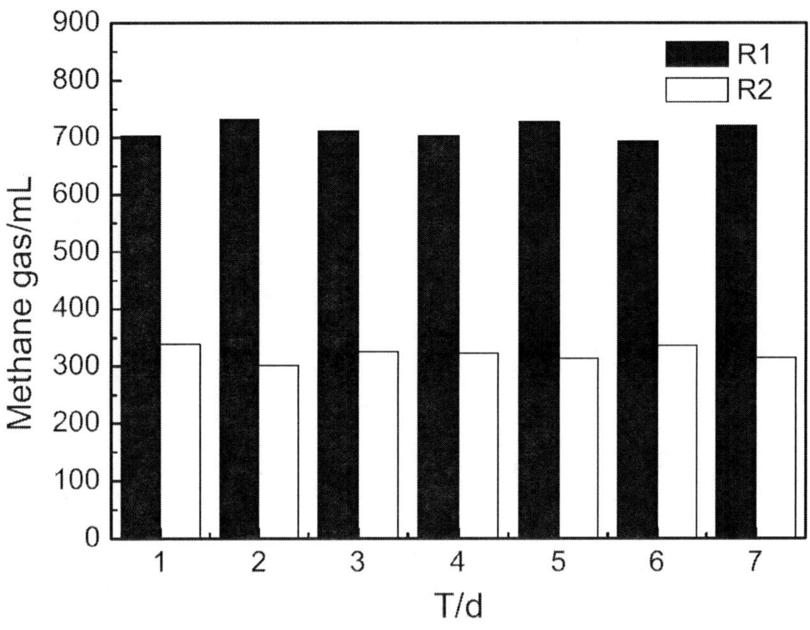

Figure. 7 Methane gas yield in the two reactors (R1, R2)

3.6 Optimization of HRT in the aerobic reactor

Anaerobic affluently moved into the aerobic reactor, the COD variation of aerobic effluent is shown in Fig. 8. The effluent COD removal significantly reduced in the aerobic process within 12 h and reached to 40.9 % at 12 h. After 12 h, the decreasing trend of COD was not obvious. So the optimal HRT of 12 h was determined.

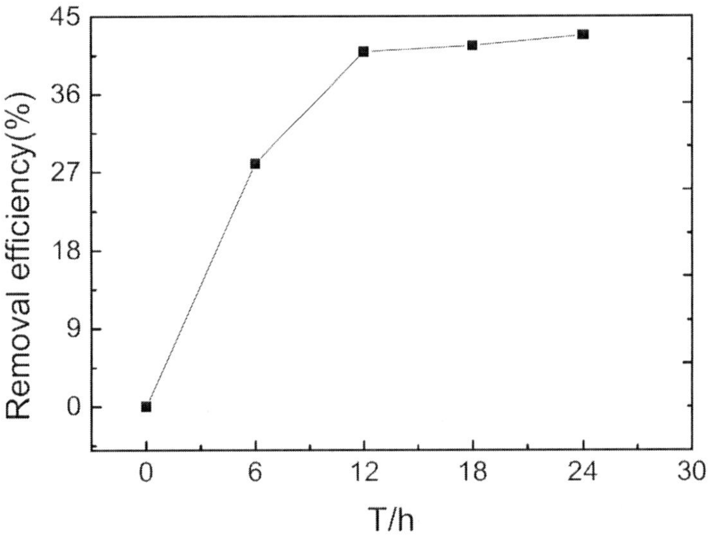

Figure. 8 Removal efficiency of COD at different HRT of the aerobic reactor

3.7 Influence of current density during electrolysis

The reaction was running under the current density of 200, 300 and 400 A/m^2 for the aerobic treated effluent. The explanation of why the initial pH of feed less than 7 was that PAC was added into the effluent of aerobic to remove microorganisms.

After the electrolysis process performed for 3 h, the COD removal efficiency significantly were 36, 73 and 78 %, as shown in Fig. 9. It was probably because Cl^- ions in the water turned into active chlorine in the electric field, resulting in direct degradation of organic pollutants. The removal rate proved that with the increase of current density, electrolysis process effect was improved naturally. During the test, the water samples would be heated as the reaction progress and the temperature rose quickly at current density of 400 A/m². To make efficient use of energy, the optimum current density was designed as 300 A/m², and all subsequent trials selected this density.

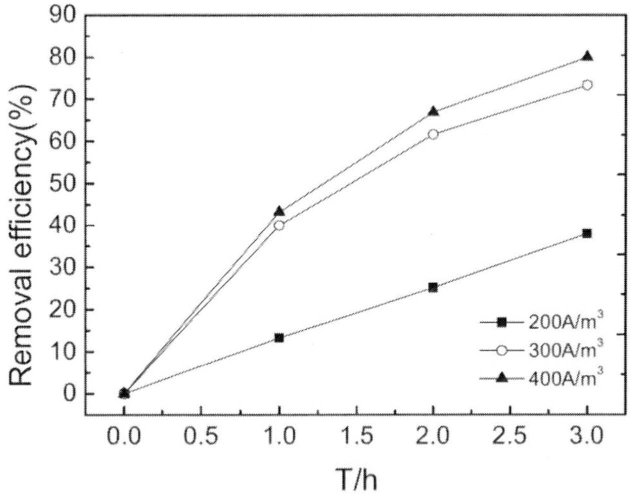

Figure. 9 Removal efficiency of COD at different current density

Salt content in wastewater was high so that sodium hypochlorite and [O], which had strong oxidizing with organic matters (Zhou et al. 2014), were produced by electrolysis of salt. Then the organic matters were degraded by electrolysis process. The mechanism displays as follows.

Electrode reaction:

$$2Cl^{-} \rightarrow Cl_2 + 2e$$

$$2H^{+} + 2e \rightarrow H_2$$

$$2NaOH + Cl_2 \rightarrow NaCl + NaClO + H_2$$

$$ClO^{-} \rightarrow Cl^{-} + [O]$$

$$ClO^{-} + 2H^{+} + 2e \rightarrow Cl^{-} + H_2O$$

The concentration of chlorine ions and available chlorine were shown in Fig. 10. In fact, during electrolysis, pH value of the electrolyzed-solution increases due to the production of hydroxide ions at the cathode electrode.

The concentration of chloride ion decreased from 22,855.5 to 4787.6 mg/L and active chlorine rose to 15,719 mg/L. Theoretical yield based on the reaction formula of active chlorine showed that, the remaining available chlorine concentration accounted for 60 % of the theoretical yield. Considering the incomplete conversion factors and 30–40 % active chlorine in the electrolysis process involved in the degradation of organic matters, both production of sodium hypochlorite and degradation of organic matters reacted in the electrolytic process. Residual chlorine in water samples could be collected after treatment and reused.

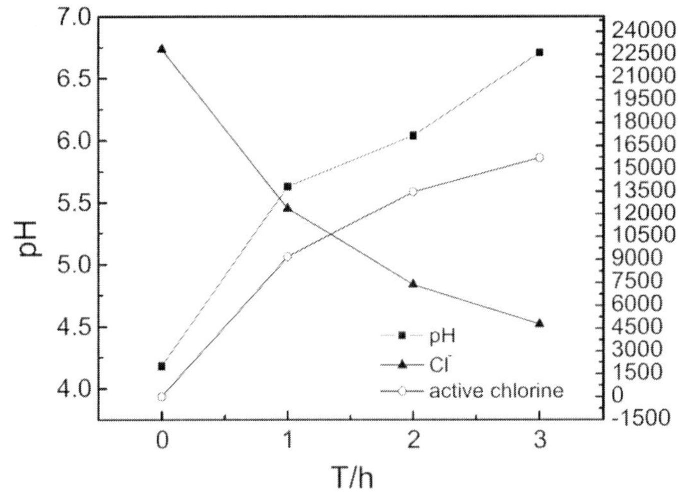

Figure. 10 The concentration of chlorine ions and available chlorine

3.8 Reuse of treated water

This study treated oxidized modified starch wastewater with high salt by micro electrolysis, two-phase anaerobic, aerobic and electrolysis. The treated effluent quality was shown in Table 2.

Table 2 Effluent quality

Parameters	Value
COD (mg/L)	220
Cl^- (mg/L)	4790
Active chlorine (mg/L)	15,720
pH	6.8

The treated water by this treatment was used in the production process of corn starch. The quality tests for the production were shown in Table 3.

Table 3 The quality testing of oxidized starch

(%)	Value	Standard
Starch content	93	≥85
Water	9	≤12.5
Acidity	15	≤18
Ash content	0.03	≤0.1
Fiber	0.09	≤0.5
Protein	0.2	≤0.4
SO_2	0.0009	≤0.004
Fats	0.02	≤0.05
Fineness	99.7	≥99.5

Because of the lack of sodium hypochlorite in the treated water, the productive process needs appropriate amount to add. But the added amount was far less than the amount in the previous production. As apparently seen from Table 3, it successfully applied the treated water to the production, and the oxidized starch product quality testing standards can be achieved.

4. CONCLUSIONS

Water quality was greatly improved and effectively reused in the production process of modified starch after the compound process of micro-electrolysis, two-phase anaerobic, aerobic and electrolysis. Thus, we can conclude that the combined process was an effective method to improve water quality of oxidized modified starch wastewater with high salt concentration. The removal rate of COD was 97.3 % in the whole process.

Micro electrolysis is a feasible method for the treatment and remediation of wastewater with high salt. After pretreatment, the removal efficiency of COD reached 46 %, and the ratio of BOD/COD in the effluent increased from 0.12 to 0.34, which provided a good foundation for the follow-up bioprocess. The arrangement of iron-carbon fillers reinforced the effect of anaerobic treatment at a best dosage of 200 mg/L.

The optimum hydraulic retention time of micro-electrolysis and acidification reactor, methane reactor and aerobic process were 5, 24 and 12 h, respectively. The most effective current density of electrolysis was 300 A/m^2 with reaction time of 3 h. The removal rates of COD in anaerobic process, aerobic treatment and electrolysis process were 45, 62, 40, 70 %, respectively.

The additional Fe^{x+} enhance methane production and 0.71 L/day methane gas was collected from anaerobic biological process. Moreover, 15,800 mg/L active chlorine was produced by electrolytic process. These resources can be recycled to produce economic benefits.

REFERENCES

1. Abeling U, Seyfried CF (1993) Anaerobic-aerobic treatment of potato-starch wastewater. Water Sci Technol 28(2):165–176
2. Abou-Elela SI, Kamel MM, Fawzy ME (2010) Biological treatment of saline wastewater using a salt-tolerant microorganism. Desalination 250(1):1–5
3. APHA, AWWA (1998) Standard methods for the examination of water and waste water, 20th edn. WPCF, New York
4. Cancino-Madariaga B, Aguirre J (2001) Combination treatment of corn starch wastewater by sedimentation, microfiltration and reverse osmosis. Desalination 279(1):285–290
5. Chavalparit O, Limpaseni W (1995) Pollutant load from modified starch factories in Thailand. The Engineering Institute of Thailand Under HM The King's Patronage (EIT), Annual Conference 77–89
6. Coates JD, Cole KA, Michaelidou U et al (2005) Biological control of hog waste odor through stimulated microbial Fe(III) reduction. Appl Environ Microbiol 71(8):4728–4735
7. Deng S, Bai R, Hu X et al (2003) Characteristics of a bioflocculant produced by *Bacillus mucilaginosus* and its use in starch wastewater treatment. Appl Microbiol Biotechnol 60(5):588–593
8. Duan J, Wu S, Zhang X et al (2008) Corrosion of carbon steel influenced by anaerobic biofilm in natural seawater. Electrochem Acta 54(1):22–28
9. Fusuwankaya K, Chaiprasert P, Suraraksa B (2009) Treatment of modified starch wastewater with high sodium chloride (NaCl) concentration using an anaerobic hybrid reactor. Desalin Water Treat 4(1–3):224–228
10. Huang L, Sun G, Yang T et al (2013) A preliminary study of anaerobic treatment coupled with micro-electrolysis for anthraquinone dye wastewater. Desalination 309:91–96
11. Kim DG, Kim WY, Yun CY et al (2013) Agro-industrial wastewater treatment by electrolysis technology. Int J Electrochem Sci 8(7):9835–9850
12. Kokabian B, Bonakdarpour B, Fazel S (2013) The effect of salt on the performance and characteristics of a combined anaerobic–aerobic biological process for the treatment of synthetic wastewaters containing Reactive Black 5. Chem Eng J 221:363–372
13. Lai B, Zhou Y, Yang P (2012) Passivation of sponge iron and GAC in Fe0/GAC mixed-potential corrosion reactor. Ind Eng Chem Res 51(22):7777–7785
14. Li G, Guo S, Li F (2010) Treatment of oilfield produced water by anaerobic process coupled with micro-electrolysis. J Environ Sci 22(12):1875–1882

15. Li G, Ma X, Tu J (2011) Treatment of starch wastewater using UASB reactor with hydraulic circulation and biomass screen. Water Resource and Environmental Protection (ISWREP), 2011 International Symposium IEEE 2:834–837
16. Lide DR (ed) (2012) CRC handbook of chemistry and physics. CRC Press, Boca Raton
17. Lin SH, Shyu CT, Sun MC (1998) Saline wastewater treatment by electrochemical method. Water Res 32(4):1059–1066
18. Qin L, Zhang G, Meng Q et al (2012) Enhanced MBR by internal micro-electrolysis for degradation of anthraquinone dye wastewater. Chem Eng J 210:575–584
19. Rajasimman M, Karthikeyan C (2007) Aerobic digestion of starch wastewater in a fluidized bed bioreactor with low density biomass support. J Hazard Mater 143(1):82–86
20. Shi X, Lefebvre O, Ng KK et al (2014) Sequential anaerobic–aerobic treatment of pharmaceutical wastewater with high salinity. Bioresour Technol 153:79–86
21. Sun L, Wan S, Yu Z et al (2012) Anaerobic biological treatment of high strength cassava starch wastewater in a new type up-flow multistage anaerobic reactor. Bioresour Technol 104:280–288
22. Tao M, Kang TT (2014) Experimental study on high salt acid pretreatment of refinery wastewater. Adv Mater Res 859:369–372
23. Toprak H (1995) Temperature and organic loading dependency of methane and carbon dioxide emission rates of a full-scale anaerobic waste stabilization pond. Water Res 29(4):1111–1119
24. Zaviska F, Drogui P, Pablo G (2012) Statistical optimization of active chlorine production from a synthetic saline effluent by electrolysis. Desalination 296:16–23
25. Zhang H, Zhang D, Zhou J (2006) Removal of COD from landfill leachate by electro-Fenton method. J Hazard Mater 135(1):106–111
26. Zhang W, Chen L, Chen H, Xia SQ (2007) The effect of Fe0/Fe^{2+}/Fe^{3+} on nitrobenzene degradation in the anaerobic sludge. Hazard Mater 143:57–64
27. Zhou Z, Hu DL, Zhang CQ et al (2014) Study on treatment of ammonium nitrogen in desulfurization and denitrification wastewater with sodium hypochlorite. Adv Mater Res 864:438–441

CHAPTER 14

Mitigating Environmental Risks of Wastewater Reuse for Agriculture

Ahmed Al-Busaidi* and Mushtaque Ahmed

Department of Soils, Water and Agricultural Engineering, College of Agricultural & Marine Sciences, Sultan Qaboos University, Muscat, Oman

ABSTRACT

The study was aimed to maximize and optimize treated wastewater reuse in conjunction with surface and ground waters resources. Moreover, environmental, agronomic and economic components were also considered. The project was funded by USAID and implemented in three countries (Oman, Tunisia and Jordan). In Oman, the study was done at Sultan Qaboos University experimental station field. Four types of waters (A: 50% of treated wastewater with 50% of groundwater, B: 100% of groundwater, C: 25% of groundwater with 75% of treated wastewater, and D: 100% of treated wastewater) were used to grow three different crops (okra, maize and sweet corn). Results showed no significant differences in soil physical and chemical properties with treatments irrigated with treated wastewater as compared to groundwater. On other hand, some chemical properties significantly increased ($p<0.05$) when treated wastewater was applied such as soil total carbon and some major elements (N, K, Mg). Crop physical analysis showed significant increases in plant productivity when plants were irrigated with treated wastewater and values of chemical properties were within the international standards. Crop biological analysis showed no effect on crop quality and all tested crops were free from any microbial contamination.

Keywords: treated wastewater, soil, fruits, yield, heavy metals

1. INTRODUCTION

Drought and overexploitation of conventional water resources present a critical problem in many regions of the world, especially the Middle East [1]. Therefore, water resources including nonconventional water should be well managed. Usage of treated wastewater (TWW) on agriculture can save fresh water resources and minimize the applications of chemical fertilizers. In many parts of the world, treated wastewater has been successfully used for irrigation, and many researchers have recognized its benefits [2, 3]. The continuous use of treated wastewater in irrigation increases the total soluble salts in the soil. The cation exchange capacity values are increased by increasing the period of using treated wastewater for irrigation, especially in the surface layer (0–30 cm). Moreover, Fe, Zn, Cu, Mn, Pb, and Co were increased by irrigation using treated wastewater as compared to virgin soil [4]. The use of treated wastewater for irrigation increased the mitotic index of divided cells, chromosome abnormality, and contents of individual amino acids. However, no differences in the profile of protein bands were observed between control and treated wastewater irrigation plants [5]. Whereas the accumulations of heavy metals in the edible part of some plants were detected which adversely affect human and animal health through the food chain [6].

Many countries have included wastewater reuse as an important component of water resources planning. Some countries like Oman have a national policy to reuse all treated wastewater effluents and have already made considerable progress toward this end [7]. Sultanate of Oman is one of the Middle East countries that is considered as the driest or semidriest region in the world with rapidly developing economy and a high population growth [8]. Soil and groundwater (GW) resources of good quality irrigation water have become limited. Rainfall is scanty to support crop production with annual mean rainfall of 100 mm. Therefore, its agriculture is almost fully dependent on groundwater [9]. Water resources augmentation together with conservation has been adopted by the government to combat the water shortage problem. The rapid development of Oman's urbanization, increase in population, and increase in agricultural production has led to high demand for water and urgent need to use treated wastewater as an alternative source of freshwater in agriculture. However, treated wastewater may contain high concentrations of salts, heavy metals, pathogens, and emerging pollutants with unknown effects on the ecological system [10]. High concentrations of heavy metals in plant fruits could affect human health and cause many environmental problems. However, the conjunctive use of treated wastewater and groundwater resources could be employed, helping to safeguard farmer's income and sustain food production. Despite this promising option, more research and education efforts are needed to ensure proper use of treated wastewater for agricultural production. Therefore, the study aimed to optimize treated wastewater reuse in conjunction with groundwater by taking into consideration their quantity and quality, in addition to the agronomic, environmental, and economic components.

2. MATERIALS AND METHODS

The field work was done in plots at the Agricultural Experiment Station, Sultan Qaboos University. Twelve plots (2.5 × 3.5 m each) were designed and sweet corn, okra, and maize crops were grown during the study. The plots were irrigated with four types of waters (A: 50% groundwater and 50% treated wastewater; B: 100% groundwater; C: 75% treated wastewater and 25% groundwater; and D: 100% treated wastewater) as shown in Figure 1. Plants were daily irrigated based on evapotranspiration (ETc). Soil samples were taken before and at the end of the study at a depth of 0–30 cm. Whereas plant samples were taken when the crop was mature and ready for analysis.

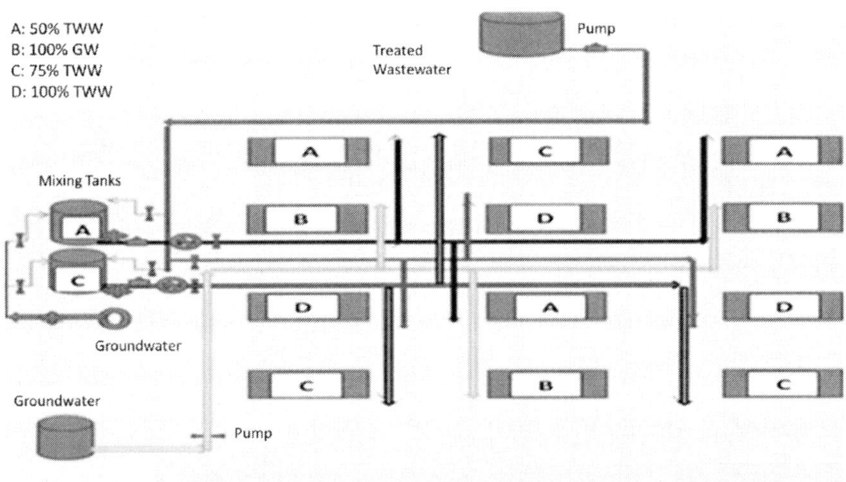

Figure 1. Field experimental design with different treatments.

Plants growth and yield of each crop irrigated by different waters were monitored. Fruits quality and quantity were assessed. Samples from soil and plants were taken for different physical, chemical, and biological analyses. All physicochemical analysis for soil and plants were done in soil and water labs (SQU) following standard methods [11] and using inductively coupled plasma (ICP) instrument for metals analysis. Soil and plant nitrogen (N) were analyzed in Rumais Research Laboratory (Ministry of Agriculture and Fisheries). Whereas biological analyses for crop samples were done in Muscat Municipality Laboratory.

The data were analyzed statistically using the analysis of variance (ANOVA) and the means were compared at the probability level of 5% using the least significant difference [12].

3. RESULT AND DISCUSSION

3.1. Heavy metals in irrigation water

Growing conditions and the irrigation water are the most important parameters controlling plant life. Table 1 demonstrates heavy metal concentrations in the irrigation waters that were used in the study. Comparing the used waters with national and international standards, it can be seen that elements concentrations mentioned in Table 1 had lower values than applied standards.

However, long-term application of some waters may accumulate some harmful elements in soil and plant tissues if mismanagement occurs. In some studies, it was found that wastewaters could carry appreciable amounts of trace toxic metals [13, 14] and concentrations of trace metals in sewage effluents vary from one city to another [15].

Although the concentration of heavy metals in sewage effluents are low, long-term use of these wastewaters on agricultural lands often results in the build-up of elevated levels of these metals in soils [15]. The results of Rattan et al. [15] reported high amount of Cr, Cu, Pb, Co, Ni, Mn, Cd, Fe, Zn, and As in sewage effluents compared to groundwater.

Whereas soil organic matter was also increased in soil samples irrigated with sewage effluents compared to groundwater.

Table 1. Comparing heavy metals concentration (mg/l) in the irrigation waters with national and international standards*.

Water	Mn	Fe	Zn	Cu	Cr	Cd	Pb	Ni	B
Groundwater	0.002	0.013	0.013	0.008	<0.002	<0.001	<0.001	<0.001	0.295
Treated wastewater	0.002	0.016	0.064	0.024	<0.002	<0.001	0.066	<0.001	0.508
EPA Standard	0.200	5.000	5.000	0.500	0.100	0.010	0.100	0.100	0.750
FAO Standard	0.200	5.000	2.000	0.200	0.100	0.010	0.500	0.200	0.750
Omani Standard	0.500	5.000	5.000	1.000	0.050	0.010	0.200	0.100	0.750

*Summary of U.S. EPA guidelines for water reuse for irrigation [16].

3.2. Soil physicochemical properties

Quality of irrigation water could affect soil physicochemical properties. It could improve the soil quality by adding more nutrients or degrading the soil by adding toxic salts. Soil organic matter and total carbon are usually interconnected parameters and they are good indicators for soil fertility. In our study, some of them were found to be high in treated wastewater (TWW) compared to groundwater (GW) treatments. It is an expected result since treated wastewater is usually rich in nitrogen and other nutrients, which enrich soil and enhance plant growth (Figure 2a and b).

The presence of more nutrients (salts) in treated wastewater helps in keeping more water in plant root zone compared to groundwater (Figure 3). Nutrients as salts increased water viscosity and reduced evaporation process and as a result more water can be kept in the root zone [17].

Treated wastewater has a good amount of nutrients (salts). Therefore, it will add more salts to the irrigated soil and increase soil salinity compared to soil irrigated with freshwater (Figure 4). Salts are usually managed and reduced by leaching process.

In addition to organic matter, treated wastewaters have higher values for several nutrients compared to groundwater (Figure 5). These nutrients can improve soil fertility and later support plant growth. The variations in some elements' concentrations between treatments could be due to original nutrients concentrations in the soil and absorbance of those metals during plant growth. Mohammed and Mazahareh [10] found that treated wastewater irrigation increased soil salinity, soil phosphorous, potassium, iron, and manganese levels. They noticed that soil organic matter increased only in the topsoil.

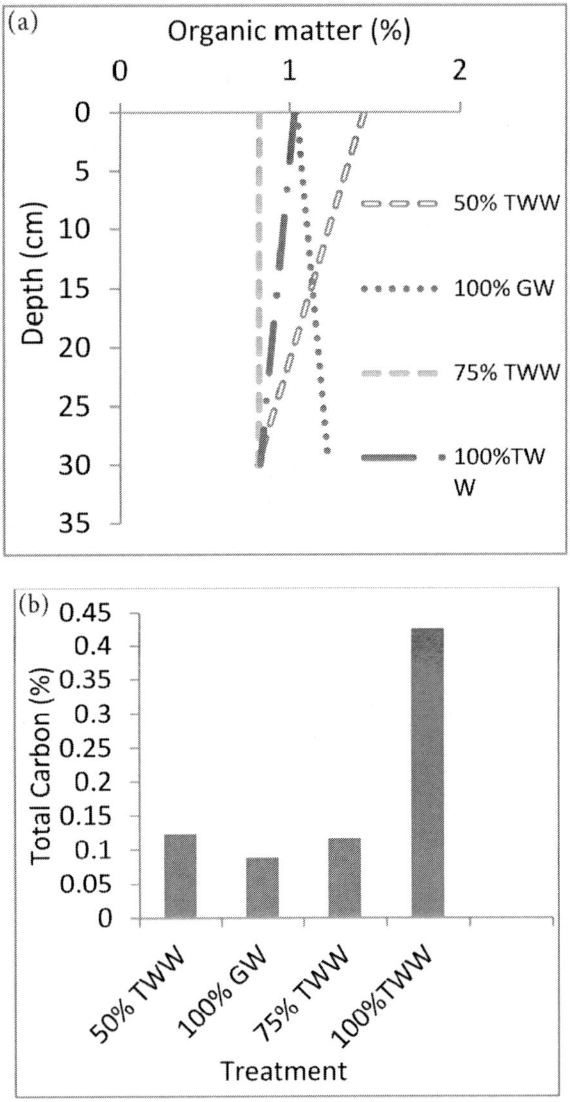

Figure 2. (a) Soil organic matter and (b) soil total carbon.

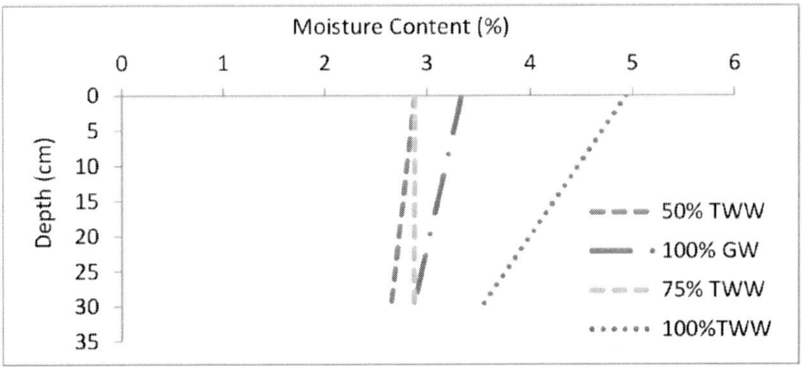

Figure 3. Soil moisture of all treatments.

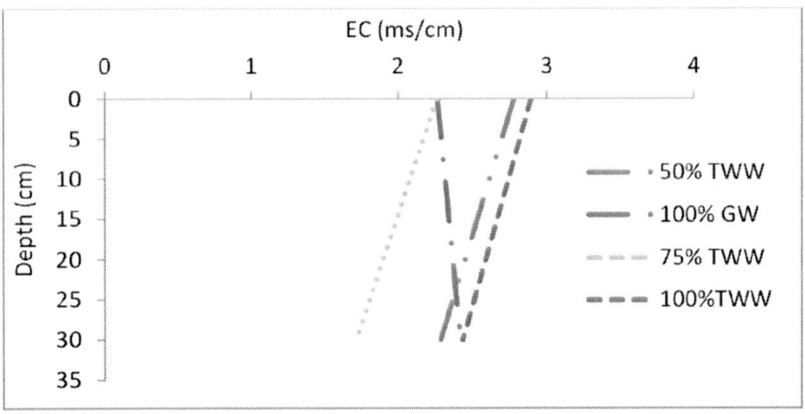

Figure 4. Soil electrical conductivity in all treatments.

Checking soil for microelements (heavy metals) concentrations, it can be seen in Table 2 that all values of heavy metals for both treatments (treated wastewater and groundwater) were very close to each other. However, some significant differences were found between some treatments which could be an indicator for long-term changes in soil chemical properties which is also found in Bansal et al. [18] and Palaniswami and Sree Ramulu [19] studies when they applied wastewater for long period. Rattan et al. [20] observed a build-up of Zn, Pb, Ni, Mn, Fe, Cu, Cr, Co, and As in the sewage-irrigated soils, over the well water-irrigated ones. Significant effect of irrigation through sewage water was observed in case of studied metals. There has been an enormous build-up in the available Fe content in the sewage-irrigated soils. Soils irrigated with groundwater and sewage water showed higher level of Cu and Zn. However, some sewage-irrigated soils accumulated more than 70 mg kg^{-1}total Zn, which could cause a phytotoxicity problem [20]. Whereas Berry et al. [21] found that soil zinc and copper were not significantly affected by wastewater irrigation.

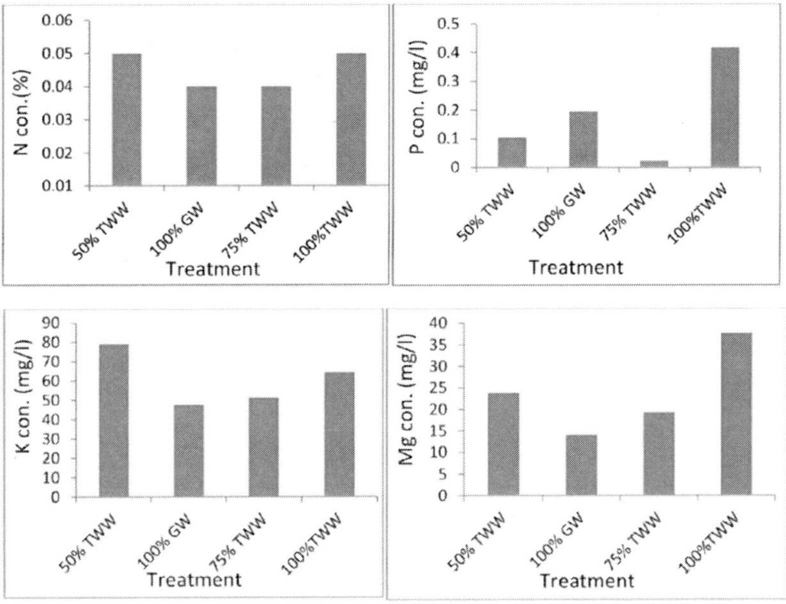

Figure 5. Mean concentration of some elements in soil samples.

Table 2. Mean concentration of heavy metals (mg/l) in soil samples*.

Treatme nt	Mn	Cd	Fe	Zn	B	Ba	Cr	Co	Pb	Ni
50%TW W	0.01 8a	0.00 1a	0.33 0a	0.02 6a	0.16 6c	0.11 8a	0.04 3a	0.05 8a	0.19 6b	0.00 5a
100%G W	0.01 8a	0.00 1a	0.33 1a	0.00 1b	0.17 1b	0.12 3a	0.03 9a	0.06 0a	0.21 9a	0.01 1a
75%TW W	0.01 6a	0.00 1a	0.33 4a	0.00 1b	0.08 8d	0.08 7b	0.04 1a	0.06 1a	0.22 0a	0.00 5a
100%T WW	0.01 8a	0.00 1a	0.34 5a	0.00 3b	0.30 9a	0.11 0a	0.03 9a	0.06 2a	0.23 4a	0.00 1a

*Means in the column with same letter indicate no difference at Duncan's Multiple Range Test at $p < 0.05$.

3.3. Crop physicochemical analysis

From Table 3, it can be seen that treated wastewater gave the best yield for all three crops compared to groundwater. The good supply of different nutrients from treated wastewater enhanced plant growth and improved plant productivity. Abohassan et al. [22] and Stewart et al. [23] have identified the beneficial effects of treated sewage water on some trees grown in Saudi Arabia and Australia. Shafiq et al. [24] found an increase of 24, 45, and 68% in maize total fresh biomass, dry yield, and grain yield irrigated by treated wastewater compared to groundwater. Same finding was also reported by Harati [25] in maize plants.

Table 3. Average weight (kg) of some crops grown in the study.

Treatment	Sweet corn	Okra	Maize
50% TWW	0.091	12.500	1.273
100% GW	0.141	11.091	1.193
75% TWW	0.085	10.556	1.160
100%TWW	0.090	13.958	1.593

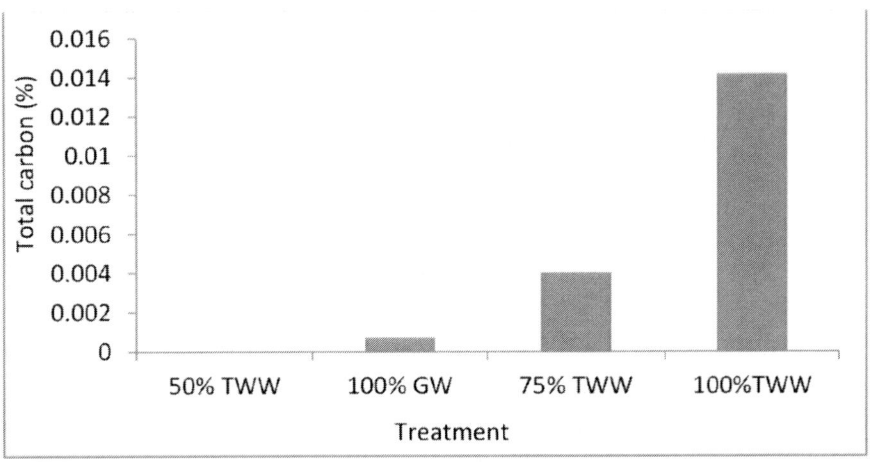

Figure 6. Percentage of total carbon in maize plant leaves.

Maize leaves were the best indicator for carbon content. Therefore, it can be seen from Figure 6 that treated wastewater got the highest values compared to other treatments. It could be a reflection for what was found in water and soil samples.

Table 4. Heavy metals concentration (mg/l) in tested crops*

Treatment	Element conc. (mg/l)								
Okra	Mn	Cd	Fe	Zn	B	Cr	Co	Pb	Ni
50%TWW	0.173d	0.001a	1.224c	0.357b	0.521c	0.060a	0.069a	0.252b	0.006b
100%GW	0.190c	0.001a	1.365b	0.364b	0.336d	0.071a	0.075a	0.229c	0.013b
75%TWW	0.242b	0.001a	2.372a	0.482a	0.745b	0.083a	0.087a	0.255b	0.127a
100%TWW	0.263a	0.001a	1.177d	0.495a	0.862a	0.057a	0.073a	0.300a	0.014b
Sweet corn	Mn	Cd	Fe	Zn	B	Cr	Co	Pb	Ni
50%TWW	0.177b	0.001a	1.295b	0.329c	0.073c	0.122b	0.091b	0.222c	0.068b
100%GW	0.204a	0.001a	1.582a	0.613a	0.492a	0.215a	0.100 a	0.191d	0.104a
75%TWW	0.152c	0.001a	1.584a	0.301c	0.122b	0.061c	0.070c	0.240b	0.011c
100%TWW	0.127d	0.001a	0.889c	0.400b	0.062c	0.064c	0.072c	0.444a	0.003d
Maize	Mn	Cd	Fe	Zn	B	Cr	Co	Pb	Ni
50%TWW	0.457b	0.001a	2.365b	0.256a	0.903a	0.136d	0.064a	0.210c	0.037a
100%GW	0.463a	0.001a	2.362b	0.219c	0.717c	0.146c	0.074a	0.213c	0.047a
75%TWW	0.366d	0.001a	2.279d	0.189d	0.454d	0.151b	0.075a	0.280a	0.040a
100%TWW	0.393c	0.001a	2.483a	0.226b	0.832b	0.181a	0.073a	0.241b	0.052a

*Means in the column with same letter indicate no difference at Duncan's Multiple Range Test at $p < 0.05$.

In a similar study done by Abd-Elfattah et al. [6], they found significant differences in metal content of plant leaves grown in soils irrigated with treated wastewater and plant leaves grown in soils irrigated with Nile water of both seasons.

Table 5. Guideline for safe limits of heavy metals in plants (mg/kg).

Standards/elements	Cd	Cu	Pb	Zn	As	Ni	Cr
WHO/FAO (2007)	0.2	40	5	60	–	–	–
European Union (EU 2006)	0.2	–	0.3	–	0.4	–	2.3
Indian Standard (Awashthi, 2000)	1.5	30	2.5	50	–	1.5	20

Source: CPCB [30].

For soil, usually there is a direct relationship between salts found in the irrigation water and irrigated land. Whereas, for plants, root selectivity and present of salts in different forms could play a role in elements movement and translocation from soil to plant. From Table 4, it can be seen that concentrations of many elements were significantly ($p < 0.05$) different from one treatment to other. However, microelements in the edible parts of all crops grown in the field were not that high and they were within the international standards (Table 5). Same results were reported by Abdelrahman et al. [26] when they observed no significant difference between fresh and treated wastewater with regards to heavy metals accumulation in grown crops. Moreover, this finding was supported by Pescod [27] study, when he concluded that the concentrations of heavy metals in seeds were within normal level when treated wastewater effluent was used. Such results make it clear that heavy metal in soil are not readily bio-available for crop uptake and do not represent a threat to quality of crop consumption.

In general, treated wastewater contains variable amounts of nutrient elements and heavy metals. Availability and translocation of these elements to and within the plant tissues is highly dependent on the environmental conditions as well as their concentration and ratios in the plant organs [28]. Same results were also found by Mahdi et al. [29] when they reported that concentration of nutrient elements of different crops indicated that the crop nutrient uptake is affected by tree age and species. Longer exposure to treated wastewater did not indicate major effects on fruit minerals, including heavy metals. Sampling over longer period of time is needed to confirm the changes in nutrient composition over time. Therefore, in the present study it can be seen that treated wastewater treatment sometime got the highest values for heavy metals which could be an indication for heavy metal accumulation with long-term application if treated wastewater is used without proper management. This prediction could be similar to Abd-Elfattah et al. [6] findings when they found a significant difference in fruit contents of heavy metals and trace elements (Pb, Cd, Ni, Cu, Mn, Fe, Zn) between fruits produced by treated wastewater compared with Nile water in both seasons. The accumulation of heavy metals in the edible part of plant was detected which adversely affects human and animal health through the food chain [6].

Finally, the findings of this study are supported by many researches. As such, Omran et al. [31] found no significant problems with orange trees when they were irrigated with treated sewage water. Furthermore, in Hamad et al. [32] study, toxicity problems for some metals (Cd, Hg, Cr, Pb) in tested crops due to irrigation with treated wastewater was not observed. In the Sultanate of Oman it was found that treated sewage water did not cause any phyto-toxicity symptoms in date palm leaves and fruits [33]. Therefore, it can be concluded that proper management of wastewater irrigation and periodic monitoring of soil and plant quality parameters are required to ensure successful safe long-term wastewater irrigation [34].

3.4. Crop biological analysis

Usually microbial analyses are the direct indicators for microbial contaminations in different crops. In this study, the edible part of grown crops was checked by Muscat Municipality laboratory and different microbes were analyzed such as coliform bacteria, *Escherichia coli*, and *Salmonella* spp. All tested samples were free from any microbial contamination. This finding was supported by Mexican and Tunisian studies where sewage effluent at different levels of treatment has been employed to irrigate various crops. It has been used with no serious effect on man and plants [35].

4. CONCLUSION

The use of treated wastewater for irrigation is increasingly being considered as a technical solution to save fresh groundwater, minimize soil degradation, and improve soil fertility. In this study, usage of treated wastewater irrigation as compared to groundwater did not affect significantly some soil physical and chemical properties. Whereas some chemical properties such as major elements (N, K, Mg) and total carbon were significantly increased when treated wastewater was applied. Concentrations of heavy metals were increased in soils irrigated with treated wastewater compared to groundwater. The differences in heavy metals concentrations of all treatments were small and data of all treatments was close to each other.

Treated wastewater is a rich source of nutrients and provides most nutrients that are necessary for crop growth. Therefore, treated wastewater improved significantly plant productivity compared to groundwater treatments. Whereas small increase was noticed with some chemical properties of plants irrigated with treated wastewater compared to groundwater. However, all measured values were within the international standards. Biologically, all tested crops were free from any microbial contaminations. In general, most crops gave higher yield with wastewater irrigation and reduced the need for chemical fertilizers, resulting in net cost savings to farmers. Therefore, it can be concluded that treated wastewater is an important source of water for agricultural production and to avoid any health or environmental problems, quality of treated wastewater should be monitored with time.

ACKNOWLEDGEMENTS

The authors like to thank the staff from Sultan Qaboos University and Ministry of Agriculture and Fisheries for their support in collecting and analyzing soil, water, and plant samples. Special thanks to USAID (FABRI) for their financial support.

REFERENCES

1. World Bank: Renewable Energy Desalination: An Emerging Solution to Close the Water Gap in the Middle East and North Africa. Washington, DC: World Bank. 2012.
2. Mujeriego R: Salal Golf course irrigation with reclaimed waste water. Sci. Technol. 1991;24 (9): 161–172.
3. Levine A, Sanot A: Recovering sustainable water from waste water. Environ sci.Technol. 2004; 38 (11): 201A–208A.
4. Selem MM, Amir S, Abdel-Aziz SM, Kandil MF, Mansour SF: Effect of irrigation withtreated waste water on some chemical characteristics of soils and plants. Egypt. J. SoilSci. 2000; 40(1/2): 49–59.
5. Hossni YA, El-Tarras A: Safety of treated waste water in the irrigation of plants. Egypt.J. Hortic. 1997; 24 (2): 261–270.
6. Abd- E lfattah A, Shehata SM, Talab AS: Evaluation of irrigation with either rawmunicipal treated waste or river water on elements up take and yield of lettuce andpotato plants. Egyptian J. Soil Sci. 2002; 42 (4): 705–714.
7. Al-Ajmy I: Wastewater in the Sultanate and its effects on environment. In Internationalconference on wastewater management and its effect on the environment in hot andarid countries, Ministry of Regional Municipalities, Environment and Water Resources,Sultanate of Oman. 2002.

8. MAF (Ministry of Agriculture & Fisheries): General soil of the Sultanate of Oman,Directorate General of Agriculture, Ministry of Agriculture and Fisheries. Sultanate ofOman. 1990. p. 50.
9. MRMWR (Ministry of Regional Municipalities and Water Resources) annual report:Environment in the Sultanate of Oman. Muscat, Oman. 2004; Vol. XIV, Issue No. 53.
10. Mohammed MJ, Mazahareh N: Changes in soil fertility parameters in response toirrigation of forage crops with secondary treated waste water. Commun. Soil Sci. PlantAnal. 2003; 34: 1281–1294.
11. Gee GW, Bauder JW: Particle-size analysis. In: Klute, A. (ed), Methods of Soil analysis,Physical and Mineralogical Methods, Part 1. 2nd ed. Agronomy series No. 9. AmericanSociety of Agronomy and Soil Science Society of America Inc. Publ., Madison, Wisconsin,USA. 1986; pp. 383–404.
12. SPSS Inc: SPSS Base 8.0 for Windows User's Guide. SPSS Inc., Chicago. 1998.
13. Brar MS, Mahli SS, Singh AP, Arora CL, Gill KS: Sewer water irrigation effects on somepotentially toxic trace elements in soil and potato plants in northwestern India. Can. J.Soil Sci. 2000; 80: 465–471.
14. Yadar RK, Goyal R, Sharma RK, Dubey SK, Minchas RS: Post irrigation impact ofdomestic sewage effluent on composition of soils, crops and ground water—a casestudy. J. Environ. Int. 2002; 28 (6): 481–486.
15. Rattan RK, Datta SP, Chandra S, Sahran N: Heavy metals and environmental quality:Indian scenario. Fertile. News. 2002; 47 (11): 21–40.
16. U.S Environmental Protection Agency: Guidelines for Waste Water Reuse, Municipalsupport office of wastewater management. Washington, DC, EPA/625/R-04/ 108. 2004.
17. Al-Busaidi A, Cookson P: Leaching potential of sea water. J. Sci. Res. 2005; 9: 27–30.Mitigating Environmental Risks of Wastewater Reuse for Agriculturehttp:// dx. doi.org/10.5772/65871
18. Bansal RL, Nayyar VK, Takkar PN: Accumulation and bioavailability of Zn, Cu, Mnand Fe in soils polluted with industrial waste water. J. Indian Soc. Soil Sci. 1992; 39: 795–799.
19. Palaniswami C, Sree Ramulu US: Effects of continuous irrigation with paper factoryeffluent on soil properties. J. Indian. Soc. Soil Sci. 1994; 42: 139–140.
20. Rattan RK, Datta SP, Chhankar PK, Suribabu K, Singh AK: Long term impact ofirrigation with sewage effluents on heavy metal contents in soils, crops and groundwater—a case study. Agric. Ecosyst. Environ. 2004; 109: 310–322.
21. Berry WL, Wallace A, Lunt OR: Ultilization of municipal wastewater for culture ofhorticultural crops. Hort. Sci. 1980; 15: 169–171.
22. Abohassan AA, Kherallah IE, Kandeel SA: Effect of sewage effluent irrigation regimeson wood quality of *Presopi juliflera* grown in Riyadh Region. Arab Gulf J. Sci. Res. 1988;136 (l): 45–53.
23. Stewart HTL, Hopmans P, Flinn DW: Nutrient accumulation in trees and soil followingirrigation with municipal effluent. Australian Environ. Pollut. 1990; 63(2): 155–177.
24. Shafiq M, Hassan I, Hassan Z: Influence of irrigation methods on the productivity ofsummer maize under saline/sodic environment. Asian Journal of Plant Sciences. 2002;1 (6): 678–680.
25. Harati M: Study on heavy metal accumulation in different parts of corn irrigated bysewage in south of Tehran. MSc Thesis. Tehran University. 2003.
26. Abdelrahman HA, Alkhamisi SA, Ahmed M, Ali H: Effects of Treated WastewaterIrrigation on Element Concentrations in Soil and Maize Plants. Conference Proceeding.Sultan Qaboos University, Oman. 2011.
27. Pescod M: Wastewater treatment and use in agriculture. Bull. FAO #47 (125) (Rome).1992.
28. Izzo K, Scagnozzi A, Belligno A, Navari-Izzo F: Influence of NaCl treatment or Ca, K,Na interrelations in maize shoots. In: M.A.C. Frageso and M.L. Beusichem (Eds.),Optimization of Plant Nutrition. Kluwer Academic Publishers, Dordrecht, The Netherlands. 1993, pp. 577–582.

29. Mahdi OE, Salama SB, Consolacion EC, Al-Solomi M: Effect of treated sewage wateron the concentration of certain nutrient elements in date palm leaves and fruits.Commun. Soil Sci. Plant Anal. 1998; 29: 5–6.
30. CPCB (Central Pollution Control Board): Parivesh, Newsletter from CPCB. 2002.Available at /http://www.cpcb.delhi.nic.in/legislation/ch15dec02a.htms
31. Omran MS, Waly TM, Abo-Eleinain EM, Elnashar BMB: Effect of sewage irrigation onyield, tree components, and heavy metals accumulation in navel orange trees. Biol.Wastes BIWAED. 1988; 23: 17–24.
32. Hamad I, Nizam AA, Suleiman MS: Biological detection of contamination in vegetablesirrigated with brackish and underground water. In: Proceedings of the Arab RegionalWorkshop on the Use of Saline, Brackish, and Treated Sewage Effluent Water inIrrigated Agriculture, Muscat, Sultanate of Oman. 1995. pp. 481–531.
33. El Mardi MO, Salama SB, Consolacion EC, Al-Shabibi MS: Effect of treated sewagewater on vegetative and reproductive growth of date palm. Commun. Soil Sci. PlantAnal. 1995; 26 (11&12): 1895–1904.
34. Rusan M, Hinnawi S, and Rousan L: Long term effect soil and plant quality parameters.Desalination. 2007; 215: 143–152.
35. Strauss M: Sewage Treatment for Wastewater Reuse. Wastewater Reuse Seminar, Institute of Health Sciences, Wattyah, Muscat, Sultanate of Oman. 1987.

CHAPTER 15

When Bioelectrochemical Systems Meet Forward Osmosis: Accomplishing Wastewater Treatment and Reuse through Synergy

Yaobin Lu [1], Mohan Qin [1], Heyang Yuan [1], Ibrahim M. Abu-Reesh [2],* and Zhen He [1],*

[1] Department of Civil and Environmental Engineering, Virginia Polytechnic Institute and State University, Blacksburg, VA 24061, USA

[2] Department of Chemical Engineering, College of Engineering, Qatar University, P.O. Box 2713 Doha, Qatar

ABSTRACT

Bioelectrochemical systems (BES) and forward osmosis (FO) are two emerging technologies with great potential for energy-efficient water/wastewater treatment. BES takes advantage of microbial interaction with a solid electron acceptor/donor to accomplish bioenergy recovery from organic compounds, and FO can extract high-quality water driven by an osmotic pressure. The strong synergy between those two technologies may complement each other and collaboratively address water-energy nexus. FO can assist BES with achieving water recovery (for future reuse), enhancing electricity generation, and supplying energy for accomplishing the cathode reactions; while BES may help FO with degrading organic contaminants, providing sustainable draw solute, and stabilizing water flux. This work has reviewed the recent development that focuses on the synergy between BES and FO, analyzed the advantages of each combination, and provided perspectives for future research. The findings encourage further investigation and development for efficient coordination between BES and FO towards an integrated system for wastewater treatment and reuse.

Keywords: bioelectrochemical systems; microbial fuel cells; microbial electrolysis cells; microbial desalination cells; forward osmosis; wastewater treatment and reuse; bioenergy; draw solution

1. INTRODUCTION

Water and energy are two crucial factors for human and sustainable socio-economic development. According to a recent report [1], 768 million people lack access to fresh water, and another 2.5 billion suffer from sanitation problems. In addition to the deterioration in ecosystems and hydrological cycle, global water consumption is estimated to increase by 55% by year 2050, resulting in significantly increased freshwater demand [1]. On the other hand, energy especially electricity is not available to more than 1.3 billion people, and roughly 2.6 billion of the population are using fossil fuels, which causes various environmental issues such as global warming and environmental pollution. It is estimated that global electricity demand will grow by 70% over the period to year 2035 [1]. A continuous increase in energy demand and serious problems regarding water shortage have called for alternative renewable energy sources and new water treatment technologies with less energy demand. Various technologies have been proposed and investigated [2,3,4,5], and among them, bioelectrochemical systems (BES) and forward osmosis (FO) are two state-of-the-art technologies that may play important roles in addressing the issues of water-energy nexus.

As a novel concept for sustainable energy production and energy-efficient wastewater treatment, BES has attracted great attention in the past decade [6]. Assisted by the exoelectrogens (electrochemically-active microorganisms) growing on the anode electrode of a BES, the organic pollutant in wastewater can be degraded anaerobically and the electrons released by exoelectrogens flow spontaneously from the anode electrode to the cathode electrode through an external circuit to produce electrical current (Figure 1A).

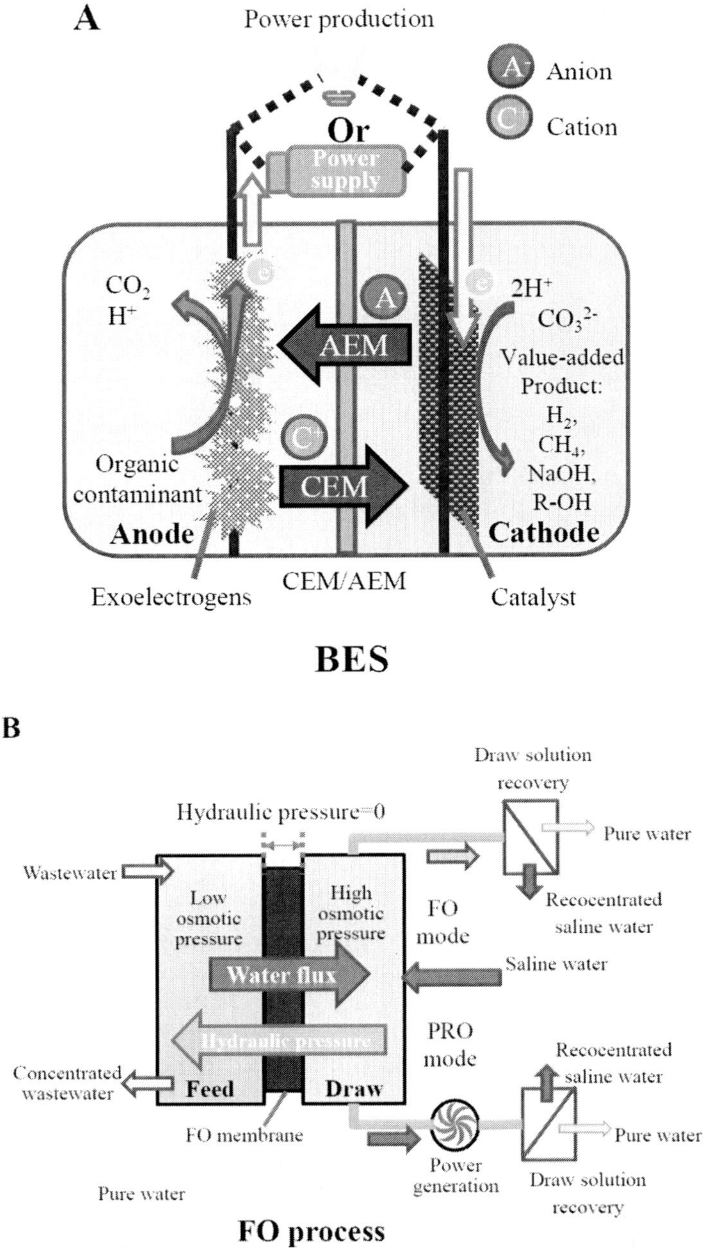

Figure 1. Schematics of bioelectrochemical system (BES) (**A**) and forward osmosis (FO) process (**B**). CEM: cation exchange membrane; AEM: anion exchange membrane.

The first concept of BES was proposed in 1931 [7], and in the past decade the BES has been greatly advanced from both fundamental research and technological development. For wastewater treatment, BES is studied mainly in three aspects [8]: (1) capturing the electrical power directly from organic contaminants in microbial fuel cells (MFCs) [9,10]; (2) harvesting the value-added products (e.g., H_2, CH_4, high-quality water) in microbial electrolysis cells (MECs), microbial electrosynthesis cells (MESs), or microbial desalination cells (MDCs) [11,12,13,14]; and (3) removing specific contaminants (heavy metal, perchlorate, *etc.*), for instance, microbial remediation cells (MRCs) [15,16]. Despite the rapid development, there are still some challenges for BES that need to be addressed before practical applications. For example, as an anaerobic wastewater treatment process, the anode effluent of a conventional BES, which contains microbes, organics, nitrogen, phosphorus, *etc.*, cannot be directly reused or discharged without further treatment. In addition, to drive some reactions in MEC or MES, extra energy is required, thereby increasing the energy demand of the whole system.

Forward osmosis (FO) is based on the natural phenomenon of osmotic processes, and can extract clean water and energy from wastewater (e.g., in pressure retarded osmosis—PRO) [17]. FO refers to the water molecule movement across a semi-permeable membrane driven by osmotic pressure difference between the solutions with high (draw) and low (feed) solute concentrations (Figure 1B). FO can achieve effective rejection of a wide range of contaminants due to the small pore radius (0.25–0.37 nm) [18,19,20], and is cost-saving and energy-efficient resulted from the low membrane fouling of the naturally spontaneous process; thus, it is advantageous over conventional wastewater treatment technologies, especially hydraulic pressure-driven membrane processes [21,22,23].

FO can be engineered for producing potable water from natural waters, extracting reusable water from complex wastewater (industrial streams, landfill leachate, digester centrate, *etc.*) [24,25,26,27,28], and dewatering of waste activated sludge [29,30,31]. However, it should be noted that FO is actually concentrating wastewater instead of degrading contaminants. The remaining concentrates from the feed side of FO still require post-treatment and the energy stored in organic contaminants are not recovered [32].

Moreover, pure water cannot be obtained directly from FO processes because the product water needs to be separated from the draw solutes, which consumes a large amount of energy. Therefore, the selections of draw solutes and the regeneration method are of essential importance to FO processes [30,33].

Because of inherent problems, FO membrane cannot prevent all dissolved draw solute moving from draw solution into feed solution [34], which is called reverse solute leakage [35], another critical shortcoming of FO processes that may decrease the flux and exert negative effects to the post-treatment of wastewater. PRO is a modified FO process with a hydraulic pressure applied in the draw side, and generates the osmotic energy stored in the salinity gradient between saline and fresh waters by releasing the pressurized, diluted salt water through a hydro-turbine [17,36,37,38]. As estimated by the Norwegian state power company, the amount of osmotic energy may reach 2000 TWh worldwide [36].

BES and FO, two distinctly different technologies, have some features that may complement each other to achieve improved wastewater treatment and reuse. Since the first introduction of an osmotic MFC that integrated FO into an MFC in 2011 [39], there has been some advancement in this new synergy. This paper has reviewed the work that links BES to FO in the past four years and provided perspectives on how the synergy between those two technologies can be further developed.

2. SYNERGY BETWEEN BES AND FO

The synergy between those two technologies lies in the complementing functions in energy recovery and water/wastewater treatment. With the advantage of effective water extraction, an FO process can improve water recovery from the treatment by BES. PRO process may provide BES with sufficient and sustainable energy via water flux. In return, BES can achieve degradation of contaminants in wastewater, provide source of treated wastewater for recovery, and harvest valuable products to improve the sustainability of the whole system. A proper combination of the two may also solve some key issues in each process, such as the supply of draw solute, reverse solute leakage in the FO, and energy requirement for carrying out certain reactions in the BES. By far, there have been two approaches to achieve the synergy between BES and FO, direct integration of FO into BES (osmotic bioelectrochemical systems, OsBES) and external linkage between the two (BES-FO/PRO). In an OsBES, FO membrane is used in place of ion exchange membrane as a separator between different compartments, for example osmotic microbial fuel cells (OsMFCs) [39,40] and osmotic microbial desalination cells (OsMDCs) [41]. For the BES-FO/PRO combined systems, the FO and BES can be integrated through hydraulic and/or electrical connection [42,43]. Those approaches result in two types of functional benefits, FO assisting BES and BES helping FO.

2.1. FO Assisting BES

2.1.1. Extracting High-Quality Water

Extraction of high-quality water via forward osmosis is a distinct feature of an OsBES compared with a conventional BES. Because of the higher water permeability coefficient of FO membrane compared with AEM [44], in the presence of an FO membrane and high salinity catholyte, high-quality water can move from the anode (wastewater) into the cathode in an OsBES. For example, an OsMFC (Figure 2) achieved water flux of 3.94 ± 0.22 liters per square meter per hour (LMH) with a catholyte containing 116 g NaCl/L, while there was no obvious water flux in a conventional MFC [39]. The extracted water can be recovered after further purification to remove draw solute in a process such as reverse osmosis (RO), electrodialysis (ED) or MDC. In this regard, the catholyte of the OsMFC functions as a "catalyst" for clean water extraction. Water flux also results in the dilution of the catholyte; the conductivity of the OsMFC catholyte decreased by 8%–35% after 10–12 h operation [39,40], implying that this system could act as a desalination system. Based on that finding, an OsMDC was developed, in which the conductivity of the artificial seawater was reduced by more than 60% during three-day treatment through combined water dilution and salt removal [41]. Unlike an AEM that allows chloride to move through, an FO membrane in the OsMDC can reject the chloride [41], which may potentially reduce the inhibition of chloride accumulation to microbes [45]. It should be noted that the water flux in an OsBES may gradually decrease due to membrane fouling and reduction of osmotic driving force, resulted from the dilution of catholyte and the concentration of anolyte by water transportation and reverse salt permeation [40,46]. In addition, the increase of salt concentration in the anode due to a concentrating effect may inhibit the microbial growth and reduce the anode performance of OsBES [32,47,48], and this adverse effect could be reduced through increasing anolyte flow rate or adding additional desalination chambers (e.g., concentrating chambers to avoid anion moving into the anode). The periodic cleaning of FO membrane, reconcentration of catholyte and replacement of anolyte are necessary for achieving continuing water extraction. Because an FO process is generally faster than biological treatment process in a BES, the unbalanced treatment capacity between the two can result in a short hydraulic retention time (HRT) in the BES anode, which will lead to low treatment efficiency for contaminants. Proper coordination between FO and BES treatment capacity (e.g., increasing the size of BES) may accommodate a longer HRT and improved treatment performance in the BES.

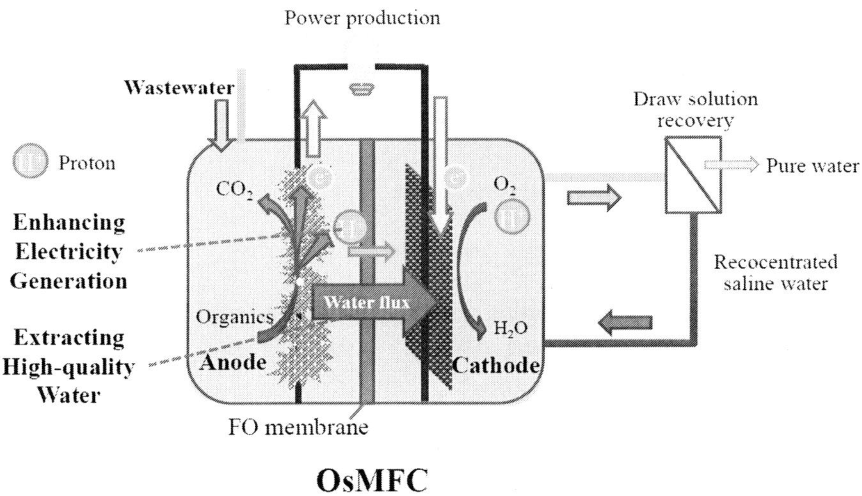

Figure 2. Schematic of an osmotic microbial fuel cell (OsMFC).

2.1.2. Enhancing Electricity Generation

It is found that by replacing ion exchange membrane with FO membrane, greater power production can be achieved in the OsBES compared to the conventional BES. According to polarization curves, the maximum power density of an OsMFC with an aeration-cathode (catholyte: 58 g/L NaCl) was 4.74 W/m^3, 36% higher than that of the MFC with cation exchange membrane (CEM) [39]. In an air-cathode OsMFC, the power density with a catholyte of 35 g/L NaCl was 8% and 87% higher than that of the MFCs containing anion exchange membrane (AEM) and CEM, respectively [40].

The reasons for this enhancement are still under investigation, and there are some preliminary thoughts/hypotheses. In general, the performance of MFCs can be assessed in terms of open circuit voltage and internal losses (including ohmic loss, activation loss, bacterial metabolic loss and concentration loss) [9]. With the same reactor configuration and electrolyte composition, the open circuit voltage of the OsMFC is comparable to that of the conventional MFCs. It is thus reasonable to hypothesize that the enhancement of electricity generation in the OsMFC can be mainly attributed to the reduction in internal losses, e.g., lower membrane resistance, less variation of pH in catholyte and/or anolyte, and higher oxygen rejection performance. It was reported that the air-cathode OsMFC had a relatively lower internal resistance (54 Ω), and the resistance to the flow of ions through the FO membrane was 9 Ω and 19 Ω lower than that through AEM and CEM, which might be due to the acceleration of ion transport [40]. Promoting proton diffusion by water flux may also contribute to the enhanced performance in the OsMFC. The efficient proton transportation supports the reactions on the electrodes and buffers the increased pH of the catholyte. In an air-cathode OsMFC the decreased anolyte pH (0.86 pH unit) was 24% of the theoretical value and 57% of that in the AEM-equipped MFC, whilst the increased catholyte pH (4.06 pH unit) was only 63% of the theoretical value and 70% of that in a regular MFC [40]. Similar results were obtained in another OsMFC, in which the pH of the catholyte reached 9.76 after 10-h operation, lower than 10.90 in a conventional MFC [39]. These results indicate that the OsMFC may exert positive effect on pH stabilization and thus effectively reduce overpotential [49,50,51]. Additionally, the use of the high-conductivity catholyte can also help reduce the resistance of the catholyte, resulting in greater power production [46,52].

2.1.3. Supplying Renewable Energy

When a BES is used for harvesting value-added products (e.g., hydrogen or synthesized organic compounds), an extra voltage (thus energy) was usually required to overcome thermodynamic barriers [53]. For example, an external voltage >0.2 V will be required to drive the hydrogen evolution reaction (HER, −0.41 V *vs.* normal hydrogen electrode, pH = 7) with a cathode electrode, when acetate is used as an electron donor in an MEC [9,12,13]. Thus, energy consumption by an external power source is one of the major challenges for the practical application of this type of BES. To make the BES technology economically viable, several sustainable electricity sources (e.g., MFC, solar, wind, waste heat, and geothermal) have been investigated as alternative power supply for BES cathode reactions [53,54,55,56,57]. PRO, a pressurized FO process, may also provide energy to drive BES reaction, and this has been demonstrated in one of our recent studies, in which we applied osmotic energy produced by a PRO to drive hydrogen production in an MEC [43].

In this system, wastewater firstly flowed through the feed side of the PRO for generating osmotic energy via water flux, and then into the anode of the MEC for degradation of organic contaminants. The diluted saline water from the draw side of the PRO functioned as a catholyte in the MEC. Based on the simulation of PRO model, a maximal energy of 106.3 J obtained during the 2-h PRO process was applied to the MEC, which removed 24.9% of COD and produced 11.5 ± 0.8 mL of hydrogen gas with a production rate of 0.031 ± 0.002 $m^3 \cdot m^{-2} \cdot d^{-1}$ [43]. Those results demonstrate the promise of linking PRO to BES via energy supply for addressing water and energy issues.

2.2. BES Helping FO

2.2.1. Enhancing Water Treatment

BES can help FO with improving the treatment of contaminant in water/wastewater. The FO processes generate concentrated feed solution (wastewater) and diluted draw solution (saline water), both of which need further treatment to meet the standard for discharge and reuse. BES can achieve organics degradation in the anode or salt removal driven by electricity. The studies of the OsMFCs showed that overall organics removal in the anode could reach 80%–90% [40,58]. In a PRO-MEC system, the concentrated feed solution from the PRO was treated in the MEC as the anolyte, which removed 20%–60% of chemical oxygen demand (COD) and the organics removal efficiency was improved by increasing the salt concentration in the draw solution of the PRO [43]. With the advantage in low energy consumption [59], MDCs can be applied to reconcentrate the draw solution and/or recover desalinated water. For example, an OsMDC was developed to recover the desalinated draw solution through both FO and salt removal processes and achieved salt removal of 17%–66% [41]. In another study, an OsMFC was hydraulically coupled with an MDC via sharing the anolyte (wastewater) and salt solution [60]. This coupled system achieved about 85% of total COD removal and more than 95% of conductivity reduction in saline water through dilution in the cathode chamber of the OsMFC and then desalination in the middle chamber of the MDC.

2.2.2. Providing Suitable Draw Solute

A key element in an FO process is draw solution, and the features of an ideal draw solution include high osmotic pressure, easy regeneration, non-toxicity and low cost [33]. The commonly used draw solution based on inorganic salts requires external supply (due to reverse salt flux) and energy-intensive regeneration processes. Ammonium bicarbonate has been demonstrated to be an efficient and recyclable draw solute because of its high osmotic efficiency, high circulation capacity and relatively simple recovery by using moderate heat [21,61,62,63]. Considering that some wastewaters such as digester centrate and landfill leachate contain a high concentration of ammonium [64], recovering ammonia from wastewater may provide a sustainable source of draw solute for FO treatment. This concept was presented in a recent study that linked an ammonium-recovery MEC to an FO process (Figure 3) [42]. In this system, the MEC was used for ammonia recovery driven by electricity generated from organics oxidation, and the recovered ammonia (0.8 M) was then used for preparing the draw solution of the following FO treatment. Carbon dioxide produced from the MEC anode was also collected to make ammonium bicarbonate. The prepared ammonium bicarbonate draw solute was successfully applied to the FO process and achieved the extraction of half of the MEC anode effluent with a maximum water flux of 2.9 ± 0.1 LMH. The diluted draw solution could be regenerated using the waste heat of combustion processes [65], and resupplied with continuing MEC operation.

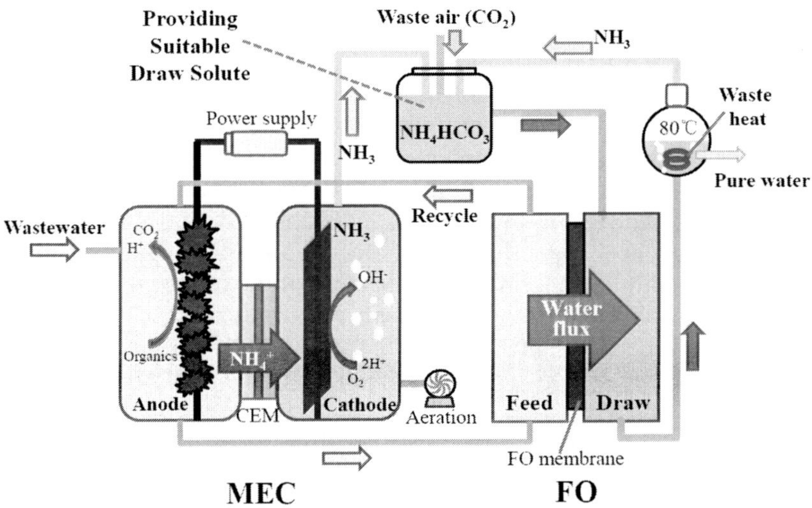

Figure 3. A coupled system consisting of an MEC providing ammonium bicarbonate draw solute for the FO.

2.2.3. Stabilizing Water Flux

Water flux is a key performance parameter for FO and it decreases with FO operation because of the decreased osmotic pressure due to draw solution dilution/feed solution concentration and reverse salt flux [66]. This is especially important to wastewater treatment when FO is applied in the form of osmotic membrane bioreactor (OsMBR) [67,68], as decreased water flux lowers the treatment capacity and increased salt concentration in wastewater inhibits biodegradation [32,47,48]. To alleviate the salt effects on wastewater treatment and maintain a low salinity in wastewater, an MDC can be linked to an OsMBR as a desalination unit. In such a system, the energy in the organic contaminants in wastewater can be extracted by the MDC, and is applied as the driving force for desalination. The desalinated effluent and biodegraded wastewater of the MDC are further treated in an OsMBR, thereby maintaining a stable water flux through decreasing salt concentration.

3. CHALLENGE AND PERSPECTIVES

The synergy between BES and FO has created great opportunities for achieving sustainable energy production, and efficient wastewater treatment and reuse simultaneously. This is a cutting-edge concept and warrants further investigation of system optimization and identifying proper application niches. To advance the research in this area, we must also understand the challenges associated with system development:

1) As a membrane-based process, the BES-FO system involves both ion exchange membranes and FO membranes, and thus membrane fouling will be a critical and inevitable problem, which is related to the operational cost [40,46]. More interestingly, the interaction between the fouling of different membranes, which is related to the interaction between BES and FO, is worth further investigation.

2) Reverse salt flux during the FO and how the leaked salt affects wastewater treatment should be further studied. The leakage of draw solute into the feed solution can decline the water quality of the treated effluent, and on the other hand, the movement of organic contaminants from feed solution to draw solution may decline the quality of recovered water.

3) BES and FO have different treating speeds; therefore, it is necessary to coordinate the treatment capacity between the two. In general, an FO is faster, compared with biological treatment in a BES. As a result, to maintain a coordinated treatment, the liquid volume of the FO will be much smaller than that of the BES. Practically, there could be multiple BES reactors linked to a single FO.

4) The energy issues should be better understood. Both BES and FO are considered as low-energy processes; however, the exact energy benefits through coupling those two are not clear and require detailed analysis through comparing with the existing technologies that can achieve similar treatment goals.

5) System scaling up remains as a great challenge, mainly for BES, as FO has been scaled up to a pilot scale [69,70]. BES scaling up involves the issues such as low-cost materials, scalability of the bench-scale systems, and operation of a scaled system. We are currently working on BES at a scale >100 L and expect to obtain experiences that are helpful with further scaling up.

4. CONCLUSIONS

The research has discovered strong synergy between BES and FO that can be used to achieve an integrated system for simultaneous wastewater treatment and sustainable resource recovery. The two processes are functionally complementary. With appropriate integration through hydraulic and electrical connection, BES and FO may gain the additional function/benefits for recovering high-quality water and improving biological wastewater treatment. Future research and development of the integrated BES-FO systems must address several key challenges including membrane fouling, reverse salt flux, scaling up, and coordination for operation capacity between the two technologies.

ACKNOWLEDGMENTS

This work was supported by funding through the National Priorities Research Program, Qatar National Research Fund (Award NPRP 6-289-2-125).

REFERENCES

1. The United Nations World Water Development Report—n° 5—2014: Water and energy—Vol. 1; facing the challenges—Vol. 2; United Nations Educational, Scientific and Cultural Organization: Paris, France, 2014.
2. Logan, B.E.; Rabaey, K. Conversion of wastes into bioelectricity and chemicals by using microbial electrochemical technologies. Science **2012**, 337, 686–690.
3. Logan, B.E.; Elimelech, M. Membrane-based processes for sustainable power generation using water. Nature **2012**, 488, 313–319.
4. McCarty, P.L.; Bae, J.; Kim, J. Domestic wastewater treatment as a net energy producer—Can this be achieved? Environ. Sci. Technol. **2011**, 45, 7100–7106.
5. Chan, Y.J.; Chong, M.F.; Law, C.L.; Hassell, D. A review on anaerobic–aerobic treatment of industrial and municipal wastewater. Chem. Eng. J. **2009**, 155, 1–18.
6. Li, W.-W.; Yu, H.-Q.; He, Z. Towards sustainable wastewater treatment by using microbial fuel cells-centered technologies. Energy Environ. Sci. **2014**, 7, 911–924.
7. Cohen, B. Thirty-second annual meeting of the society of american bacteriologists. J. Bacteriol. **1931**, 21, 1–60.
8. Arends, J.B.; Verstraete, W. 100 years of microbial electricity production: Three concepts for the future. Microb. Biotechnol. **2012**, 5, 333–346.
9. Logan, B.E.; Hamelers, B.; Rozendal, R.; Schröder, U.; Keller, J.; Freguia, S.; Aelterman, P.; Verstraete, W.; Rabaey, K. Microbial fuel cells: Methodology and technology. Environ. Sci. Technol. **2006**, 40, 5181–5192.
10. Liu, H.; Logan, B.E. Electricity generation using an air-cathode single chamber microbial fuel cell in the presence and absence of a proton exchange membrane. Environ. Sci. Technol. **2004**, 38, 4040–4046.

11. Cheng, S.; Xing, D.; Call, D.F.; Logan, B.E. Direct biological conversion of electrical current into methane by electromethanogenesis. Environ. Sci. Technol. **2009**, 43, 3953–3958.
12. Logan, B.E.; Call, D.; Cheng, S.; Hamelers, H.V.; Sleutels, T.H.; Jeremiasse, A.W.; Rozendal, R.A. Microbial electrolysis cells for high yield hydrogen gas production from organic matter. Environ. Sci. Technol. **2008**, 42, 8630–8640.
13. Rabaey, K.; Rozendal, R.A. Microbial electrosynthesis—Revisiting the electrical route for microbial production. Nat. Rev. Microbiol. **2010**, 8, 706–716.
14. Luo, H.; Xu, P.; Roane, T.M.; Jenkins, P.E.; Ren, Z. Microbial desalination cells for improved performance in wastewater treatment, electricity production, and desalination. Bioresour. Technol. **2012**, 105, 60–66.
15. Gregory, K.B.; Lovley, D.R. Remediation and recovery of uranium from contaminated subsurface environments with electrodes. Environ. Sci. Technol. **2005**, 39, 8943–8947.
16. Butler, C.S.; Clauwaert, P.; Green, S.J.; Verstraete, W.; Nerenberg, R. Bioelectrochemical perchlorate reduction in a microbial fuel cell. Environ. Sci. Technol. **2010**, 44, 4685–4691.
17. Chung, T.-S.; Li, X.; Ong, R.C.; Ge, Q.; Wang, H.; Han, G. Emerging forward osmosis (FO) technologies and challenges ahead for clean water and clean energy applications. Curr. Opin. Chem. Eng. **2012**, 1, 246–257.
18. Coday, B.D.; Xu, P.; Beaudry, E.G.; Herron, J.; Lampi, K.; Hancock, N.T.; Cath, T.Y. The sweet spot of forward osmosis: Treatment of produced water, drilling wastewater, and other complex and difficult liquid streams. Desalination **2014**, 333, 23–35.
19. Xie, M.; Nghiem, L.D.; Price, W.E.; Elimelech, M. Comparison of the removal of hydrophobic trace organic contaminants by forward osmosis and reverse osmosis. Water Res. **2012**, 46, 2683–2692.
20. Fang, Y.; Bian, L.; Bi, Q.; Li, Q.; Wang, X. Evaluation of the pore size distribution of a forward osmosis membrane in three different ways. J. Membr. Sci. **2014**, 454, 390–397.
21. Klaysom, C.; Cath, T.Y.; Depuydt, T.; Vankelecom, I.F. Forward and pressure retarded osmosis: Potential solutions for global challenges in energy and water supply. Chem. Soc. Rev. **2013**, 42, 6959–6989.
22. Cath, T.Y.; Childress, A.E.; Elimelech, M. Forward osmosis: Principles, applications, and recent developments. J. Membr. Sci. **2006**, 281, 70–87.
23. Thompson, N.A.; Nicoll, P.G. Forward Osmosis Desalination: A Commercial Reality; IDA World Congress–Perth Convention and Exhibition Centre (PCEC): Perth, Western Australia, 2011; pp. 4–9.
24. Cath, T.Y.; Gormly, S.; Beaudry, E.G.; Flynn, M.T.; Adams, V.D.; Childress, A.E. Membrane contactor processes for wastewater reclamation in space: Part I. Direct osmotic concentration as pretreatment for reverse osmosis. J. Membr. Sci. **2005**, 257, 85–98.
25. Cath, T.Y.; Adams, D.; Childress, A.E. Membrane contactor processes for wastewater reclamation in space: II. Combined direct osmosis, osmotic distillation, and membrane distillation for treatment of metabolic wastewater. J. Membr. Sci. **2005**, 257, 111–119.
26. Holloway, R.W.; Childress, A.E.; Dennett, K.E.; Cath, T.Y. Forward osmosis for concentration of anaerobic digester centrate. Water Res. **2007**, 41, 4005–4014.
27. Cornelissen, E.R.; Harmsen, D.; Beerendonk, E.F.; Qin, J.J.; Oo, H.; de Korte, K.F.; Kappelhof, J.W. The innovative osmotic membrane bioreactor (OMBR) for reuse of wastewater. Water Sci. Technol. **2011**, 63, 1557–1565.
28. Zhang, S.; Wang, P.; Fu, X.; Chung, T.-S. Sustainable water recovery from oily wastewater via forward osmosis-membrane distillation (FO-MD). Water Res. **2014**, 52, 112–121.
29. Zhu, H.; Zhang, L.; Wen, X.; Huang, X. Feasibility of applying forward osmosis to the simultaneous thickening, digestion, and direct dewatering of waste activated sludge. Bioresour. Technol. **2012**, 113, 207–213.
30. Chung, T.-S.; Zhang, S.; Wang, K.Y.; Su, J.; Ling, M.M. Forward osmosis processes: Yesterday, today and tomorrow. Desalination **2012**, 287, 78–81.
31. Cornelissen, E.; Harmsen, D.; Dekorte, K.; Ruiken, C.; Qin, J.; Oo, H.; Wessels, L. Membrane fouling and process performance of forward osmosis membranes on activated sludge. J. Membr. Sci. **2008**, 319, 158–168.
32. Achilli, A.; Cath, T.Y.; Marchand, E.A.; Childress, A.E. The forward osmosis membrane bioreactor: A low fouling alternative to mbr processes. Desalination **2009**, 239, 10–21.

33. Ge, Q.; Ling, M.; Chung, T.-S. Draw solutions for forward osmosis processes: Developments, challenges, and prospects for the future. J. Membr. Sci. **2013**, 442, 225–237.
34. Phillip, W.A.; Yong, J.S.; Elimelech, M. Reverse draw solute permeation in forward osmosis: Modeling and experiments. Environ. Sci. Technol. **2010**, 44, 5170–5176.
35. Hancock, N.T.; Cath, T.Y. Solute coupled diffusion in osmotically driven membrane processes. Environ. Sci. Technol.**2009**, 43, 6769–6775.
36. Aaberg, R.J. Osmotic power: A new and powerful renewable energy source? Refocus **2003**, 4, 48–50.
37. Skilhagen, S.E.; Dugstad, J.E.; Aaberg, R.J. Osmotic power—Power production based on the osmotic pressure difference between waters with varying salt gradients. Desalination **2008**, 220, 476–482.
38. Kim, Y.C.; Elimelech, M. Adverse impact of feed channel spacers on the performance of pressure retarded osmosis. Environ. Sci. Technol. **2012**, 46, 4673–4681.
39. Zhang, F.; Brastad, K.S.; He, Z. Integrating forward osmosis into microbial fuel cells for wastewater treatment, water extraction and bioelectricity generation. Environ. Sci. Technol. **2011**, 45, 6690–6696.
40. Werner, C.M.; Logan, B.E.; Saikaly, P.E.; Amy, G.L. Wastewater treatment, energy recovery and desalination using a forward osmosis membrane in an air-cathode microbial osmotic fuel cell. J. Membr. Sci. **2013**, 428, 116–122.
41. Zhang, B.; He, Z. Integrated salinity reduction and water recovery in an osmotic microbial desalination cell. RSC Adv.**2012**, 2, 3265–3269.
42. Qin, M.; He, Z. Self-supplied ammonium bicarbonate draw solute for achieving wastewater treatment and recovery in a microbial electrolysis cell-forward osmosis coupled system. Environ. Sci. Technol. Lett. **2014**, 1, 437–441.
43. Yuan, H.; Lu, Y.; Abu-Reesh, I. M.; He, Z. Hydrogen production in a microbial electrolysis cell biased by osmotic energy. J. Power Sources **2015**. unpublished work.
44. Sata, T. Ion Exchange Membranes: Preparation, Characterization, Modification and Application; Royal Society of Chemistry: Cambridge, UK, 2004.
45. Mehanna, M.; Saito, T.; Yan, J.; Hickner, M.; Cao, X.; Huang, X.; Logan, B.E. Using microbial desalination cells to reduce water salinity prior to reverse osmosis. Energy Environ. Sci. **2010**, 3, 1114–1120.
46. Ge, Z.; He, Z. Effects of draw solutions and membrane conditions on electricity generation and water flux in osmotic microbial fuel cells. Bioresour. Technol. **2012**, 109, 70–76.
47. Yap, W.J.; Zhang, J.; Lay, W.C.L.; Cao, B.; Fane, A.G.; Liu, Y. State of the art of osmotic membrane bioreactors for water reclamation. Bioresour. Technol. **2012**, 122, 217–222.
48. Alturki, A.; McDonald, J.; Khan, S.J.; Hai, F.I.; Price, W.E.; Nghiem, L.D. Performance of a novel osmotic membrane bioreactor (OMBR) system: Flux stability and removal of trace organics. Bioresour. Technol. **2012**, 113, 201–206.
49. Zhao, F.; Harnisch, F.; Schröder, U.; Scholz, F.; Bogdanoff, P.; Herrmann, I. Challenges and constraints of using oxygen cathodes in microbial fuel cells. Environ. Sci. Technol. **2006**, 40, 5193–5199.
50. Harnisch, F.; Schröder, U.; Scholz, F. The suitability of monopolar and bipolar ion exchange membranes as separators for biological fuel cells. Environ. Sci. Technol. **2008**, 42, 1740–1746.
51. Rozendal, R.A.; Hamelers, H.V.; Buisman, C.J. Effects of membrane cation transport on ph and microbial fuel cell performance. Environ. Sci. Technol. **2006**, 40, 5206–5211.
52. Gil, G.-C.; Chang, I.-S.; Kim, B.H.; Kim, M.; Jang, J.-K.; Park, H.S.; Kim, H.J. Operational parameters affecting the performannce of a mediator-less microbial fuel cell. Biosens. Bioelectron. **2003**, 18, 327–334.
53. Zhang, Y.; Angelidaki, I. Microbial electrolysis cells turning to be versatile technology: Recent advances and future challenges. Water Res. **2014**, 56, 11–25.
54. Sun, M.; Sheng, G.-P.; Zhang, L.; Xia, C.-R.; Mu, Z.-X.; Liu, X.-W.; Wang, H.-L.; Yu, H.-Q.; Qi, R.; Yu, T.; et al. An mec-mfc-coupled system for biohydrogen production from acetate. Environ. Sci. Technol. **2008**, 42, 8095–8100.

55. Chae, K.-J.; Choi, M.-J.; Kim, K.-Y.; Ajayi, F.F.; Chang, I.-S.; Kim, I.S. A solar-powered microbial electrolysis cell with a platinum catalyst-free cathode to produce hydrogen. Environ. Sci. Technol. **2009**, 43, 9525–9530.
56. Tartakovsky, B.; Manuel, M.F.; Wang, H.; Guiot, S.R. High rate membrane-less microbial electrolysis cell for continuous hydrogen production. Int. J. Hydrog. Energy **2009**, 34, 672–677.
57. Aelterman, P.; Rabaey, K.; Pham, H.T.; Boon, N.; Verstraete, W. Continuous electricity generation at high voltages and currents using stacked microbial fuel cells. Environ. Sci. Technol. **2006**, 40, 3388–3394.
58. Ge, Z.; Ping, Q.; Xiao, L.; He, Z. Reducing effluent discharge and recovering bioenergy in an osmotic microbial fuel cell treating domestic wastewater. Desalination **2013**, 312, 52–59.
59. Kim, Y.; Logan, B.E. Microbial desalination cells for energy production and desalination. Desalination **2013**, 308, 122–130.
60. Zhang, B.; He, Z. Improving water desalination by hydraulically coupling an osmotic microbial fuel cell with a microbial desalination cell. J. Membr. Sci. **2013**, 441, 18–24.
61. McCutcheon, J.R.; McGinnis, R.L.; Elimelech, M. A novel ammonia—Carbon dioxide forward (direct) osmosis desalination process. Desalination **2005**, 174, 1–11.
62. Trypuć, M.; Kiełkowska, U. Solubility in the NH_4HCO_3 + $NaHCO_3$ + H_2O system. J. Chem. Eng. Data **1998**, 43, 201–204.
63. McGinnis, R.L.; McCutcheon, J.R.; Elimelech, M. A novel ammonia–carbon dioxide osmotic heat engine for power generation. J. Membr. Sci. **2007**, 305, 13–19.
64. Renou, S.; Givaudan, J.; Poulain, S.; Dirassouyan, F.; Moulin, P. Landfill leachate treatment: Review and opportunity. J. Hazard. Mater. **2008**, 150, 468–493.
65. Cusick, R.D.; Kim, Y.; Logan, B.E. Energy capture from thermolytic solutions in microbial reverse-electrodialysis cells. Science **2012**, 335, 1474–1477.
66. Lutchmiah, K.; Verliefde, A.R.; Roest, K.; Rietveld, L.C.; Cornelissen, E.R. Forward osmosis for application in wastewater treatment: A review. Water Res. **2014**, 58, 179–197.
67. Wang, X.; Yuan, B.; Chen, Y.; Li, X.; Ren, Y. Integration of micro-filtration into osmotic membrane bioreactors to prevent salinity build-up. Bioresour. Technol. **2014**, 167, 116–123.
68. Chen, L.; Gu, Y.; Cao, C.; Zhang, J.; Ng, J.W.; Tang, C. Performance of a submerged anaerobic membrane bioreactor with forward osmosis membrane for low-strength wastewater treatment. Water Res. **2014**, 50, 114–123.
69. Hancock, N.T.; Xu, P.; Roby, M.J.; Gomez, J.D.; Cath, T.Y. Towards direct potable reuse with forward osmosis: Technical assessment of long-term process performance at the pilot scale. J. Membr. Sci. **2013**, 445, 34–46.
70. Hancock, N.T.; Xu, P.; Heil, D.M.; Bellona, C.; Cath, T.Y. Comprehensive bench- and pilot-scale investigation of trace organic compounds rejection by forward osmosis. Environ. Sci. Technol. **2011**, 45, 8483–8490.

CHAPTER 16

Comparative bacterial degradation and detoxification of model and kraft lignin from pulp paper wastewater and its metabolites

Amar Abhishek[1,2], Ashish Dwivedi[3], Neeraj Tandan[4] Urwashi Kumar[5]

[1] *Environmental Microbiology Section, CSIR-Indian Institute of Toxicology Research, Post Box 80, M. G. Marg, Lucknow 226001, Uttar Pradesh, India*

[2] *Department of Urology, King George's Medical University, Lucknow 226003, Uttar Pradesh, India*

[3] *Phototoxicity Division, CSIR-Indian Institute of Toxicology Research, Post Box No. 80, M. G. Marg, Lucknow 226001, Uttar Pradesh, India*

[4] *Scientific and Applied Research Center, Post Box No. 2241, Meerut 250001, Uttar Pradesh, India*

[5] *International Centre for Genetic Engineering and Biotechnology, P. O. Box 10504, Aruna Asaf Ali Marg, New Delhi 110067, India*

ABSTRACT

Continuous discharge of lignin containing colored wastewater from pulp paper mill into the environment has resulted in building up their high level in various aquatic systems. In this study, the chemical texture of kraft lignin in terms of pollution parameters (COD, TOC, BOD, etc.) was quite different and approximately twofold higher as compared to model lignin at same optical density (OD 3.7 at 465 nm) and lignin content (2000 mg/L). For comparative bacterial degradation and detoxification of model and kraft lignin two bacteria *Citrobacter freundii* and *Serratia marcescens* were isolated, screened and applied in axenic and mixed condition. Bacterial mixed culture was found to decolorize 87 and 70 % model and kraft lignin (2000 mg/L), respectively; whereas, axenic culture *Citrobacter freundii* and *Serratia marcescens* decolorized 64, 60 % model and 50, 55 % kraft lignin, respectively, at optimized condition (34 °C, pH 8.2, 140 rpm). In addition, the mixed bacterial culture also showed the removal of 76, 61 % TOC; 80, 67 % COD and 87, 65 % lignin from model and kraft lignin, respectively. High pollution parameters (like TOC, COD, BOD, sulphate) and toxic chemicals slow down the degradation of kraft lignin as compared to model lignin. The comparative GC–MS analysis has suggested that the interspecies collaboration, i.e., each bacterial strain in culture medium has cumulative enhancing effect on growth, and degradation of lignin rather than inhibition. Furthermore, toxicity evaluation on human keratinocyte cell line after bacterial treatment has supported the degradation and detoxification of model and kraft lignin.

Keywords: Bacteriam, Kraft lignin, GC–MSm, Toxicitym, Cell line

1. INTRODUCTION

The increasing scarcity of water in the world caused by the steady growth of the population and the industrial development raises serious concerns about the need for appropriate water management practices. The pulp paper mills are water-intensive industries generating several types of wastewater with different characteristics, depending on the production process and its raw material. Most significant sources of pollutants in pulp paper mills are wood preparation, pulping, bleaching and coating processes. The color of pulp paper wastewater is mainly due to lignin and its derivatives, which causes various problems along with aesthetically unacceptable intense coloring of water bodies, which block the passage of light to the lower depths of the aquatic system resulting in cessation of photosynthesis, leading to anaerobic conditions, which in turn result in the death of aquatic life causing foul smelling and toxicity of waters bodies (Ali and Sreekrishnan 2001). Thus, it is obligatory to treat the wastewater adequately before final disposal into the environment. Despite the fact that, several physical and chemical methods are available for the treatment of wastewater, they are less desirable than biological process because of cost-ineffectiveness and residual effect (Singhal and Thakur 2009).

Microorganisms are nature's innovative recyclers, converting toxic organic compounds to innocuous species. Among biological treatment methods tried so far, most of the literature confined to a few genera of white rot fungi because of their powerful enzymatic system. But, the stability of fungi is not good in practical treatment under extreme environmental and substrate conditions, such as higher pH, oxygen limitation, and high lignin concentrations. On the other hand, the bacterial ability to use low molecular weight fraction of lignin indicated that bacteria have many unique and specific enzymes with the ability to catalyze the production of various useful compounds (Masai et al. 1999; Raj et al. 2007). In lignin bioremediation, synthetic/model lignin has been widely used as an experimental lignin (Forney and Reddy 1979; Perestelo et al. 1989; Morii et al. 1995; Raj et al. 2007). But, kraft lignin (KL) which is present in pulp paper wastewater differs from model lignin (ML) as it undergoes a variety of reactions including aryl–alkyl cleavages, strong modification of side chains, and various ill-defined condensation reactions causing the polymer to fragment into smaller water/alkali-soluble fragments (Chakar and Ragauskas 2004). However, the above studies provide indirect evidence for the degradation of lignin or its model compounds that cannot be directly explain the degradation of KL. Therefore, there is a need for the comparative study of bacterial removal of ML and KL simultaneously, to explore the inhibitory factors responsible for poor/less degradation of lignin. For this reason, there is still a need for continued search for more efficient ligninolytic bacterial strains for the lignin bioremediation.

Due to the diversity of pollutants in pulp paper mill wastewater, their control is very difficult and presence in water resources poses a high risk to the environment as well as human health. Several studies reported that the pollutants present in pulp paper mill wastewater are responsible for the oxidative stress, DNA damage and lipid peroxidation which leads to apoptosis. Moreover, terpenoids, phytosterols, 2-methoxyphenol and related compounds have been identified as a potential source of cytotoxicity caused by paper mill wastewaters (Chamorro et al. 2013). Therefore, the in present investigates a potential bacterial strain was isolated and screened from pulp and paper mill sludge and examined its ability to decolorize/degrade lignin, which is a major challenge for the pulping industry and characterization of its metabolic products. Furthermore, the study has been extended to assess the cytotoxicity and intracellular reactive oxygen species (ROS) generation using human cell line for evaluation whether toxicity increases or reduces.

2. METHODS

2.1 Sampling site, sample collection and analysis

The wastewater and sludge samples were collected from M/s Yash paper mill Ltd., Darsannagar, Faizabad (UP), India, having co-ordinates 26°44′20″N latitude and 82°13′42″E longitude. This industry came into operation in 1982 and uses eucalyptus, wheat straw and sugarcane bagasse as a raw material and adopts kraft process for pulping followed by multistage chlorine bleaching for making white paper. The wastewater was collected from pulping stage in sterile plastic containers, brought to laboratory and stored at 4 °C for further study. The physico-chemical parameters were accomplished as described in standard methods for the examination of water and wastewaters (APHA 2005). The pH of the sample was analyzed with the selective ion electrode (9172 BN) of Thermo Orion (Model 960). Model lignin (ML) was purchased from Sigma Aldrich (USA), whereas the kraft lignin (KL) was extracted from pulp paper mill wastewater by precipitation method (Raj et al. 2007).

2.2 Media composition and source of microorganism

The medium used throughout this study was lignin-amended mineral salt medium (MSM), which contained (mg/L): [lignin, 2000; K_2HPO_4, 85; KH_2PO_4, 17; $MgSO_4$, 30; $FeSO_4 \cdot 7H_2O$, 30; $CaSO_4$, 30; $MnSO_4 \cdot H_2O$, 30; $(NH_4)_2SO_4$, (Chandra et al. 2011)]. For comparative study both lignins (2000 mg/L, w/v) were added separately to the liquid medium in such a way to give an optical density 3.7 at 465 nm with the reference of distilled water as a blank and pH was adjusted to 8.2.

For degradation study, potential bacteria were isolated from the sludge of main drainage of pulp paper industry with regular discharge of their wastewater. A stable bacterial community was obtained after enrichment of microbial cell. Ten purified isolated bacterial strains designated as SU1–SU10 were screened on the basis of their COD, color and lignin-removal potential (Chandra and Abhishek 2011). Two potential bacterial strains (SU1 and SU5) showing rapid growth and reduction of pollution parameters were selected for further study. For the construction of effective bacterial co-culture, the screened strains were tested in mixed condition. Further, the identification of strains was established on the basis of morphological and biochemical characteristics (Barrow and Feltham 1993).

2.3 Removal of model and kraft lignin by axenic and mixed bacterial culture

For degradation study, 2 % (v/v) overnight-grown suspension of SU1 and SU5 having inoculum size 22 × 10^4 and 30 × 10^4, respectively, was transferred aseptically to 250-mL flask containing 98 mL MSM medium containing 2000 mg/L lignin, amended with 1 % glucose and 0.3 % peptone (w/v) at pH 8.2. Similarly, in mixed condition, 1 % (v/v) from each culture was added to obtain final inoculum size 2 % (v/v) in same condition. The inoculated flasks were incubated at 34 ± 2 °C under shaking flask condition (140 rpm) for 6 days. The bacterial growth was measured at 620 nm using uninoculated medium as blank. For the measurement of color and residual lignin, bacterial degraded and undegraded samples were centrifuged at 8000*g* for 30 min. Supernatant (1 mL) was diluted by adding 3 mL phosphate buffer (pH 7.6) and absorbance measured at 465 nm for color reduction and at 280 nm for lignin degradation on a UV–visible spectrophotometer. To confirm whether the decolorization/degradation process was biological or adsorption phenomenon, NaOH extraction method was adopted (Jiranuntipon et al. 2008). The residual pellets of bacteria were resuspended in equal volume of NaOH (0.1 M) to extract color substances adsorbed to the cell surface. The extracts were centrifuged and absorbance was measured at 465 nm.

2.4 Metabolite characterization

Bacterial undegraded and degraded samples (50 mL) were centrifuged at 3000*g* for 20 min. The supernatant obtained was acidified to pH 2.0 using 1 N HCl and extracted thrice with the equal volume (50 mL) of ethyl acetate in a separating funnel (500 mL) by intermittent shaking. The organic layer was separated and evaporated to dryness under vacuum at 40 °C. Further, the residues were dried under a stream of nitrogen gas and dissolved in acetonitrile for GC–MS analysis and de-ionized water for toxicity evaluation.

An aliquot of 1 μl of silylated compounds was injected into the GC–MS equipped with a PE auto system XL gas chromatograph interfaced with a turbomass mass spectrometric mass selective detector. The analytical column connected to the system was a PE-5MS capillary column (20 m × 0.18 mm i.d., 0.18 μm film thickness). Helium gas with flow rate of 1 mL/min was used as carrier gas. The column temperature was programmed as 50 °C (5 min); 50–300 °C (10 °C/min, hold time: 5 min) (Chandra and Abhishek 2011). The metabolic products were identified by comparing their mass spectra with that of NIST library available with instrument and by comparing the retention time with those of available authentic organic compounds.

2.5 Toxicity evaluation

Acute toxicity test was performed to evaluate the reactive oxygen species (ROS) generation and cytotoxicity by using human keratinocyte (HaCaT) cell line. The cell line was grown in DMEM F-12 HAM media supplemented with 10 % FBS, antibiotics and antimycotic solution (1.5 %) at 5 % CO_2 and 95 % relative humidity at 37 °C (Agrawal et al. 2007).

2.6 Determination of intracellular ROS production

Cells were grown in 96-multiwell black plates (2×10^4 cells/well) and treated with four types of samples (50 μl) extracted from undegraded and degraded samples of ML and KL, respectively. Cells were then incubated for 30 min at 37 °C with 10 μM carboxy H2-DCFDA (dichlorofluorescin diacetate) in HBSS (Hank's balanced salt solution) and exposed. Culture medium was used as a control (C). After treatment the cells were incubate for 24 h. The fluorescence of DCF was measured by using 485 nm excitation and 520 nm emission wavelengths. The generation of intracellular ROS was further substantiated by the administration of NAC (*N*-acetyl cysteine) (10 and 100 μM) as a specific quencher (Valencia and Kochevar 2006).

2.7 Cell viability assay by MTT

In brief, cells (2×10^4) were seeded per well in 96-well plates and kept in CO_2 incubator for 48 h. The treatments were performed with four different sample extracts (50 μL) by complete medium (200 μL) and incubated for 24 h. Further, the media were replaced by 3-(4,5-dimethylthiozolyl-2)-2,5-diphenyl tetrazolium bromide (MTT) (5 mg/mL). The absorbance was recorded at 530 nm by using multiwell micro plate reader (Agrawal et al. 2007).

3. RESULTS

3.1 Characterization of the isolated bacterial strain

The chemical texture of ML and KL at same optical density and lignin concentration was quite different in nature, which was noted as pH: 8.2,

8.2; sulphate: 4.7, 211; color: 6700, 6700 Co.pt; COD: 8068, 16,700 mg/L; BOD: 3018, 6080 mg/L; TOC: 3130, 6250; lignin: 2000, 2000 mg/L, respectively. This was indicated that KL found in paper mill effluent was very different in nature. In this study, two bacterial strains SU1 and SU5 were found to be very efficient and could reduce 45, 41 and 33, 31 %, COD and 52, 57 and 48, 52 % color in ML and KL, respectively, during 6 days of incubation period. Further, on the basis of biochemical tests both the strains were found, gram-negative rods and belong to Enterobacteriaceae family.

Table 1 Bio-chemical characterization of screened bacterial strains

Characteristics	**SU1**	**SU5**
Gram reaction	−ve	−ve
Shape	Rod	Rod
Culture characteristic on agar plate	Translucent, moist, medium, round	Opaque, round, and produces red pigment
Growth of MacConkey agar	–	–
Urea	–	–
MR/VP	+/–	–/+
Gelatin	–	+
Oxidase	–	–
Catalase	+	+
Starch hydrolysis	–	–
Citrate utilization	+	+
Motility	+	+
Indole	–	–
H_2S	–	–
Carbohydrate fermentation		
Glucose	A/G	A
Sucrose	A	A
Lactose	A	–

– negative, + positive, *A* acid, *G* gas

In addition, strain SU1 was motile, gram-negative rod-shaped bacteria and gave positive test for oxidase and catalase; whereas, strain SU5 was differentiated from SU1 produces a reddish-orange pigment. Finally, these two bacteria (SU1 and SU5) were identified as *Citrobacter freundii* and *Serratia marcescens*, respectively, (Table 1).

3.2 Comparative removal of model and kraft lignin

Bacterial growth, measured by absorbance at 620 nm showed the optimum growth at 4th day of incubation period during lignin degradation (Fig. 1a, b). The pH decline tendency of the medium during degradation was recorded at 2nd day; which subsequent increases to a level of neutral at the end of experiment (Fig. 1a, b). Results revealed that mixed bacterial culture was found to be more effective than axenic culture for lignin degradation as well as remove 76, 61 % TOC; 80, 67 % COD and 87, 65 % lignin from ML and KL, respectively (Fig. 2a, b). In addition, to confirm that lignin degradation is a biological phenomenon not adsorption, suspended bacterial pellet was dissolved in NaOH solution. It showed that the fraction of NaOH-extractable color substances were negligible (Fig. 2c). Hence, the results clearly indicated that lignin degradation process was a completely biological phenomenon.

3.3 GC–MS analysis

Results obtained for comparative degradation of ML and KL through GC–MS analysis of the ethyl extracted products showed the massive consumption of compounds as compared to their respective controls (Fig. 3). Many organic compounds were detected in control sample of ML such as guaiacol (RT-12.9); benzoic acid (RT-13.8); phthalic anhydride (RT-15.5); butylated hydroxytoluene (RT-17.1); 1,2,3-trihydroxy propane (RT-19.2); vanillin acid (20.5); dibutyl phthalate (21.5); octadecanoic acid (26.4) and benzyl butyl phthalate (RT-28.1) were diminished after bacterial treatment (Table 2); whereas, in KL control sample some different but lignin-related fragments were identified such as acetic acid (RT-9.55); phthalate (RT-18.2); butanoic acid (RT-25.4); propyl benzene (RT-30.4) were identified. In addition to lignin fragments, some chlorophenols such as 3,4,5-trichlorophenol (RT-16.0); 2,3,4,6-tetrachlorophenol (RT-18.8) and pentachlorophenol (RT-21.8) were found in kraft lignin undegraded (control) and degraded samples by axenic culture, whereas in mixed condition these chlorophenols were not identified using NIST database. These data indicated that, the developed mixed bacterial culture showed better overall removal efficiencies than axenic culture. While on the other hand, some compounds such as hexadecanoic acid and phthalate in case of model lignin and kraft lignin, respectively, remain unchanged (Table 3).

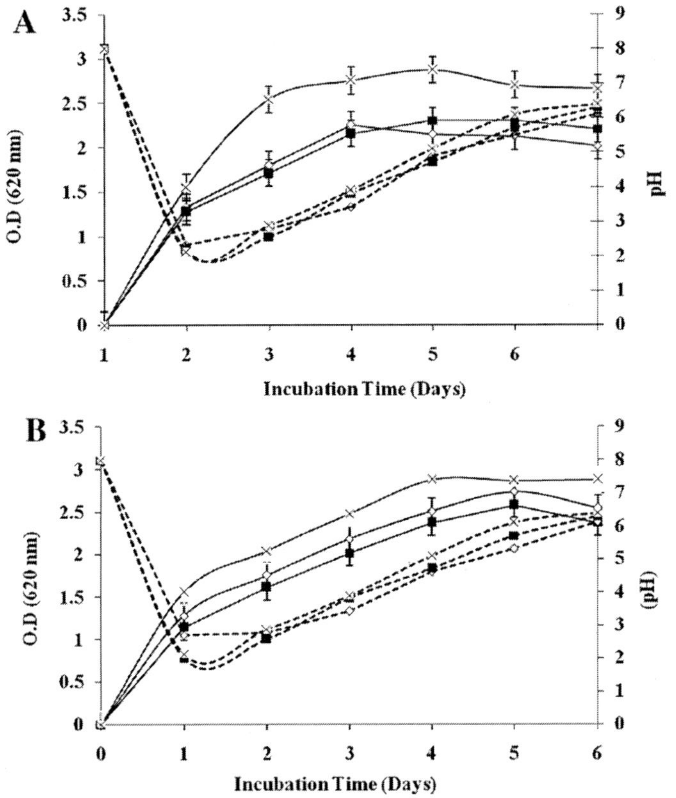

Figure. 1 Bacterial growth and pH pattern of model lignin (ML) (**a**) and kraft lignin (KL) (**b**). *Error bar* represents standard deviation. *COD* chemical oxygen demand, *TOC* total organic carbon

3.4 Intracellular ROS generation

Intracellular ROS levels generated by different sample extract were measured by using DCF assay (Fig. 4). The undegraded sample of KL (control) revealed the maximum generation of ROS than other samples. The generation of ROS is following order, medium control < degraded sample of ML < degraded sample of KL < undegraded sample of ML < undegraded sample of KL (Fig. 4a). Further, quenching of intracellular ROS generation of undegraded sample of KL by different concentrations of NAC (10 and 100 μM) showed that DCF fluorescence was decreased as the concentration of NAC increased (Fig. 4c). Figure 4b, d exhibited fluorescent photographs of different conditions, which also support the above results.

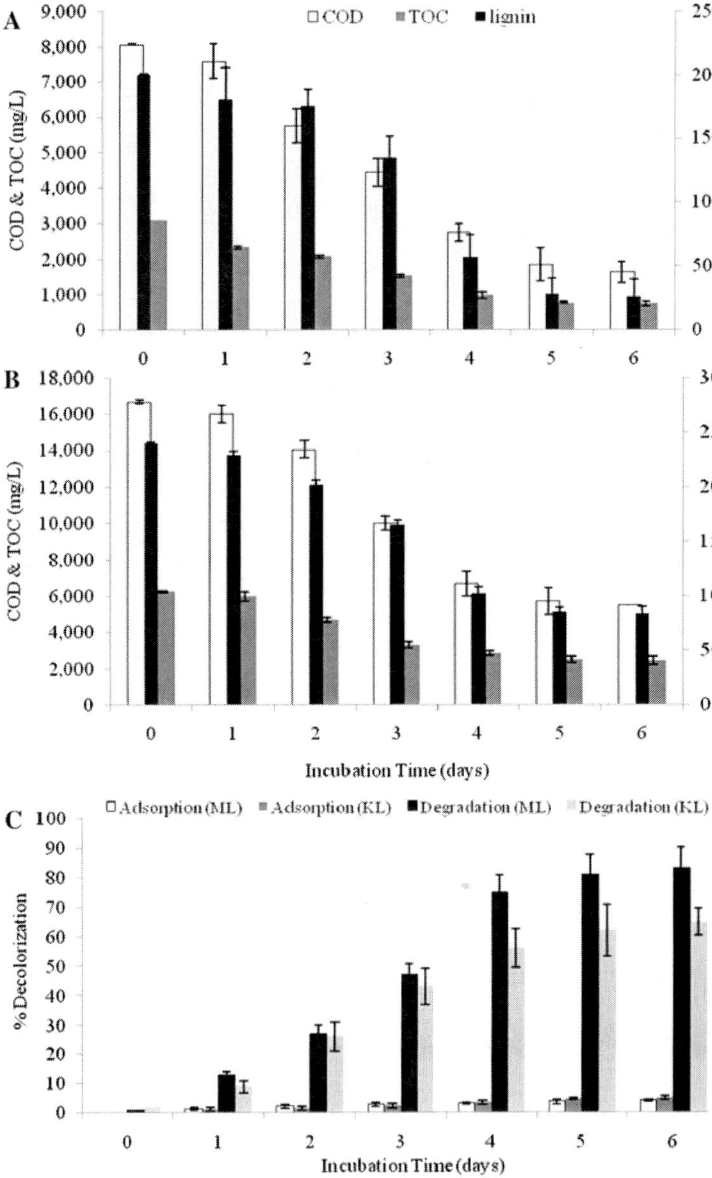

Figure. 2 Reduction of pollution parameters of ML (model lignin) (**a**) and KL (kraft lignin) (**b**); and adsorption vs degradation during bacterial degradation by mixed culture

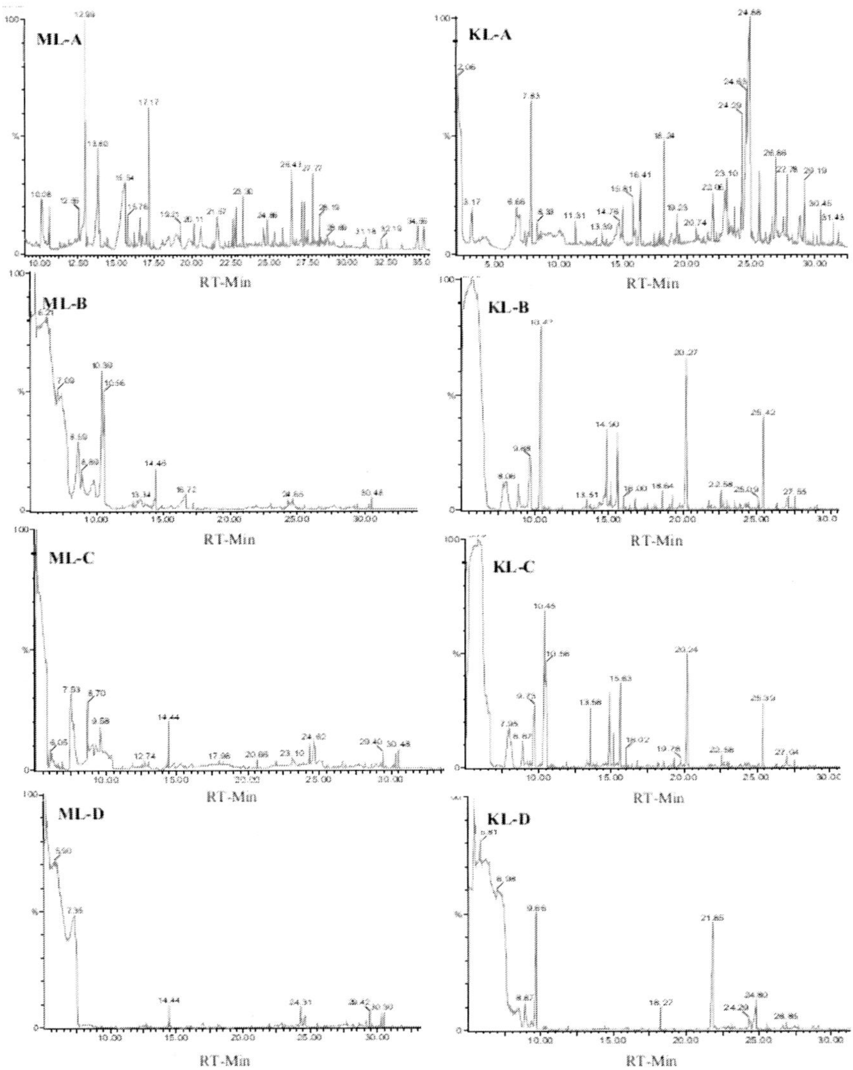

Figure. 3 Total ion chromatogram (TIC) of ethyl acetate extracts from undegraded (**a**) and degraded samples of model (ML) and kraft lignin (KL) with SU1 (**b**); SU5 (**c**) and mixed bacterial culture (**d**)

Table 2 Compounds identified as TMS derivatives in ethyl acetate extract from control and bacterial degraded samples of model lignin

S. no.	Compounds	Model lignin				
		R.T	a	b	c	d
1	Propanoic acid	10.3	–	+	–	–
2	Guaiacol	12.9	+	–	+	–
3	Phenyl acetic acid	13.3	–	+	–	–
4	Benzoic acid	13.8	+	–	–	–
5	Benzyl benzoate	14.4	–	+	+	+
6	Phthalic anhydride	15.5	+	–	–	–
7	Homovanillic acid	16.7	–	+	–	–
8	Butylated hydroxytoluene	17.1	+	–	–	–
9	Gloxylic acid	17.9	–	–	+	–
10	1,2,3-Trihydroxy propane	19.2	+	–	–	–
11	Vanillin acid	20.5	+	–	+	–
12	Dibutyl phthalate	21.5	+	–	–	–
13	Tetradecanoic acid	23.1	–	–	+	–
14	Hexadecanoic acid	24.8	+	+	+	+
15	Octadecanoic acid	26.4	+	–	–	–
16	Unidentified compound	27.7	+	–	–	–
17	Benzyl butyl phthalate	28.1	+	–	–	–
18	1-Phenanthrene carboxylic acid	29.4	–	–	+	+
19	Bis (2-ethylhexyl) phthalate	30.4	–	+	+	+

a control (untreated), *b* SU1, *c* SU5, *d* mixed culture, *RT* retention time in minutes, + present, – absent

Table 3 Compounds identified as TMS derivatives in ethyl acetate extract from control and bacterial degraded samples of kraft lignin

S. no.	Compounds	Kraft lignin				
		R.T	a	b	c	d
1	Ethanedioic acid	7.83	+	–	–	–
2	Acetic acid	9.55	–	+	+	+
3	Propanoic acid	10.4	+	+	–	+
4	3,4,6-Trichloro phenol	11.5	+	–	–	–
5	Unidentified compound	14.9	+	+	–	–
6	Phthalic anhydride	15.6	+	+	+	–
7	3,4,5 Trichloro phenol	16.0	+	+	+	–
8	Phthalate	18.2	+	+	+	+
9	2,3,4,6-Tetrachloro phenol	19.7	+	–	+	
10	Vanillin acid	20.2	+	+	+	–
11	Pentachlorophenol	21.8	+	–	+	–
12	Tetradecanoic acid	22.5	+	+	+	–
13	Unidentified compound	23.1	+	–	–	–
14	Hexadecanoic acid	24.2	+	–	+	–
15	Unidentified compound	24.8	+	–	–	–
16	Butanoic acid	25.4	–	+	–	+
17	Octadecanoic acid	27.0	+	+	+	–
18	Propyl benzene	30.4	+	–	–	–

a control (untreated), *b* SU1, *c* SU5, *d* mixed culture, *RT* Retention Time in minutes, + present, – absent

3.5 Cytotoxicity

The cytotoxicity of degraded and undegraded lignin sample extracts on HaCaT cell line was performed through MTT assay. The result of MTT showed highest reduction in cell viability (60 %) by undegraded sample of KL extract than control. However, undegraded sample of ML, degraded sample of KL and degraded sample of ML showed reduction in cell viability in 40, 30 and 10 % than control, respectively (Fig. 5).

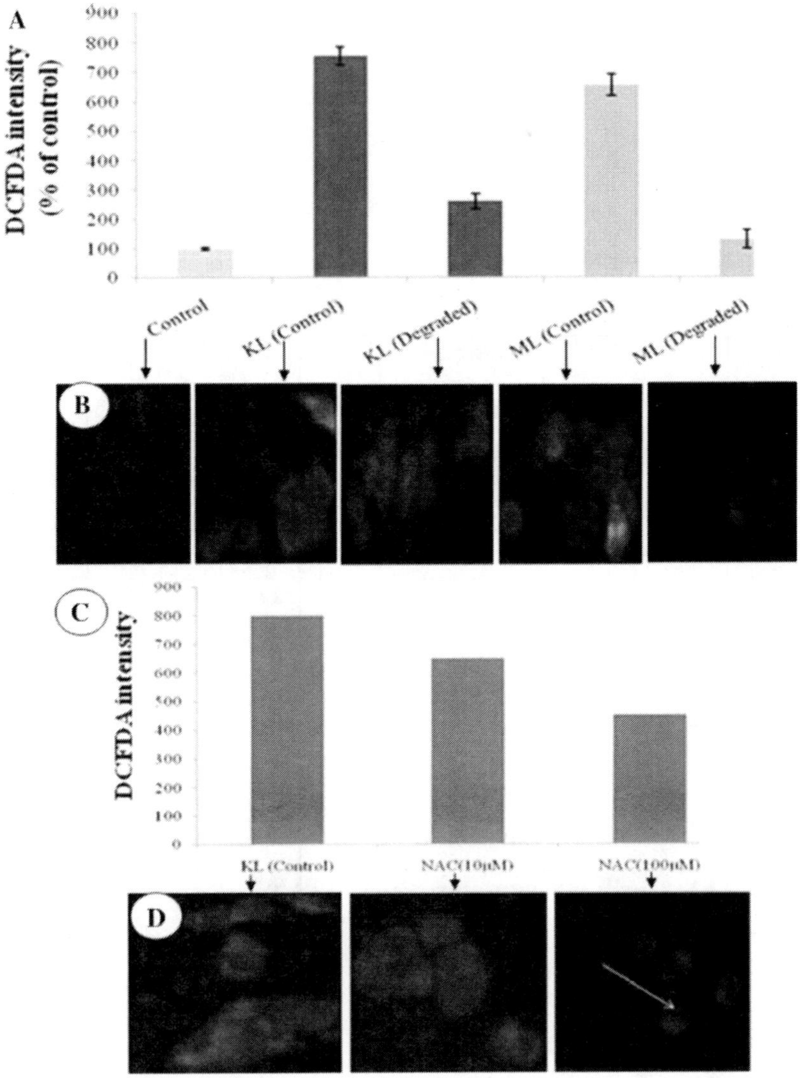

Figure. 4 Intracellular reactive oxygen species generation on HaCaT cell line after 24 h incubation by extracts of undegraded and degraded sample of kraft lignin (KL) and model lignin (ML) (**a**). Quenching of intracellular ROS by different concentrations of *N*-acetyl-L-cysteine (10–100 μM) (**c**). Qualitative analysis of DCF fluorescence under different conditions (**b**, **d**)

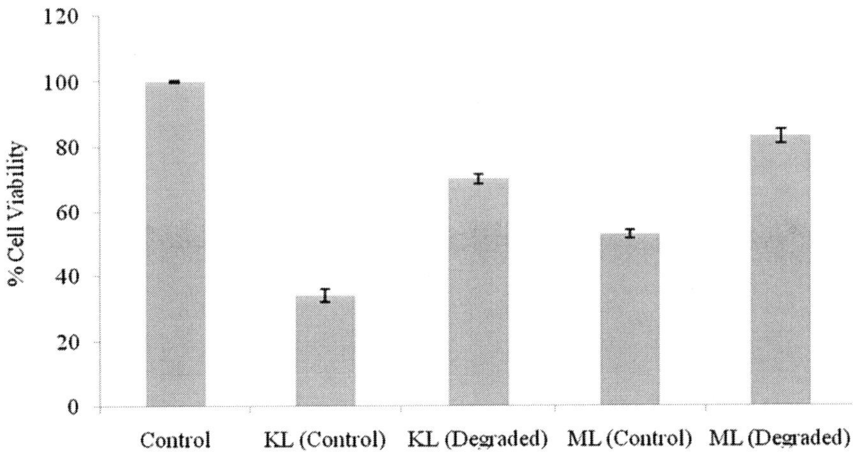

Figure. 5 Cytotoxicity of control and degraded sample of kraft lignin (KL) and model lignin (ML) extracts was performed on HaCaT cell line by MTT assay after incubation of 24 h

4. DISCUSSION

Pulp paper industry has been considered as one of the biggest consumers of natural resources (wood, water) and energy (fossil fuels and electricity). These industries also considered as a significant taxpayer of discharge of pollutants to the environment. The high pH and sulphate value of kraft lignin was due to the presence of sodium hydroxide and sodium sulphide which is used in pulping process. Besides this, the compounds with high molecular weight, such as lignin, chlorophenols and its derivatives do not affect BOD, but would contribute high COD that why these values were higher in KL compared to ML (Ugurlu et al. 2008; Chandra et al. 2011). In addition, the presence of other pollution parameters like high TOC and sulphate values in kraft lignin has made more complex than model lignin. Many authors have reported that the *Citrobacter* sp. and *Serratia marcescens* showing good potential for the degradation of various azo, anthraquinone and other toxic pollutants (Wang et al. 2009; Khalid et al. 2009). Despite the good degrading potential, these organisms have not been used for the treatment of pulp paper wastewater.

During the bacterial degradation of ML and KL, the axenic and mixed bacterial culture attained maximum growth at 4th day of incubation period in terms of optical density at 620 nm. Results revealed that the mixed bacterial culture was found to be more effective for the removal of lignin. Low decolorization at initial phase in spite of fast bacterial growth might be possible due to the utilization of simpler form of carbon (glucose) and nitrogen source (peptone) available in growth media, but depletion of these sources from media compelled bacteria to utilize lignin as a co-substrate (Das et al. 2012). Similar to this study, a co-metabolism mechanism in bacteria and fungi during lignin degradation of kraft pine lignin and model lignin (DHP) has been reported (Perestelo et al. 1989; Kirk et al. 1978; Raj et al. 2007). During bacterial degradation the shift in pH towards acidic condition was noted within initial 2 days of bacterial growth which indicated the formation of acidic compounds through tricarboxylic acid (TCA) cycle, utilizing the simpler form of carbon sources present in medium by bacteria (Yang et al. 2008). As the supplementary nutritional source depleted, the pH of medium shifted towards higher value facilitating the lignin degradation, as lignins are uniformly soluble at high pH which is why more reduction in pollution parameters were noted during late stage of degradation. Similar observation during the microbial degradation of alkali lignin and pulp paper mill wastewater has also been reported (Yang et al. 2008; Singhal and Thakur 2009; Chandra et al. 2011). The treatment with mixed bacterial culture has revealed that presence of each bacterial strain in culture medium has cumulative enhancing effect on growth, and degradation of lignin rather than inhibition.

Pulp paper wastewater is a complex aqueous system; it is difficult to comprehensively investigate the changes during the degradation process. Many studies only focused on optimizing the processes itself while only few reports emphasized on the compositional changes during the bio-treatment process. In the present work, phenolic units derived from lignin and some natural compounds were detected by GC/MS. The low molecular weight compounds like vanillin acid, 1-phenanthrene carboxylic acid, bis (2-ethylhexyl) phthalate, etc., were derivatives of phenolic unit of lignin and could be consumed by mixed bacterial culture to a large extent (Raj et al. 2007; Ko et al. 2009). The massive consumption of these compounds were found in degraded sample of model lignin and kraft lignin as compared with its control indicated that the mixed bacterial culture have strong ability to utilize its constituents as sole source of carbon, nitrogen and energy. Coincident with earlier reports (Gupta et al. 2001), apart from these aromatic compounds, more acid-type compounds like propanoic acid, benzoic acid, gloxylic acid, hexadecanoic acid, etc., were detected than aldehyde and ketone-type due to degradation of lignin in both ML and KL. The phthalate derivative [bis (2-ethylhexyl) phthalate] had been detected from fungal peroxidase degradation of lignosulfonate (Shin and Lee 1999) and also from photodegradation of black liquor lignin (Ksibi et al. 2003). On the other hand, compounds such as hexadecanoic acid in case of model lignin remain undegraded. Unfortunately, there were also many compounds like RT-27.7 in model lignin and RT-14.9, 19.7 and 23.3 in kraft lignin that could not be identified by mass spectrometry. Along with lignin fragments some chlorophenols such as 3,4,5-trichlorophenol (RT-16.0); 2,3,4,6-tetrachlorophenol (RT-18.8) and pentachlorophenol (RT-21.8) found in kraft lignin control and degraded samples by axenic culture. However, these chlorophenols are not found in mixed bacterial culture because these might be utilized as carbon and energy source. Chlorophenols are not the fragment of lignin; it is a priority pollutant that has been used widely as a general biocide in commercial wood treatment (Lee and Wang 2006) and it probably comes during the extraction of lignin from pulp paper wastewater because at low pH, lignin and chlorophenol both are precipitated. These findings revealed that the slow degradation of KL compared to ML due to the high pollution load in terms of COD, TOC, sulphate, etc., and toxic chemicals (organochlorines).

The discharge of this pulp paper mill wastewater may promote skin diseases in human beings who are directly or indirectly contact with this aquatic system (Das et al. 2012). Therefore, it is important to assess the toxic response of these pollutants by using human skin cell line as a model. The cytotoxicity assay of different lignin sample extracts on HaCaT cell line showed highest reduction in cell viability of undegraded sample of KL extract than control, which indicates the maximum pollution load in kraft lignin which was extracted from effluent released by pulp paper mill and is more hazardous for the people.

However, degraded samples of ML and KL showed a significant reduction in cell viability than control illustrating that the toxicity was decreased. Further, to evaluate the mechanism of pollutants inside the cell, intracellular ROS were performed to observe the generation of ROS. Generally, high pollution load increase the reactive oxygen species level inside the cell, which damage different cell organelles which further promote apoptosis (Lee et al. 2002). The DCFDA fluorescence is a very sensitive method to detect the ROS level inside the cell. Here our results of DCFDA on HaCaT cell line showed that as the concentration of pollutants increases, the intracellular ROS is also increases thereby increasing the fluorescence. The extract of undegraded sample of KL showed maximum fluorescence due to high pollution and degraded samples showed lesser. Further, pretreatment with NAC significantly attenuates the fluorescence intensity of DCF, which confirmed the involvement of ROS in cytotoxicity.

5. CONCLUSION

The bacterial cultures were able to co-metabolically remove lignin, using glucose and peptone as primary source of nutrients. The developed mixed bacterial culture showed better removal efficiency compared to axenic culture. Further, the GC–MS analysis of bacterial degraded samples of ML and KL has shown that presence of each bacterial strain in culture medium has cumulative enhancing effect on growth, lignin degradation and detoxification rather than inhibition. It is our view, the above information would be useful for modeling and designing the unit for the treatment of lignin containing wastewater.

ACKNOWLEDGMENTS

The authors are thankful and gratefully acknowledge the Director, CSIR-Indian Institute of Toxicology Research, for providing constant support for this work.

REFERENCES

1. Agrawal N, Ray RS, Farooq M, Pant AB, Hans RK (2007) Photosensitizing potential of ciprofloxacin at ambient level of UV radiation. Photochem Photobiol 83:1226–1236
2. Ali M, Sreekrishnan TR (2001) Aquatic toxicity from pulp mill effluents: a review. Adv Environ Res 5:175–196
3. APHA (2005) Standard methods for the examination of water and wastewater, 21st edn. American Public Health Association, Washington, DC
4. Barrow GI, Feltham RKA (1993) Cowan and Steel's Manual for the Identification of Medical Bacteria, 3rd edn. Cambridge University Press, Great Britain
5. Chakar SF, Ragauskas JA (2004) Review of current and future softwood kraft lignin process chemistry. Ind Crop Prod 20:131–141
6. Chamorro S, Monsalvez E, Piña B, Olivares A, Hernández V, Becerra J, Vidal G (2013) Analysis of aryl hydrocarbon receptor ligands in kraft mill effluents by a combination of yeast bioassays and CG–MS chemical determinations. J. Environ Sci Health A 48:1–7
7. Chandra R, Abhishek A (2011) Bacterial decolorization of black liquor in axenic and mixed condition and characterization of metabolites. Biodegrade 22:603–611
8. Chandra R, Abhishek A, Sankhwar M (2011) Bacterial decolorization and detoxification of black liquor from rayon grade pulp manufacturing paper industry and detection of their metabolic products. Bioresour Technol 102:6429–6436
9. Das MT, Budhraja V, Mishra M, Thakur IS (2012) Toxicological evaluation of paper mill sewage sediment treated by indigenous dibenzofuran-degrading *Pseudomonas* sp. Bioresour Technol 110:71–78
10. Forney LJ, Reddy CA (1979) Bacterial degradation of kraft lignin. Develop Ind Microbiol 20:163–175
11. Gupta VK, Minocha AK, Jain N (2001) Batch and continuous studies on treatment of pulp mill wastewater by *Aeromonas formicans*. J Chem Technol Biotechnol 76:547–552
12. Jiranuntipon S, Chareornwattana S, Damronglerd S, Albasi C (2008) Decolorization of synthetic melanoidins-containing wastewater by a bacterial consortium. J Ind Microbiol Biotechnol 38:1313–1321
13. Khalid A, Arshad M, Crowley DE (2009) Biodegradation potential of pure and mixed bacterial cultures for removal of 4-nitroaniline from textile dye wastewater. Water Res 43:1110–1116
14. Kirk TK, Schulz E, Connor WL, Lorenz LF, Zeikus JG (1978) Influence of cultural parameters on lignin metabolism by *Phanerochaete chrysosporium*. Arch Microbiol 117:227–287

15. Ko JJ, Shimizu Y, Ikeda K, Kim SK, Park CH, Matsui S (2009) Biodegradation of high molecular weight lignin under sulfate reducing condition: lignin degradability and degradation by-products. Bioresour Technol 100:1622–1627
16. Ksibi MM, Amor SB, Elaloui Cherif S, Houas EA, Elaloui M (2003) Photodegradation of lignin from black liquor using UV/TiO_2 system. J Photochem Photobiol A: Chem 154:211–218
17. Lee YCM, Wang CC (2006) Isolation and physiological characterization of the pentachlorophenol degrading bacterium *Sphingomonas chlorophenolica*. Chemosphere 62:709–714
18. Lee J, Koh D, Andijani M, Saw SM, Munoz C, Chia SE, Wong ML, Hong CY, Ong CN (2002) Effluents from a pulp and paper mill: a skin and health survey of children living in upstream and downstream villages. Occup Environ Med 59:373–379
19. Masai E, Shinohara S, Hara H, Nishikawa S, Katayama Y, Fukuda M (1999) Genetic and biochemical characterization of a 2-pyrone-4, 6-dicarboxylic acid hydrolase involved in the protocatechuate 4, 5-cleavage pathway of *Sphingomonas paucimobilis* SYK-6. J Bacteriol 181:55–62
20. Morii H, Nakamiya K, Kinoshita S (1995) Isolation of lignin decolourising bacterium. J Ferment Bioeng 80:296–299
21. Perestelo F, Falcon MA, Perez ML, Roig EC, de la Fuente Martin G (1989) Bioalteration of kraft pine lignin by *Bacillus megaterium* isolated from compost piles. J Ferment Bioeng 68:151–153
22. Raj A, Reddy MMK, Chandra R (2007) Identification of low molecular weight aromatic compounds by gas chromatography–mass spectrometry (GC–MS) from kraft lignin degradation by three *Bacillus* sp. Int Biodeterior Biodegrade 59:292–296
23. Shin KS, Lee YJ (1999) Depolymerization of lignosulfonate by peroxidase of the white-rot basidiomycetes, *Pleurotus ostreatus*. Biotechnol Lett 21:585–588
24. Singhal A, Thakur IS (2009) Decolourization and detoxification of pulp and paper mill effluent by *Cryptococcus* sp. Biochem Eng J 46:21–27
25. Ugurlu M, Gurses A, Dogar C, Yalcın M (2008) The removal of lignin and phenol from paper mill effluents by electro-coagulation. J Environ Manage 87:420–428

CHAPTER 17

Carbon Turnover during Effluent Application to the Land: A Potential Role for Vegetation?

Vasileios A. Tzanakakis [1,*], Giannis Vagiakis [2], Myrto Tsiknia [2], Andreas N. Angelakis [3] and Nikolaos V. Paranychianakis [2]

[1]*Region of Crete, Directorate of Agricultural Economy, Iraklion 71201, Greece*

[2]*School of Environmental Engineering, Technical University of Crete, Chania 73132, Greece*

[3]*National Agricultural Research Foundation (N.AG.RE.F.), Institute of Iraklion, Iraklion 71307, Greece*

ABSTRACT

This work investigates the effect of plant species (*Eucalyptus camaldulensis vs. Arundo donax*) on carbon (C) turnover during wastewater application to the land. The study was carried out in 40-liter pots under field conditions and plant species were treated either with pre-treated municipal wastewater or freshwater. Plant species had a strong effect on soil organic matter with pots planted with *E. camaldulensis* showing greater values than pots planted with *A. donax*. In accordance, greater respiration rates were measured in *E. camaldulensis* pots compared to those planted with *A. donax*. The respiration rate followed a decreasing trend with the progress of the season for both species. These findings suggest differences in soil microbial community composition and/or activity in the rhizosphere of plant species. Minor effects of plant species or effluent were observed in dissolved organic carbon, protein, and hexoses content. In conclusion, the results of the present study reveal an important role of plant species on C cycling in terrestrial environments with potential implications on the sequestration of C and release of nutrients and pollutants.

Keywords: land application; land treatment; wastewater treatment; *E. camaldulensis*; *A. donax*; carbon cycling; respiration rate; carbon mineralization; dissolved organic carbon (DOC)

1. INTRODUCTION

Effluent application to the land constitutes a widespread practice and it contributes to the further treatment of effluents and the beneficial recycling of water and nutrients [1]. Depending on the pre-treatment level and composition, effluent application onto land may result in relatively high amounts of carbon (C). Previous studies have revealed the great potential of soil for C assimilation provided through effluent [2], but contrasting effects have been reported in the literature with the regard to its effect on soil organic matter (SOM) [2,3,4]. These findings indicate that environmental factors and management practices affect the turnover of the organic-C applied to the soil through effluent. Given the interest on the factors regulating C sequestration in soils and the release of nutrients, especially nitrogen (N), concern for the fate of C in land treatment systems has been renewed. Thus, a better understanding of the processes regulating C-cycling in effluent-irrigated lands is required to improve their potential for C and N assimilation and to reduce N losses to the environment.

Earlier studies dealing with effluent application showed enhancement of soil respiration rates, attributable to easily decomposable organic compounds in the effluent [5] which in turn stimulate microbial growth and activity [6,7]. However, such an influence is not always the case [8,9,10]. Many factors have been reported to affect the turnover of C in soils including the composition of organic substrate, soil properties and management, climatic conditions, and vegetation [11,12]. With regard to the effect of vegetation, its role has not been investigated in land treatment systems although indirect impacts have been proposed through its effect on effluent application rate and litter deposition [1]. Recent studies have provided evidence that plant species may affect the turnover of C in the soil through its effects on litter composition and rhizodeposition as well as microbial community composition and activity [5,13,14,15]. Given the high rates of effluent (and C) application, especially in areas with (semi)arid climates, it remains unclear whether the influence of plant species, if any, will be maintained or if it will be masked.

Up to date studies dealing with the role of vegetation in the treatment performance of land treatment systems have been mainly focused on its adaptation in salinity, biomass production, nutrient and heavy metal removal, water use, and soil (bio)chemical properties [1,2,16,17,18]. Thus, the purpose of this study is to investigate the influence of plant species (*Eucalyptus camaldulensis* and *Arundo donax*) on the turnover of C applied to the soil through effluent. To achieve this goal soil respiration rates, dissolved organic carbon (DOC) and SOM were monitored throughout the growing season.

2. MATERIALS AND METHODS (OR EXPERIMENTAL)

2.1. Site Description and Experimental Set Up

The study was carried out in an experimental station close to Iraklion city (35°16′51″ N–25°10′53″ E). The climate is semi-arid with relatively humid winters and dry and warm summers. In April 2010 one-year-old eucalyptus plants (*E. camaldulensis*) and rhizomes of giant reed (*A. donax*) were transplanted into separate pots (40 cm in diameter × 40 cm in height), filled with surface soil taken form a fallow field. Pots were placed at 1.5 m distance within a row and at the same distance between rows. The chemical properties of the soil were pH: 7.59, electrical conductivity (EC) (dS/m): 0.35, SOM (%): 1.15, total kjeldahl nitrogen (TKN) (%): 0.162, Olsen-P (mg/kg): 13.5, and $CaCO_3$ (%): 55%.

Twelve replications per treatment (plant species) were used, while six additional pots planted with both species and irrigated with fresh water were also included to serve as controls. Pots were irrigated with primary treated wastewater transferred from the Wastewater Treatment Plant of Iraklion city. Effluent application started at the beginning of May and continued until the end of November. The average composition of the effluent, with regard to the parameters of interest, was chemical oxygen demand (COD): 617 ± 120 mg/L, biochemical oxygen demand (BOD_5): 247 ± 45 mg/L, TKN: 108 ± 30 mg/L, NH_4^+–N: 44 ± 10 mg/L, NO_3^-–N: 19 ± 5 mg/L, EC: 3.5 ± 0.3 dS/m, and pH: 7.5 ± 0.2. Plants were irrigated according to evapotranspiration demands determined by the use of tensiometers located at 20 cm depth. The frequency of irrigation varied from one irrigation per three days, early and late in the season (May to mid-June and October) to one irrigation per day or two days from July to September following climatic conditions and leaf area development. The cumulative hydraulic loading approached 2000 mm. Hydraulic loading and reference evapotranspiration (ETo) are shown in Figure 1. No effect of plant species on water requirements was observed throughout the study period.

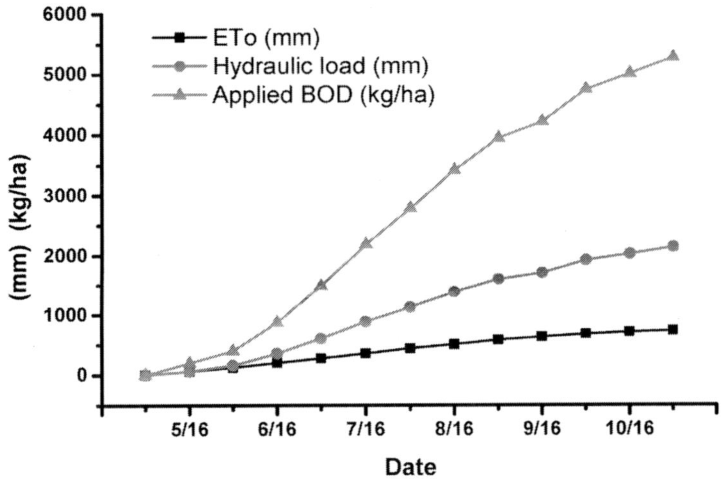

Figure 1. Cumulative reference evapotranspiration (ETo) and hydraulic and BOD_5loading of experimental pots throughout the study period.

2.2. Sampling and Chemical Analyses

Representative soil samples of 200 g weight were collected from 0–13, 13–26, and 26–40 cm of soil profile by using a metallic soil sampler at fortnightly intervals during May and June and monthly during the rest period (July to October). The soil samples were prepared and analyzed according to the Soil Sampling and Methods of Analysis for SOM, DOC, exozes, and proteins [19]. Since no differentiation was found with soil depth the averaged values are presented.

2.3. Respiration Rate

The respiration rate assays were carried out under field conditions on 28 June, 25 August and 6 October. The respiration rates were estimated by capturing CO_2 emissions from the soil within 24 h. More detailed, after the uniform application of the irrigation dose at the pot surface, a PVC cylinder (6.3 cm in diameter and 15 cm in height) with a gas-tight screw lid was immediately inserted at a depth of 3 cm. To capture the generated CO_2 a vial containing 10 mL of 1 M·NaOH was placed inside the cylinder. The NaOH traps were replaced by new ones at 0.5 (only in the measurement of 25 August), 1, 3, 6, and 24 h after irrigation (either with fresh water or wastewater). The concentration of CO_2 was assessed by titration according to the Methods of Soil Analysis [20].

2.4. Statistical Analysis

Analysis of variance ANOVA was used for treatment comparison (plant species, irrigation water quality, and soil depth) on CO_2, SOM, DOC, hexoses, and proteins. These statistical analyses were done using SPSS 19.0 software [21].

3. RESULTS

3.1. Organic Matter Loading

The organic matter loading rate increased gradually until mid-June and then remained almost constant until the end of the application period (September), following the pattern of hydraulic loading rate. The cumulative organic matter loading (kg·BOD_5/ha) applied in pots treated with effluent is shown in Figure 1.

3.2. Soil Organic Matter Content, Dissolved Organic Matter, Hexoses, and Proteins

Neither plant species nor effluent affected SOM content by 7 June. Then, SOM increased gradually in pots planted with *E. camaldulensis* until mid-July when it approached its maximum value (1.37%). An opposite trend was identified in pots planted with *A. donax* in which SOM approached its minimum value (Figure 2a) at the same time. Thereafter and by the end of the study, SOM decreased gradually in pots planted with *E. camaldulensis* and approached its background levels (1.15%). Surprisingly, SOM remained constant during the same period in pots planted with *A. donax* (Figure 2a). At the end of the study, SOM content differed significantly between plant species ($p < 0.05$). In pots irrigated with fresh water, no differences were observed in SOM content independently of the plant species (Figure 2a).

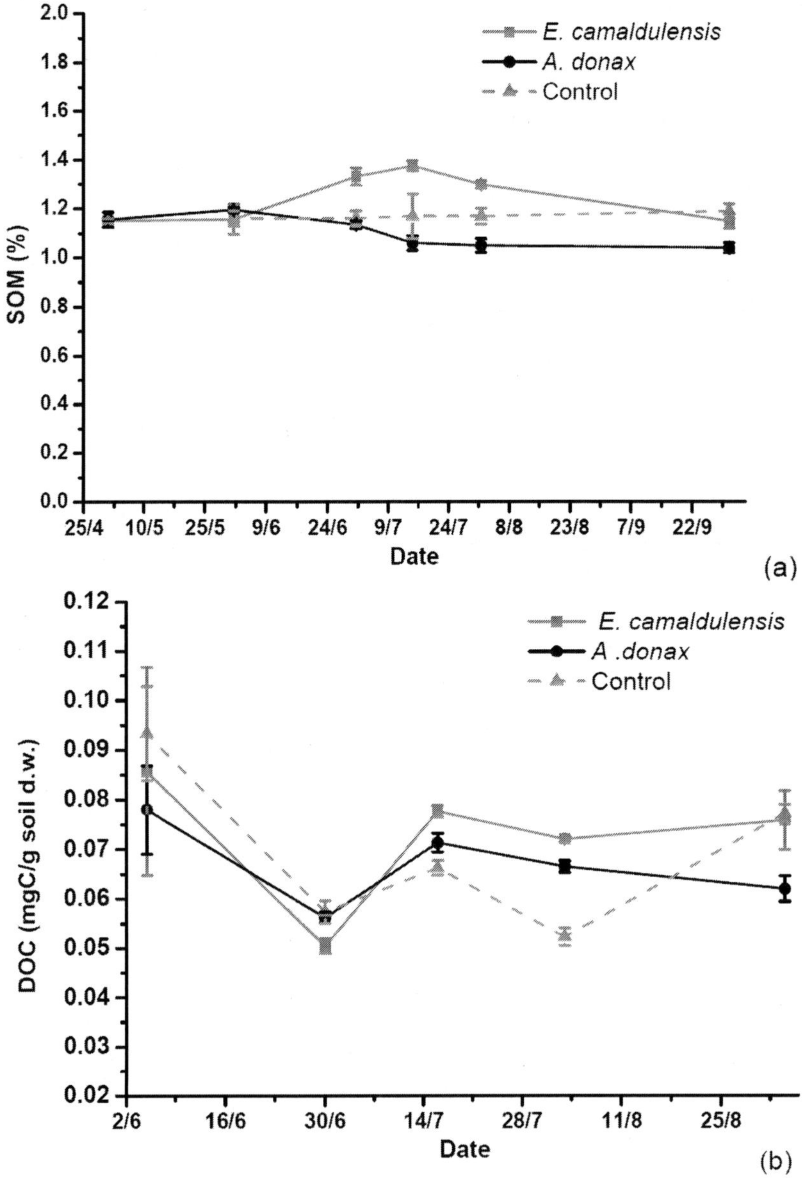

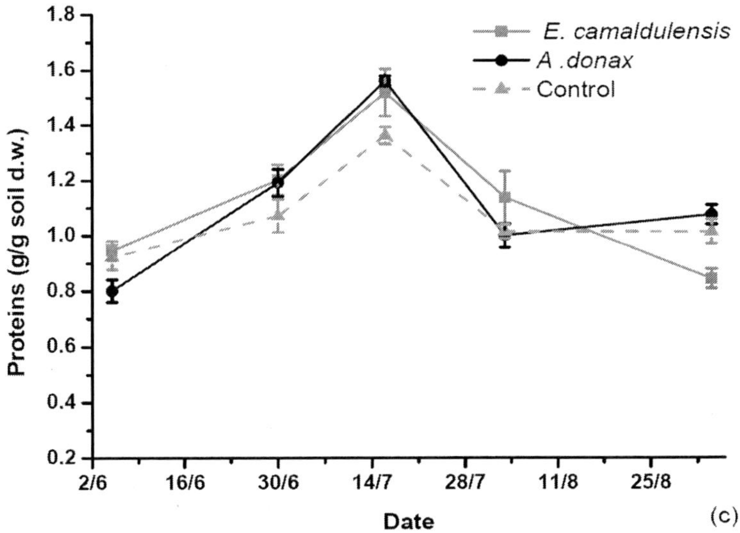

Figure 2. (a) Soil organic matter (SOM); (b) dissolved organic C (DOC); and (c) protein content in pots planted with *E. camaldulensis* and *A. donax* and irrigated with primary treated wastewater. Control refers to average of freshwater treated pots since no differences were observed between plant species.

With regard to the DOC, a slight decrease was observed until the end of June independently of water source (Figure 2b). Thereafter, DOC increased by mid-July in effluent-irrigated pots, exceeding the concentration of DOC in pots irrigated with fresh water, and then remained constant (Figure 2b).

In fresh water irrigated pots DOC remained constant until August, when a slight recovery took place (Figure 2b). A significant influence of plant species on DOC was observed from mid-July onwards, when pots planted with *E. camaldulensis* showed greater values of DOC compared to pots planted with *A. donax* (Figure 2b). At the study end (6 October) DOC content in *E. camaldulensis* pots was 22% higher compared to pots planted with *A. donax* ($p < 0.05$).

Protein concentration in soil solution samples increased in pots irrigated with fresh water until mid-July and then declined gradually to the concentrations measured at the beginning of the study (Figure 2c). A quite similar pattern was also observed for effluent-irrigated pots of *A. donax*. A differential response of protein concentration was found in effluent-irrigated pots of *E. camaldulensis* in which the concentration of proteins continued to decrease resulting finally in significant difference between plant species ($p < 0.01$) at the last sampling (Figure 2c). Hexoses concentration showed fluctuations between samplings, but overall it was greater in the rhizosphere of *E. camaldulensis* (data not shown).

3.3. Respiration Rates

The respiration rate followed an exponential pattern including an initial rapid phase followed by a slower second one. It was significantly affected by both plant species and water source, while seasonal effects were also observed (Figure 3).

In effluent-treated pots, at the first sampling when the greatest respiration rates were measured, *E. camaldulensis* showed 26% greater cumulative respiration compared to *A. donax* ($p < 0.01$) at 24 h. Moreover, effluent-treated pots showed higher emissions of CO_2 compared to pots irrigated with fresh water up to 32% for *E. camaldulensis* ($p < 0.01$) and 28% for *A. donax* ($p < 0.01$), respectively (Figure 3a).

These differences among effluent and fresh water irrigated plants became greater with the progress of the season approaching differences up to 86% and 41% for E. camaldulensis and *A. donax*, respectively at the last sampling (Figure 3b,c), suggesting that the contribution of the effluent-driven heterotrophic respiration to the overall CO2 evolution increased with progress of the season.

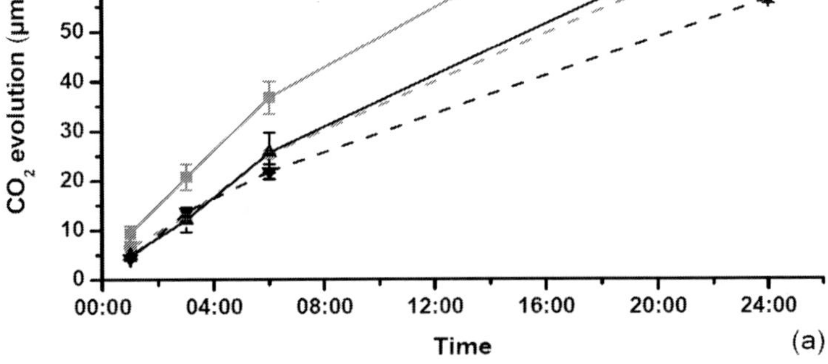

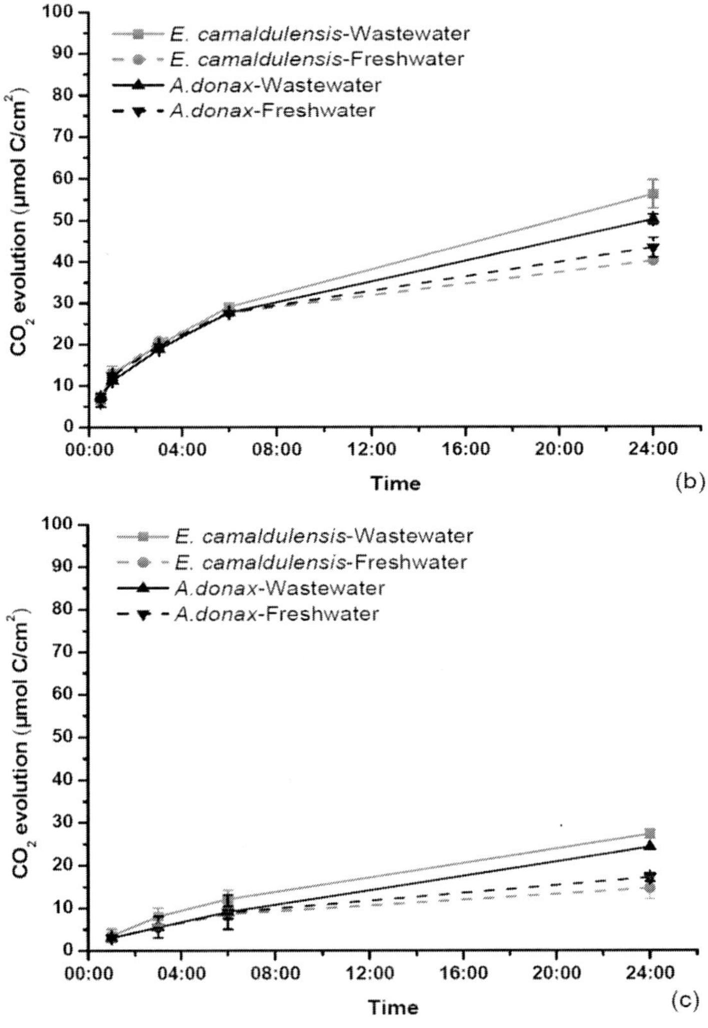

Figure 3. Emitted CO_2 from pots planted with *E. camaldulensis* and *A. donax* plants and irrigated with primary treated wastewater and fresh water on (a) 28 June; (b) 25 August; and (c) 6 October.

By subtracting the values of respiration of fresh water irrigated pots from those irrigated with effluent, to remove the autotrophic (root) and basal heterotrophic respiration [22], data supported again the seasonal pattern of decreasing CO_2, despite the fact that organic C loading did not differ significantly among CO_2 samplings (Figure 4). Moreover, this data revealed that the influence of plant species remained almost the same in absolute values over samplings. Considering this CO_2 fraction, *E. camaldulensis* had 40% higher emissions of CO_2 than *A. donax* at the first sampling, an effect that successively increased to 135% and 79% at the following samplings due to the greater contribution of respiration derived from the effluent. Finally, taking into account the applied C via effluent application and the net amount of C emission derived from effluent application, it was estimated that only 27% and 20% of the applied C was released at the first sampling at the eucalyptus and reed species, respectively (Figure 4).

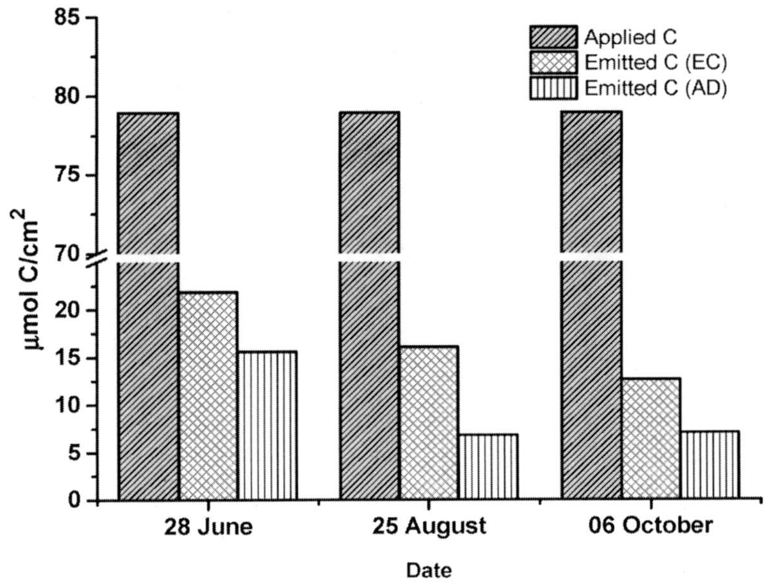

Figure 4. Estimated applied and net emitted C 24 h after effluent dose application at *E. camaldulensis* (EC) and *A. donax* (AD) pots.

4. DISCUSSION

It is well documented that plant species can shape the soil microbial community and that this effect is largely driven by root activity and liter deposition rates and composition [23,24,25]. It remains however questionable if external application of organic substrates, e.g., effluent application, could modify this influence. Differences in microbial composition and enzyme activity were found in the rhizosphere of various plant species, and among them *A. donax*, in vertical flow constructed wetlands [26]. Accordingly, plant species influenced respiration, SOM and, to a lower extent, DOC and protein concentration in soil solution during the study period and this influence was only detected or exacerbated when effluent was used as a water source.

In agreement with earlier studies [27,28] effluent application stimulated CO_2 emissions reflecting the enhancement of heterotrophic activity due to organic-C addition. Pots planted with *E. camaldulensis* showed constantly greater respiration rates than those planted with *A. donax* under effluent irrigation, but the mechanisms responsible remain to be elucidated. The greater concentrations of DOC in the rhizosphere of *E. camaldulensis* may suggest a greater release of root exudates that stimulated microbial activity, and thus CO_2 emissions. The fact that the differences among plant species remained, even when the respiration from fresh water irrigated pots was subtracted to remove the effects of autotrophic respiration and rhizodeposition, documents a strong effect of plant species on the activity of microorganisms involved in organic-C mineralization. Moreover, this finding suggests that it is the availability of substrates that express the potential of vegetation to affect C cycling. This assumption explains the lack of differences among plant species in fresh water irrigated pots which can be attributed to the lower availability of substrates.

CO_2 evolution across the treatments followed the characteristic biphasic pattern, including an initial rapid phase followed by a slower one [29,30]. It has been shown that a double exponential decay model describes better the mineralization of labile organic substrates added to the soil [29]. Application of a similar model was not possible in the present study since CO_2 monitoring was limited to 24 h. Another important limitation is that CO_2 emission measured in effluent-irrigated pots is the cumulative effect of regularly repeated applications of effluent with variable contribution of various C pools. However, our data revealed that the only a portion of organic-C applied to the soil with the effluent was emitted as CO_2 within this interval.

The normalized data of respiration (effluent respiration–fresh water respiration) suggests that the decline observed in CO_2 emissions with the season's progress could only partially be explained by a decrease in autotrophic respiration and/or rhizodeposition as a consequence of plant ageing and the cessation of growth. These data also suggest that the SOM differences between plant species irrigated with effluent may have contributed (Figure 2a) to the occurrence of this pattern of respiration. Cumulative CO_2 attributed to the effluent application from *A. donax* pots remained approximately the same in the last two samplings when SOM was constant, while in *E. camaldulensis* it continued to decline following the decrease in SOM (Figure 3b,c).

This study revealed a significant effect of plant species on SOM only in the effluent-irrigated pots. SOM declined rapidly early in the season in *A. donax* which may suggest a species-driven priming of SOM as has been shown in other studies [24,31]. The first measurement of CO_2 coincided with the initiation of SOM decline attributed to the priming event, and the lower differences in respiration rate among plant species observed during that sampling (Figure 4) provide some documentation that SOM content had indeed an effect on respiration rate. Unlike *A. donax*, pots planted with *E. camaldulensis* showed a gradual accumulation in SOM during the first half of the study period. This finding clearly suggests a greater potential of the *E. camaldulensis* rhizosphere for temporary retention and/or stabilization of organic-C supplied through effluent application and rhizodeposition. Since the soil properties were the same, the mechanisms responsible for this effect remain obscure. Long term studies are required to fully address this issue and to provide further documentation for the effect of plant species on SOM.

Moreover, the findings of the present study revealed a seasonality of the influence of plant species on SOM (Figure 2) resulting probably from differences in exudation rates and the activity of the microbial community [32,33,34]. This hypothesis is supported by the observed decline in SOM differences among plant species towards the end of the growing season. Indeed plant species exerted a minor effect on microbial community in autumn attributed to low plant activity and high nutrient and energy inputs from litter fall [35].

5. CONCLUSIONS

In summary, plant species exerted a strong effect on the processes regulating the biogeochemical cycling of C in soils treated with effluent. Greater amounts of CO_2 were emitted from the rhizosphere of *E. camaldulensis* compared to *A. donax*, an effect which is in agreement with the greater concentrations of DOC and hexoses measured. This effect of plant species remained even when the emissions of fresh water irrigated plants were subtracted from those irrigated with effluent to account for the contribution of autotrophic and basal soil respiration. A possible explanation is that *E. camaldulensis* stimulate changes in the activity and/or the composition of microorganisms involved in organic-C processing. Surprisingly, greater contents of SOM were also assessed in the rhizosphere of *E. camaldulensis*. These contrasting effects could only be explained by a priming effect on SOM induced by *A. donax* through yet unknown mechanisms. Apparently, further work is needed to elucidate the mechanisms through which vegetation affects SOM turnover as well as to test these effects in long-term trials.

ACKNOWLEDGMENTS

Funding was provided by the Technical University of Crete.

REFERENCES

1. Paranychianakis, N.; Angelakis, A.; Leverenz, H.; Tchobanoglous, G. Treatment of wastewater with slow rate systems: A review of treatment processes and plant functions. *Crit. Rev. Environ. Sci. Technol.* 2006, *36*, 187–259.
2. Tzanakakis, V.A.; Paranychianakis, N.V.; Angelakis, A.N. Nutrient removal and biomass production in land treatment systems receiving domestic effluent. *Ecol. Eng.* 2009, *35*, 1485–1492.
3. Tzanakakis, V.A.; Paranychianakis, N.V.; Londra, P.A.; Angelakis, A.N. Effluent application to the land: Changes in soil properties and treatment potential. *Ecol. Eng.* 2011, *37*, 1757–1764.
4. Jueschke, E.; Marschner, B.; Tarchitzky, J.; Chen, Y. Effects of treated wastewater irrigation on the dissolved and soil organic carbon in Israeli soils. *Water Sci. Technol.* 2008, *57*, 727–733.
5. Elifantz, H.; Kautsky, L.; Mor-Yosef, M.; Tarchitzky, J.; Bar-Tal, A.; Chen, Y.; Minz, D. Microbial activity and organic matter dynamics during 4 years of irrigation with treated wastewater. *Microb. Ecol.* 2011, *62*, 973–981.
6. Cleveland, C.; Nemergut, D.; Schmidt, S.; Townsend, A. Increases in soil respiration following labile carbon additions linked to rapid shifts in soil microbial community composition. *Biogeochemistry* 2007, *82*, 229–240.
7. Adrover, M.; Farrús, E.; Moyà, G.; Vadell, J. Chemical properties and biological activity in soils of Mallorca following twenty years of treated wastewater irrigation. *J. Environ. Manag.* 2012, *95*, 188–192.
8. Zaman, M.; Cameron, K.C.; Di, H.J.; Inubushi, K. Changes in mineral N, microbial biomass and enzyme activities in different soil depths after surface applications of dairy shed effluent and chemical fertilizer. *Nutr. Cycl. Agroecosyst.*2002, *63*, 275–290.
9. Ramirez-Fuentes, E.; Lucho-Constantino, C.; Escamilla-Silva, E.; Dendooven, L. Characteristics, and carbon and nitrogen dynamics in soil irrigated with wastewater for different lengths of time. *Bioresour. Technol.* 2002, *85*, 179–187.
10. Truu, M.; Truu, J.; Heinsoo, K. Changes in soil microbial community under willow coppice: The effect of irrigation with secondary-treated municipal wastewater. *Ecol. Eng.* 2009, *35*, 1011–1020.
11. Adrover, M.; Moyà, G.; Vadell, J. Effect of treated wastewater irrigation on plant growth and biological activity in three soil types. *Commun. Soil Sci. Plant Anal.* 2012, *43*, 1163–1180.
12. Lõhmus, K.; Truu, M.; Truu, J.; Ostonen, I.; Kaar, E.; Vares, A.; Uri, V.; Alama, S.; Kanal, A. Functional diversity of culturable bacterial communities in the rhizosphere in relation to fine-root and soil parameters in alder stands on forest, abandoned agricultural, and oil-shale mining areas. *Plant Soil* 2006, *283*, 1–10.
13. Cheng, W.; Kuzyakov, Y. Roots and Soil Management: Interactions between Roots and the Soiled. In *Root Effects on Soil Organic Matter Decomposition, in Roots and Soil Management: Interactions between Roots and the Soil*; Zobel, R.W., Wright, S.F., Eds.;

American Society of Agronomy, Crop Science Society of America, Soil Science Society of America: Madison, WI, USA, 2005; pp. 119–143.
14. Nielsen, U.N.; Ayres, E.; Wall, D.H.; Bardgett, R.D. Soil biodiversity and carbon cycling: A review and synthesis of studies examining diversity-function relationships. *Eur. J. Soil Sci.* 2011, *62*, 105–116.
15. Leff, J.; Nemergut, D.; Grandy, A.S.; O'Neill, S.; Wickings, K.; Townsend, A.; Cleveland, C. The effects of soil bacterial community structure on decomposition in a Tropical Rain Forest. *Ecosystems* 2012, *15*, 284–298.
16. Strickland, M.S.; Lauber, C.; Fierer, N.; Bradford, M.A. Testing the functional significance of microbial community composition. *Ecology* 2009, *90*, 441–451.
17. Tsiknia, M.; Tzanakakis, V.A.; Paranychianakis, N.V. Insights on the role of vegetation on nitrogen cycling in effluent irrigated lands. *Appl. Soil Ecol.* 2013, *64*, 104–111.
18. Nsanganwimana, F.; Marchand, L.; Douay, F.; Mench, M. *Arundo donax* L., a candidate for phytomanaging water and soils contaminated by trace elements and producing plant-based feedstock. A review. *Int. J. Phytoremed.* 2013, *16*, 982–1017.
19. Chantigny, M.; Angers, D.; Kaiser, K.; Kalbitz, K. Extraction and characterization of dissolved organic matter. In *Soil Sampling and Methods of Analysis*, 2nd ed.; Carter, M.R., Gregorich, E.G., Eds.; Taylor & Francis Group, LLC: Boca Raton, FL, USA, 2007; pp. 617–635.
20. Weaver, R.W., Angle, S., Bottomley, P., Bezdicek, D., Smith, S., Tabatabai, A., Wollum, A., Mickelson, S., Eds.;*Methods of Soil Analysis: Part 2, Microbiological and Biochemical Properties*; Soil Science Society of America: Madison, WI, USA, 1994.
21. IBM. IBM SPSS 19.0 for Windows, IBM Company, Armonk, NY, USA, 2010.
22. Hanson, P.J.; Edwards, N.T.; Garten, C.T.; Andrews, J.A. Separating root and soil microbial contributions to soil respiration: A review of methods and observations. *Biogeochemistry* 2000, *48*, 115–146.
23. Berg, G.; Smalla, K. Plant species and soil type cooperatively shape the structure and function of microbial communities in the rhizosphere. *FEMS Microbiol. Ecol.* 2009, *68*, 1–13.
24. Haichar, F.E.Z.; Santaella, C.; Heulin, T.; Achouak, W. Root exudates mediated interactions belowground. *Soil Biol. Biochem.* 2014, *77*, 69–80.
25. Shi, S.; Richardson, A.E.; O'Callaghan, M.; DeAngelis, K.M.; Jones, E.E.; Stewart, A.; Firestone, M.K.; Condron, L.M. Effects of selected root exudate components on soil bacterial communities. *FEMS Microbiol. Ecol.* 2011, *77*, 600–610.
26. Ge, Y.; Zhang, C.; Jiang, Y.; Yue, C.; Jiang, Q.; Min, H.; Fan, H.; Zeng, Q.; Chang, J. Soil microbial abundances and enzyme activities in different rhizospheres in an integrated vertical flow constructed wetland. *CLEAN Soil Air Water*2011, *39*, 206–211.
27. Fernández-Luqueño, F.; Reyes-Varela, V.; Cervantes-Santiago, F.; Gómez-Juárez, C.; Santillán-Arias, A.; Dendooven, L. Emissions of carbon dioxide, methane and nitrous oxide from soil receiving urban wastewater for maize (Zea mays L.) cultivation. *Plant Soil* 2010, *331*, 203–215.
28. Mapanda, F.; Mangwayana, E.N.; Nyamangara, J.; Giller, K.E. The effect of long-term irrigation using wastewater on heavy metal contents of soils under vegetables in Harare, Zimbabwe. *Agric. Ecosyst. Environ.* 2005, *107*, 151–165.
29. Boddy, E.; Hill, P.W.; Farrar, J.; Jones, D.L. Fast turnover of low molecular weight components of the dissolved organic carbon pool of temperate grassland field soils. *Soil Biol. Biochem.* 2007, *39*, 827–835.
30. Glanville, H.; Rousk, J.; Golyshin, P.; Jones, D.L. Mineralization of low molecular weight carbon substrates in soil solution under laboratory and field conditions. *Soil Biol. Biochem.* 2012, *48*, 88–95.
31. De Graaff, M.-A.; Jastrow, J.D.; Gillette, S.; Johns, A.; Wullschleger, S.D. Differential priming of soil carbon driven by soil depth and root impacts on carbon availability. *Soil Biol. Biochem.* 2014, *69*, 147–156.
32. Ucisik, A.S.; Trapp, S. Uptake, removal, accumulation, and phytotoxicity of phenol in willow trees (Salix viminalis). *Environ. Toxicol. Chem.* 2006, *25*, 2455–2460.

33. Wang, X.; Piao, S.; Ciais, P.; Janssens, I.A.; Reichstein, M.; Peng, S.; Wang, T. Are ecological gradients in seasonal Q10 of soil respiration explained by climate or by vegetation seasonality? *Soil Biol. Biochem.* 2010, *42*, 1728–1734.
34. Zhang, X.H.; Li, L.Q.; Pan, G.X. Topsoil organic carbon mineralization and CO2 evolution of three paddy soils from South China and the temperature dependence. *J. Environ. Sci. China* 2007, *19*, 319–326.
35. Thoms, C.; Gleixner, G. Seasonal differences in tree species' influence on soil microbial communities. *Soil Biol. Biochem.* 2013, *66*, 239–248.

CHAPTER 18

Prediction of Biomass Production and Nutrient Uptake in Land Application Using Partial Least Squares Regression Analysis

Vasileios A. Tzanakakis [1,*], Andy Mauromoustakos [2] and Andreas N. Angelakis [3]

[1]*Region of Crete, Directorate of Agricultural Economy, Iraklion 71201, Greece*

[2]*Agricultural Statistics Lab, University of Arkansas, Fayetteville, AR 72701, USA*

[3]*National Agricultural Research Foundation (N.AG.RE.F.), Institute of Iraklion, Iraklion 71307, Hellas*

ABSTRACT

Partial Least Squares Regression (PLSR) can integrate a great number of variables and overcome collinearity problems, a fact that makes it suitable for intensive agronomical practices such as land application. In the present study a PLSR model was developed to predict important management goals, including biomass production and nutrient recovery (*i.e.*, nitrogen and phosphorus), associated with treatment potential, environmental impacts, and economic benefits. Effluent loading and a considerable number of soil parameters commonly monitored in effluent irrigated lands were considered as potential predictor variables during the model development. All data were derived from a three year field trial including plantations of four different plant species (*Acacia cyanophylla*, *Eucalyptus camaldulensis*, *Populus nigra*, and *Arundo donax*), irrigated with pre-treated domestic effluent. PLSR method was very effective despite the small sample size and the wide nature of data set (with many highly correlated inputs and several highly correlated responses). Through PLSR method the number of initial predictor variables was reduced and only several variables were remained and included in the final PLSR model. The important input variables maintained were: Effluent loading, electrical conductivity (EC), available phosphorus (Olsen-P), Na^+, Ca^{2+}, Mg^{2+}, K^{2+}, SAR, and NO_3^--N. Among these variables, effluent loading, EC, and nitrates had the greater contribution to the final PLSR model. PLSR is highly compatible with intensive agronomical practices such as land application, in which a large number of highly collinear and noisy input variables is monitored to assess plant species performance and to detect impacts on the environment.

Keywords: land application; land treatment systems; biomass production; nutrient uptake; partial least squares regression (PLSR); JMP

1. INTRODUCTION

Plant biomass production and nutrient recovery are important management goals in land application associated with treatment efficiency, potential impacts on the environment, and economic benefits [1,2]. Plant species with high biomass potential usually achieve increased nutrient recovery resulting from high biomass yield and nutrient assimilation in plant tissues [3]. On the other hand, such plant species may receive high effluent loading, due to their high water requirements, which in turn may result in changes in soil properties and put the surrounding environment at risk as a result of the increased nutrient and/or pollutant release [4,5,6]. The latter as well as potential negative impacts on soil properties are undesirable during land application and may have negative influence on vegetation and overall system performance. Considering this close relationship between vegetation performance and soil properties in land application any potential quantitative description of this relationship in the form of a strong prediction model would be valuable and could provide useful information during system design and monitoring. Until now several statistical methods have been used to develop efficient prediction models in crop and soil science, such as principal component regression (PCR), multiple regression, and partial least squares analysis [7,8,9,10].

Partial least squares regression (PLSR) has become a popular statistical technique used widely in Chemometrics and other related areas [11,12,13]. There is also wide application in soil and crop studies providing information either for soil or plant parameters by considering spectroscopic measurements [14,15,16,17]. PLSR regression analysis construct models by linking predictors (Xs) and responses (Ys) and this is achieved via a projection procedure which reduce data dimensionality to a small number of important factors (also called *latent variables*). Characteristic of the method, unlike other projection methods (e.g., PCR) is that it integrates the compression and regression steps while selection of the orthogonal factors is carried out to achieve maximum covariance between the predictor and response variables [11]. Because of its principles, PLSR approach fits best in cases when matrix of predictors has more variables than observations and predictor variables are highly collinear [18].

PLSR models can deal effectively the following scenarios: (a) wide data (where number of input variables is much greater than the number of observations); (b) tall data; (c) square data; (d) collinear data; and (e) noisy data. This makes PLSR suitable for studies dealing with land application where usually a great number of highly collinear and noisy input variables is monitored to describe plant species performance and effects on soil properties and environment. However, PLSR and even multivariate techniques in general are either lacking or under-utilized in land application schemes. Thus, the primary objective in the present study was the application of a PLSR approach in order to develop a robust model capable of predicting important management goals in land application (*i.e.*, biomass production and N and P uptake) using as predictor variables critical soil parameters. The information provided here is expected to help in the development of the appropriate methodology during design and monitoring of intensive agronomical practices, such as land application in quest of appropriate management strategies with respect to vegetation and field practices.

2. MATERIALS AND METHODS

2.1. LTS Set Up, Sampling, and Chemical Analyses

A three-year-field trial with four different plant species (*Eucalyptus camandulensis*, *Acacia cyanophylla*, *Populus nigra*, and *Arundo donax*), each forming a separated land treatment system (LTS), was carried out at Skalani village, located approximately 5 km south of Iraklion city, Hellas (at 35°16'50.87" N, 25°10'52.61" E). Plant species received septic tank municipal effluents for three consecutive years (2001–2003) at hydraulic loading rate based on crop water requirements and evaporation losses. The soil in which LTS were established was characterized as a clay loam with relatively high calcium content (55% $CaCO_3$). Details about LTS set up, effluent loading and characteristics, soil properties, climatic conditions of the area, and methods used to determine and assess soil data were previously described (soil surface data from 0–7.5 to 55–65 cm obtained from the third irrigation period were included in PLSR as described below) [4]. In brief soil samples prepared and analyzed according to methods referred to the Methods of Soil Analysis [19]. pH, EC, soluble Na^+, Ca^{2+} and Mg^{2+} were assessed in saturation paste extracts with atomic absorption spectrometry (Ca^{2+} and Mg^{2+}) and flame photometer (Na^+). Soil organic matter (SOM) was assessed by the Walkley and Black wet-digestion method and available-P according to the Olsen method after extraction with $NaHCO_3$. Total Kjeldahl Nitrogen (TKN) was assessed by a macro-Kjeldahl device and analysis of NO_3^--N in soil solution samples was carried out using the phenol-disulfonic acid method. In the present work, additional measurements were carried out: Soil C:N ratio was determined as the quotient of organic matter and TKN contents; gravimetric moisture content (ω) was determined by oven drying a representative undisturbed ring of moist soil at 105 °C; bulk density (ρb) was determined by the weight of the soil per unit volume (g/cm^3) at 105 °C. Samplings and measurements regarding biomass and nutrient recovery across plant species are also presented in our previous study [3]. In brief, at the end of every growing season (October), one representative tree from each of four plots was harvested and separated into individual organs. The fresh weight of leaves, shoots and trunk (old wood) were recorded. For reeds, the whole plot surface was harvested each season and separated in leaves and shoots. Dry weights of vegetation were determined by drying (65 °C) to a constant weight. In 2002 and 2003 one replicate plot was harvested from each treatment and the tissue dry weights were determined. The dried samples were ground to 1-mm and used in elemental analysis. Micro-Kjeldahl N-digestion was used to determine total-N content of biomass samples [20] and P content was determined by the vanado-molybdo-phosphoric acid colorimetric method after digesting the samples with a mixture of perchloric-nitric acid.

2.2. Statistical Analysis

The effluent loading and some soil parameters were selected according to their importance in system description and ease of determination as candidate X variables in the PLSR model. This model contained plant biomass and N and P uptake as response variables (Ys). The soil parameters were SOM, dissolved organic matter (as COD), TKN, pH, EC, soil solution NH_3-N (in soil solution sampler), soil solution P, soil solution EC, NO_3^--N, C:N, Olsen-P, Na^+, Ca^{2+}, Mg^{2+}, SAR, K^+, ρb, and ω. Most of these soil input variables were measured in each location across five depth intervals over the 0.65 m of soil profile and their average values were used in the initial PLSR model to eliminate some of the noise. After the preliminary PLSR regression the important Xs variables were identified and included in the final PLSR prediction model.

The non-linear iterative partial least squares (NIPALS) algorithm was used for computing the first few factors. KFold validation was used to select the number of factors that minimize the Root Mean PRESS statistic. The variable VIP (variable importance for the projection) value measures its influence on the factors that define the model. Wold and others advocated cut-off-values for VIP to separate terms that do not make important contribution to the dimensionality reduction involved in PLSR (VIP < 8) and those that might (VIP ≥ 8). In addition, the percentage of variation explained for X variables and Y responses and the contribution of each of the important factors were assessed. Loadings were also calculated and plotted to give another way to view the relationships between the Xs the Ys and the PLSR factors. For each X variable VIP (variable importance for the projection) was calculated to assess its importance in the determination of the PLSR projection model for both predictors and responses [21]. Also, Xs coefficients in the PLSR model were calculated to assess their contribution to the prediction of the Ys. Based on PLSR model the predicted values for the responses were calculated and plotted *versus* the observed values. Also, validation of PLSR prediction model was performed based on data derived from the previous year (2002). All analyses were carried out the PLS platform of JMP® (SAS Institute Inc.: Cary, NC, USA) Pro Version 11.2.1 [22].

3. RESULTS

Preliminary PLSR regression removed several soil parameters with minor contribution to the prediction model. The remaining parameters were included in final PLSR model. These parameters were effluent loading, EC, available phosphorus (Olsen-P), Na^+, Ca^{2+}, Mg^{2+}, K^{2+}, and NO_3^--N. Based on PLSR method, the above X data set was reduced to two principal factors. The first explained the 45.4% of the variation while the second explained the 34.2%. Thus, the cumulative variation explained by two principal factors was 79.6%. The percentage of variation explained for X variables and the contribution of each of the two factors are shown in Figure 1a. Among X variables K^+, SAR, and Na^+ were the greater contributors to explained variation. Also, Mg^{2+}, EC, nitrates, effluent, Ca^{2+}, and P had distinctively higher factor loadings (X loadings) at the first factor compared to K^+, Na^+, and SAR. With regard to second factor SAR, Na^+, EC, and effluent received positive loadings and were higher than the other variables (Figure 1c). In terms of Y data (Y responses) two factors explained the 78.3% of the variation, with the first contributing with 70.4% and the second with 7.9%. Total biomass had the highest percentage of the explained variation followed by N uptake and P uptake (Figure 1b). In the first and strongest factor, plant biomass had the highest factor loading (Y loadings) and N uptake the lowest. In the second factor, N uptake received the highest factor loading over plant biomass and P uptake (Figure 1d).

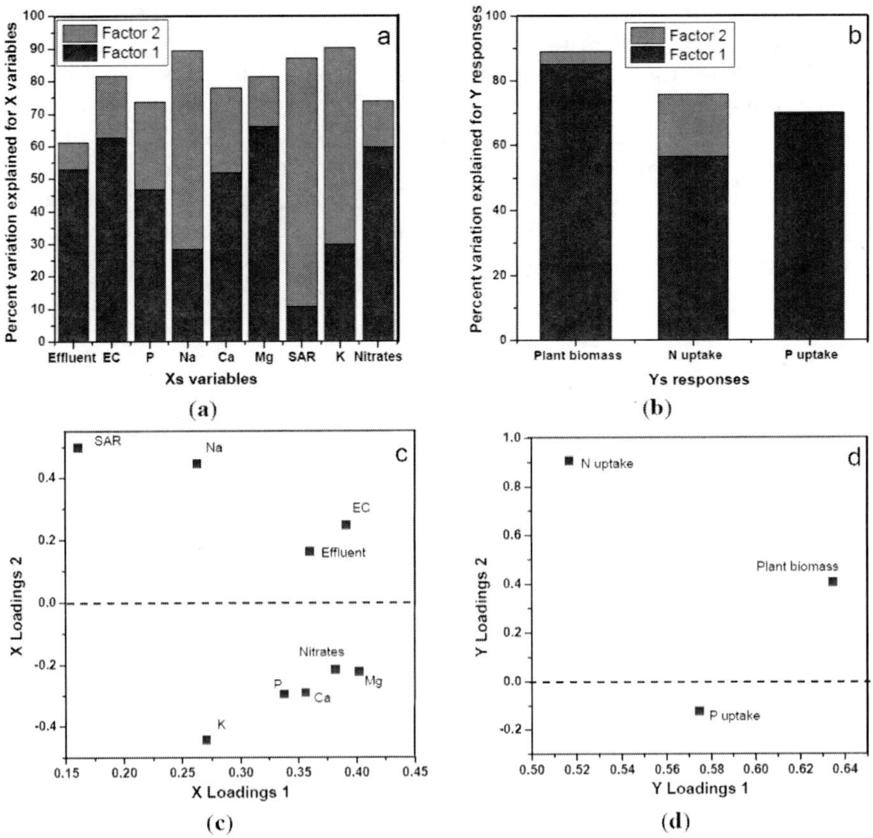

Figure 1. Explained variation and factor loadings for X variables (**a**,**c**) and Y responses (**b**,**d**).

The amount of the effluent, EC, nitrates and Mg^{2+} were among the most powerful X variables in the determination of PLSR model (Figure 2). Based on 0.8 threshold two X variables (SAR and K^{+}) were less influential in the final PLSR model (Figure 2), however, because of their importance for Xs factors (Figure 1a) and contribution in the prediction of Ys (Figure 3) were retained within the model.

Effluent had the greater coefficient values across Ys prediction models followed by EC, Na^{+}, SAR, and/or nitrates dependent on the Y response (Figure 3). Specifically, apart from the amount of effluent, EC had also high coefficient values either for the prediction of biomass produced or prediction of the nutrients. $Na^{+}$and SAR also received high values with exception of the P recovery prediction model.

Interestingly, nitrates received high values in both biomass and P recovery, but mid values were registered in terms of N recovery model. Predicted values derived from PLSR model were plotted over the observed values showing close relationship across all Y responses (Figure 4).

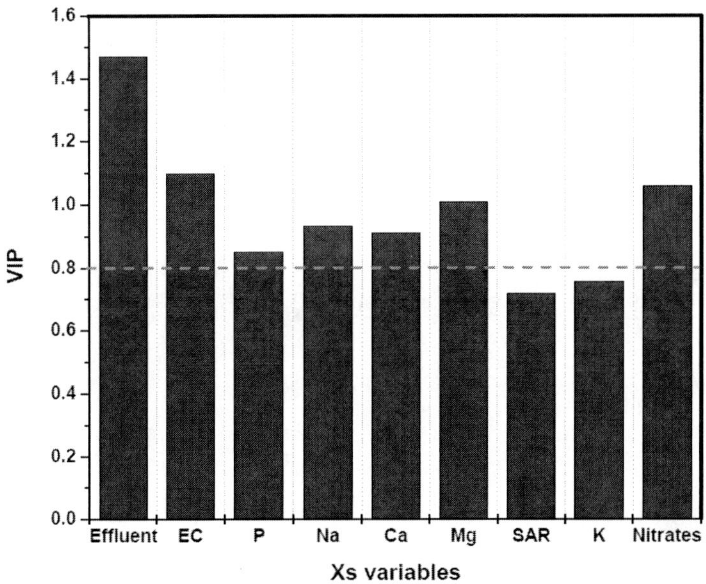

Figure 2. Important X variables in the Partial Least Squares Regression (PLSR) model.

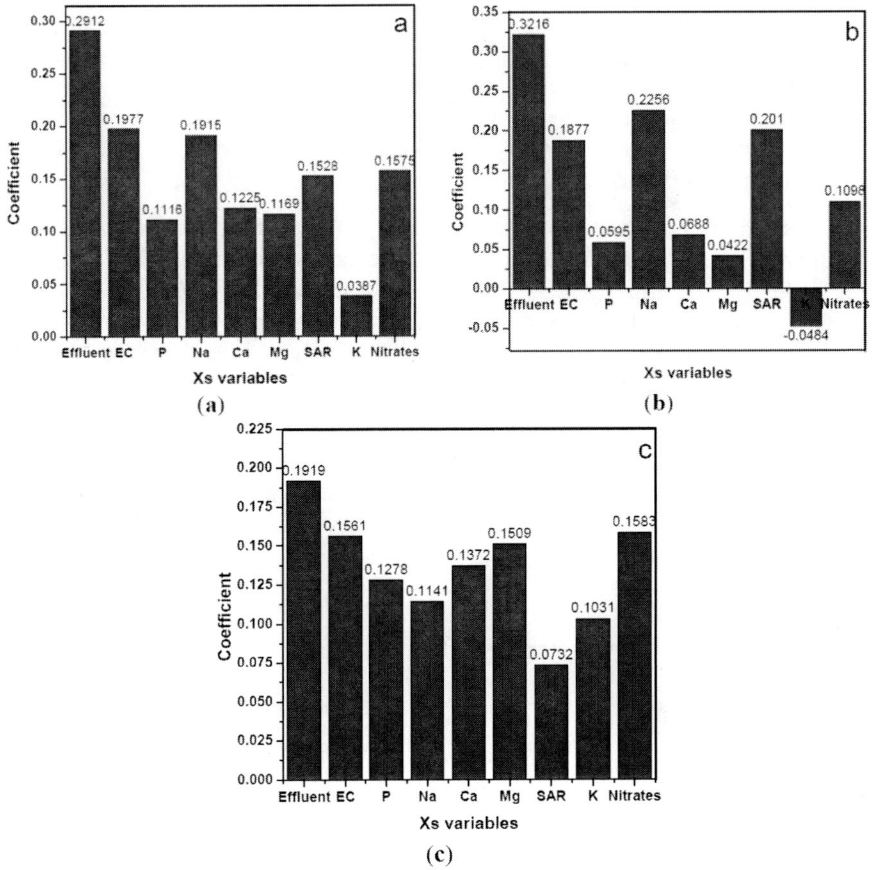

Figure 3. Coefficients of the Xs in the PLSR regression for the models of: (**a**) plant biomass; (**b**) N uptake; and (**c**) P uptake, based on the centered and scaled data.

4. DISCUSSION

The PLSR model in this study is well fitted to the experimental data and successfully predicted the established Y responses (*i.e.*, plant biomass and P and N uptake). Important X variables were effluent loading and certain soil parameters such as soil salinity and nitrates. Effluent loading is considered essential component in the design of land application associated with organics and nutrient loading, growth of plant biomass, and potential environmental impacts [2]. In the present study it was applied at rates equivalent to the evapotranspiration requirements of plant species, which reasonably justify its close relationship with the response variables.

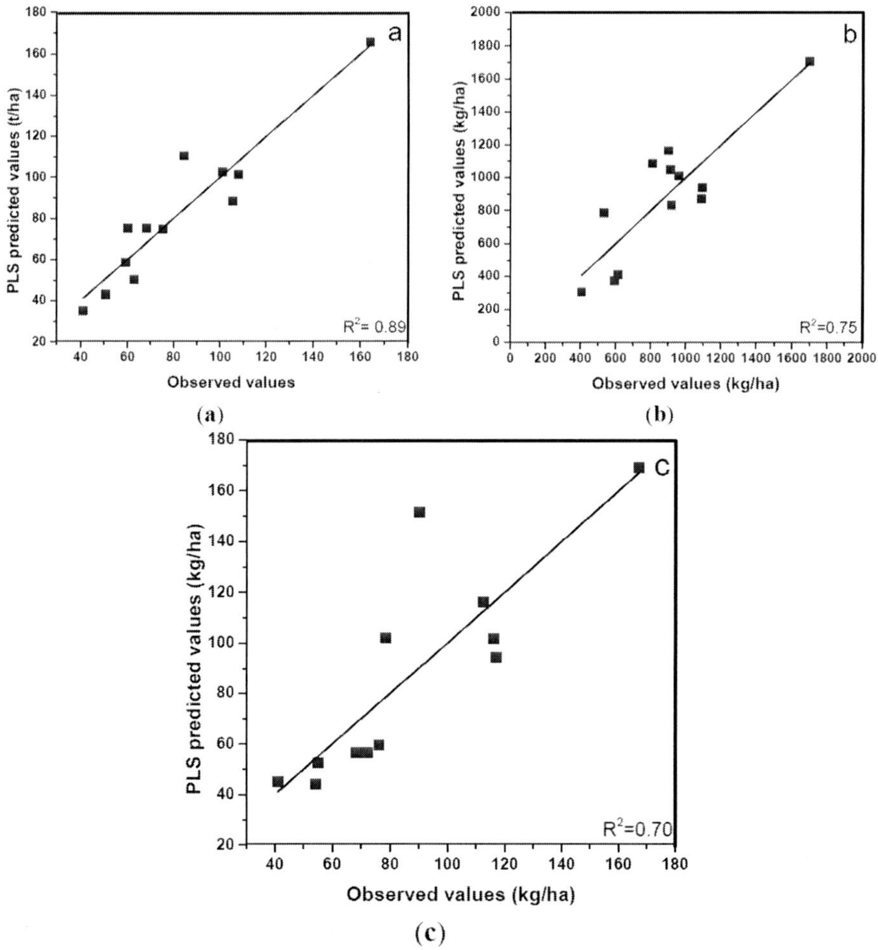

Figure 4. Observed values *versus* PLSR predicted values for: (**a**) plant biomass; (**b**) N uptake; and (**c**) P uptake.

Soil salinity, expressed by EC, was also an important contributor to PLSR model showing strong and positive relationship to the response variables. Soil salinity is closely associated with effluent loading which in turn is associated with plant biomass. For example, species with great biomass potential receive high effluent loading that may result in salts accumulation in the rhizosphere [4,6]. This effect, however, is temporary because of the effect of rainfall which removes salts from the rhizosphere reducing the possible negative impacts on soil and vegetation. EC also has been included in a previous PCR model predicting plant biomass from several soil parameters [23]. With regard to Na^+, it had also significant contribution to the prediction model, particularly for biomass and N uptake, as shown in Figure 2 and Figure 3. SAR had lower contribution than Na^+ indicated also by lower VIP and relative smaller standardized regression coefficients. Increase in Na^+, and SAR values do not necessary imply risk for soil physical properties and/or crop yield considering the high salt and organic matter contents in the effluent. Indeed, in this field experiment effluent application resulted in enhancement in soil physical properties, presented in a previous study [4], an effect that is in agreement with the findings of previous studies [24,25].

Interestingly nitrate was important predictor in our model being among the most important X variables particularly for plant biomass and P uptake. This confirms previous arguments linking nitrates with plant biomass and effluent loading [1,3]. Inorganic N availability in the soil is considered important driver of plants growth with significant effect on biomass yield, tissues nutrient content, and total nutrient/pollutants assimilation [26,27,28,29]. The lower effect of nitrates in the model compared to effluent could be attributed to the existence of additional factors with significant influence on fate of inorganic N in the soil. In this field study inconsistent results were registered between the produced biomass and nitrates in *A. donax* species compared to other species [3,4], which is in agreement with the findings of a recent study [30]. Authors suggested the potential effect of species on soil microbial communities with significant role in N cycling. Moreover, N overloading, up to a critical threshold, and the favorable environmental conditions that induced high nitrification rates may have mitigated the contribution of nitrates in the model.

In this PLSR model important, yet easily measurable, parameters were included as predictor variables. These parameters were capable of providing important information for three specified response variables. In addition, PLSR model was validated by the data of the previous year, a fact, that increases further its potential for future use. The successful development of the prediction model in this study arises from the advantages of the PLSR method. In comparison to other statistical techniques PLSR can overcame the small or large number of highly collinear and noisy input variables and still provide reasonable prediction of multiple collinear plant response variables. Moreover, PLSR has greater reliability compared to other techniques (single multiple regression or combination of multiple regression with other multivariate methods) when identifying relevant variables and their magnitudes of influence, independently of the sample size used in the analysis [10]. On the other hand, probably there are limitations with regard to the applicability of the suggested model to other areas or different plant species [31]. Because of its advantages, PLSR method could also be ideal for the prediction of important management goals involving soil bio (chemical) parameters such as those involved in C and N cycles [30,32].

5. CONCLUSIONS

PLSR is highly compatible with intensive agronomic practices, such as land application, in which a large number of highly collinear and noisy input variables is monitored to assess plant species performance and detect potential impacts on soil and surrounding environment. In the present study a robust PLSR model was developed to predict important management goals in land application (*i.e.*, plant biomass and N and P uptake), associated with treatment potential, environmental impacts, and benefits. Effluent loadings and several soil parameters commonly monitored in effluent irrigated lands, yet easily measurable, were included in the final PLSR model as predictor variables. Among these variables effluent, soil salinity (as EC), and nitrates had the greater effect on the model. PLSR model suggested in this study could help during the system design procedure and monitoring providing valuable information in terms of system performance and suggesting adjustments according to the management objectives. Monitoring of a large scale (spatial and temporal) data set of the pre-defined soil parameters can strengthen the reliability of the PLSR-drawn results, relevant recommendations, and future extrapolations.

ACKNOWLEDGMENTS

This work was financed by the EU Coretech project: ICA 3-CT 1999-00012.

REFERENCES

1. Tzanakakis, V.A.; Chatzakis, M.K.; Angelakis, A.N. Energetic environmental and economic assessment of three tree species and one herbaceous crop irrigated with primary treated sewage effluent. *Biomass Bioenergy* **2012**, *47*, 115–124.
2. Paranychianakis, N.V.; Angelakis, A.N.; Leverenz, H.; Tchobanoglous, G. Treatment of wastewater with slow rate systems: A review of treatment processes and plant functions. *Crit. Rev. Environ. Sci. Technol.* **2006**, *36*, 187–259.

3. Tzanakakis, V.A.; Paranychianakis, N.V.; Angelakis, A.N. Nutrient removal and biomass production in land treatment systems receiving domestic effluent. *Ecol. Eng.* **2009**, *35*, 1485–1492.
4. Tzanakakis, V.A.; Paranychianakis, N.V.; Londra, P.A.; Angelakis, A.N. Effluent application to the land: Changes in soil properties and treatment potential. *Ecol. Eng.* **2011**, *37*, 1757–1764.
5. Lado, M.; Ben-Hur, M. Treated domestic sewage irrigation effects on soil hydraulic properties in arid and semiarid zones: A review. *Soil Tillage Res.* **2009**, *106*, 152–163.
6. Leal, R.M.P.; Herpin, U.; Fonseca, A.F.D.; Firme, L.P.; Montes, C.R.; Melfi, A.J. Sodicity and salinity in a Brazilian Oxisol cultivated with sugarcane irrigated with wastewater. *Agric. Water Manag.* **2009**, *96*, 307–316.
7. Ruffo, M.L.; Bollero, G.A. Residue decomposition and prediction of carbon and nitrogen release rates based on biochemical fractions using principal-component regression. *Agron. J.* **2003**, *95*, 1034–1040.
8. Jabro, J.D.; Sainju, U.; Stevens, W.B.; Evans, R.G. Carbon dioxide flux as affected by tillage and irrigation in soil converted from perennial forages to annual crops. *J. Environ. Manag.* **2008**, *88*, 1478–1484.
9. Kulmatiski, A.; Beard, K.H.; Stevens, J.R.; Cobbold, S.M. Plant-soil feedbacks: A meta-analytical review. *Ecol. Lett.* **2008**, *11*, 980–992.
10. Carrascal, L.M.; Galván, I.; Gordo, O. Partial least squares regression as an alternative to current regression methods used in ecology. *Oikos* **2009**, *118*, 681–690.
11. Wold, S.; Sjöström, M.; Eriksson, L. PLS-regression: A basic tool of chemometrics. *Chemom. Intell. Lab. Syst.* **2001**, *58*, 109–130.
12. Kehimkar, B.; Hoggard, J.C.; Marney, L.C.; Billingsley, M.C.; Fraga, C.G.; Bruno, T.J.; Synovec, R.E. Correlation of rocket propulsion fuel properties with chemical composition using comprehensive two-dimensional gas chromatography with time-of-flight mass spectrometry followed by partial least squares regression analysis. *J. Chromatogr. A* **2014**, *1327*, 132–140.
13. Fonville, J.M.; Richards, S.E.; Barton, R.H.; Boulange, C.L.; Ebbels, T.M.D.; Nicholson, J.K.; Holmes, E.; Dumas, M.-E. The evolution of partial least squares models and related chemometric approaches in metabonomics and metabolic phenotyping. *J. Chemom.* **2010**, *24*, 636–649.
14. Vohland, M.; Emmerling, C. Determination of total soil organic C and hot water-extractable C from VIS-NIR soil reflectance with partial least squares regression and spectral feature selection techniques. *Eur. J. Soil Sci.* **2011**, *62*, 598–606.
15. Nocita, M.; Stevens, A.; Toth, G.; Panagos, P.; van Wesemael, B.; Montanarella, L. Prediction of soil organic carbon content by diffuse reflectance spectroscopy using a local partial least square regression approach. *Soil Biol. Biochem.* **2014**, *68*, 337–347.
16. Fu, Y.; Yang, G.; Wang, J.; Song, X.; Feng, H. Winter wheat biomass estimation based on spectral indices, band depth analysis and partial least squares regression using hyperspectral measurements. *Comput. Electron. Agric.* **2014**, *100*, 51–59.
17. Li, F.; Mistele, B.; Hu, Y.; Chen, X.; Schmidhalter, U. Reflectance estimation of canopy nitrogen content in winter wheat using optimised hyperspectral spectral indices and partial least squares regression. *Eur. J. Agron.* **2014**, *52*, 198–209.
18. Wold, S.; Ruhe, A.; Wold, H.; Dunn, W.J., III. The collinearity problem in linear regression. The partial least squares (PLS) approach to generalized inverses. *SIAM J. Sci. Stat. Comput.* **1984**, *5*, 735–743.
19. Page, A.L., Ed.; *Methods of Soil Analysis: Chemical and Microbiological Proerpties*; American Society of Agronomy: Madison, WI, USA, 1982.
20. Markus, D.K.; McKinnon, J.P.; Buccafuri, A.F. Automated analysis of nitrite, nitrate, and ammonium nitrogen in soils. *Soil Sci. Soc. Am. J.* **1985**, *49*, 1208–1215.
21. Cox, I.; Gaudard, M. *Discovering Partial Least Squares with JMP*; SAS Institute Inc.: Cary, NC, USA, 2013.
22. JMP® Pro, version 11.2.1. J. SAS Institute Inc. Cary, NC, USA.

23. Mandal, U.K.; Warrington, D.N.; Bhardwaj, A.K.; Bar-Tal, A.; Kautsky, L.; Minz, D.; Levy, G.J. Evaluating impact of irrigation water quality on a calcareous clay soil using principal component analysis. *Geoderma* **2008**, *144*, 189–197.
24. Guo, L.B.; Sims, R.E.H. Soil response to eucalypt tree planting and meatworks effluent irrigation in a short rotation forest regime in New Zealand. *Bioresour. Technol.* **2003**, *87*, 341–347.
25. Yaron, B.; Dror, I.; Berkowitz, B. Contaminant-induced irreversible changes in properties of the soil-vadose-aquifer zone: An overview. *Chemosphere* **2008**, *71*, 1409–1421.
26. Christersson, L. Biomass production of intensively grown poplars in the southernmost part of Sweden: Observations of characters, traits and growth potential. *Biomass Bioenergy* **2006**, *30*, 497–508.
27. Labrecque, M.; Teodorescu, T.I. High biomass yield achieved by Salix clones in SRIC following two 3-year coppice rotations on abandoned farmland in southern Quebec, Canada. *Biomass Bioenergy* **2003**, *25*, 135–146.
28. Adegbidi, H.G.; Volk, T.A.; White, E.H.; Abrahamson, L.P.; Briggs, R.D.; Bickelhaupt, D.H. Biomass and nutrient removal by willow clones in experimental bioenergy plantations in New York State. *Biomass Bioenergy* **2001**, *20*, 399–411.
29. Guo, L.B.; Sims, R.E.H.; Horne, D.J. Biomass production and nutrient cycling in Eucalyptus short rotation energy forests in New Zealand. I: Biomass and nutrient accumulation. *Bioresour. Technol.***2002**, *85*, 273–283.
30. Tsiknia, M.; Tzanakakis, V.A.; Paranychianakis, N.V. Insights on the role of vegetation on nitrogen cycling in effluent irrigated lands. *Appl. Soil Ecol.* **2013**, *64*, 104–111.
31. Goodhue, D.; Lewis, W.; Thompson, R. Small sample size, and statistical power in MIS research. In Proceedings of the 39th Annual Hawaii International Conference on System Sciences, Kauai, HI, USA, 4–7 January 2006; Volume 8, p. 202b.
32. Tsiknia, M.; Tzanakakis, V.; Oikonomidis, D.; Paranychianakis, N.; Nikolaidis, N. Effects of olive mill wastewater on soil carbon and nitrogen cycling. *Appl. Microbiol. Biotechnol.* **2014**, *98*, 2739–2749.

CHAPTER 19

Degrading Organic Micropollutants: The Next Challenge in the Evolution of Biological Wastewater Treatment Processes

Naresh Singhal * and Octavio Perez-Garcia

Department of Civil and Environmental Engineering, The University of Auckland, Auckland, New Zealand

Global water scarcity is driving the need for identifying new water source. Wastewater could be a potential water resource if appropriate treatment technologies could be developed. One of the barriers to obtaining high quality water from wastewater arises from the presence of organic micropollutants, which are biologically active at trace levels. Removal of these compounds from wastewater by current physico-chemical technologies is prohibitively expensive. While biological treatment processes are comparatively cheap, current systems are not capable of degrading the wide range of organic micropollutants present in wastewater. As current wastewater treatment processes were developed for treating conventional pollutants present at mg/L levels, degrading the ng/L levels of micropollutants will require a different approach to system design and operation. In this paper we discuss strategies that could be employed to develop biological wastewater treatment systems capable of degrading organic micropollutants.

Keywords: emerging contaminants, organic micropollutants, biocatalysis, metabolic diversity, biodegradation, oxidoreductases

1. WASTEWATER, A WASTED RESOURCE

The Oxford English Dictionary defines wastewater as superfluous water, or water that has served its purpose. This definition of wastewater is ingrained in our perception and attitude. Yet, this definition would appear to be outdated in light of the emerging view that wastewater represents an unutilized water resource and not a waste product to be discarded. Furthermore, promoting such use of wastewater is neither new nor radical. Virtually all urban societies already make use of water sources laden with wastewater, albeit in varying amounts. For example, consider the Mississippi, the longest river in North America, which serves as both a water source and wastewater receptor for the numerous towns and cities located along its path. As the river meanders downstream, water increasingly enriched by upstream discharges of wastewater serves as a source of potable water for downstream communities. Such observations support the view that wastewater is better defined as a potential water resource that is currently wasted. A third of the world's population lives in water-stressed countries and this proportion is expected to rise to two thirds by 2025 (Service, 2006). Finding alternate water sources is critical to the survival of these water stressed societies. Although wastewater has been overlooked as a potential water resource, it could become an option if appropriate treatment technologies can be developed.

While biological treatment processes are significantly cheaper to build and operate, current designs are not capable of delivering high quality water suitable for a variety of purposes. Despite several advances since the inception of biological wastewater treatment processes, with an initial focus on removing organic carbon through to subsequent extension to include nitrogen and phosphorous removal, these systems continue to only be able to effectively degrade substances present at mg/L concentrations.

Over the past decade the rapid increase in instrument sensitivity has led to wide detection of ng/L levels of biologically active contaminants, referred to as emerging pollutants or organic micropollutants, in the environment. These contaminants include pharmaceuticals, personal care products, steroid hormones, industrial chemicals, and pesticides (Galloway et al., 2010; Oulton et al., 2010; Vandenberg et al., 2013; Richardson and Ternes, 2014; Petrie et al., 2015; Richardson and Kimura, 2016). Discharges from wastewater treatment plants are acknowledged as a major source of these contaminants in receiving environments (Spongberg and Witter, 2008; Bartelt-Hunt et al., 2009; Loganathan et al., 2009). Various studies have highlighted the need to remove the ng/L levels of these contaminants from the treated wastewater discharge due to their negative effect on human and animal health and ecotoxicology effects (Escher et al., 2011)—for example, a recent critical review concluded that doses of bisphenol A up to 4 orders of magnitude lower than the currently prescribed lowest observed adverse effect level of 50 mg/kg/day reliably produced effects in animals (Vandenberg et al., 2013) and another study reported changes in sex hormones associated with exposure to bisphenol A in men (Galloway et al., 2010). The cost of treating impacted environments can be large. London's drinking water is reported to be tainted with synthetic estrogens from excreted contraceptive pill residues in the Thames that have resulted from the water having gone through the equivalent of six people before reaching London. The cost of cleaning up Britain's contaminated waterways is estimated to exceed £30bn (McKie, 2012). Development of appropriate technologies is needed to convert wastewater into a water resource, potentially even for meeting potable needs. This paper aims to stimulate a discussion on ways to enhance the performance of biological treatment systems to achieve this vision.

1.1 Current Wastewater Treatment Approaches

The biological treatment approach, specifically the activated sludge process, is the most commonly used wastewater treatment technology. While it was originally designed to only remove organic carbon, it was subsequently extended to remove nitrogen and phosphorous. Transitioning between aerobic, anoxic, and anaerobic conditions is key to achieving biological nutrient removal. Doing so triggers the utilization of different electron acceptors and donors, therefore promoting the transformation of C, N, and P compounds. This strategy has continued to be exploited with the development of new system configurations, such as the Anammox (Lackner et al., 2014), SHARON (Hellinga et al., 1998; van Dongen et al., 2001), and Nareda (van der Roest et al., 2011) processes. Different redox environments influence organic micropollutant removal in different ways. Various studies have reported that organic micropollutant removal occurs to varying extent in biological treatment systems via a combination of biosorption and biodegradation. Maximum achievable removal for micropollutant such estrogens, nonylphenolics and metals was recorded at the highest sludge retention times (SRT) and hydraulic retention time (HRT) studied (Petrie et al., 2014). This micropollutant removal is related to the concomitant reduction in food: microorganism ratio. Nevertheless, studies have concluded that micropollutant degradation is insensitive to different SRTs (Falås et al., 2016). Other aspects influencing micropollutant removal in biological treatment processes are heterotrophic activity (Majewsky et al., 2010), pH (Gulde et al., 2014), and suspended/attached growth configuration (Falås et al., 2013). Although the heterogeneity of micropollutants in wastewaters makes removal difficult to predict since their chemistry is so diverse. For instance micropollutants can be broadly classified as easily, moderate and poorly degradable. The reality is that the biological processes are not designed to remove these pollutants, resulting in their incomplete removal and detection in final effluents and receiving surface waters (Joss et al., 2006; Racz and Goel, 2010). No specific strategies appear to have been successfully developed to enhance micropollutant removal by biological systems. Instead, organic micropollutant removal is typically achieved using physicochemical processes. These processes include adsorption in to organic matrices, passive effluent treatment in wetlands and aquifers, advanced oxidation processes (e.g., ozonation, UV treatment, photocatalysis, and Fenton oxidation), membrane filtration (nanofiltration and reverse osmosis membranes), and membrane biological reactor (Bolong et al., 2009; Rossner et al., 2009; Oulton et al., 2010; Luo et al., 2014). The shortcomings of these treatment systems are high investment and maintenance costs, generation of toxic residuals, and complex treatment procedures (Grassi et al., 2012). Operational difficulties of these physicochemical processes also need to be considered. For instance, wetlands and aquifers require large surface areas and large HRTs, which significantly complicates its implementation in urban areas. Similarly, advanced oxidation processes require chemicals not easily available (e.g., O_3, $F+2F_2+$, H_2O_2) and the non-selective nature of these reactions can result in formation of daughter products that are more toxic than the parent micropollutant (Rosal et al., 2009). Other concerns may also need consideration: energy and chemical inputs of ozonation processes are substantial (Cañizares et al., 2009); the Fenton process requires rigorous pH control (Chong et al., 2012); specially configured reactors are needed to maximize light exposure by UV and photocatalysis based water treatment, or water turbidity could significantly influence treatment efficiency (Chong et al., 2012); and membrane filtration and membrane biological reactors treatments routinely suffer from membrane fouling and biofilm formation (Guo et al., 2012).

2. ENHANCING BIOLOGICAL WASTEWATER TREATMENT

2.1 Exploiting Microbial Diversity

Complex microbial consortia composed by bacteria-bacteria, bacteria-archea and bacteria-fungi can be developed to enhance biological degradation of micropollutants. Oxidase enzymes have great potential as biocatalysts for micropollutant and organic waste breakdown. Two of such class of enzymes are the oxygenase Cytochromes P450 (CYPs or Cyt P450), a highly efficient group of monooxygenases responsible for the destruction of drugs and toxins in organisms, and the laccases (EC 1.10.3.2), a class of copper-containing oxidase enzymes used by microorganisms to break down lignin) (Riva, 2006; Kumar, 2010). Both enzymes have been shown to efficiently degrade a vast array of organic micropollutants in pure enzyme assays (Harms et al., 2011; Lah et al., 2011). Despite the demonstrated effectiveness of these enzymes, their use in wastewater treatment is just starting to be investigated and has not been implemented in pilot or full scale (Lah et al., 2011). Current biological systems are reliant on using prokaryotic bacteria, which can rapidly oxidize organic carbon but generally do not express Cyt P450. Additionally, no applications of bacterial laccases have yet been realized due to their limited characterization (Werck-Reichhart and Feyereisen, 2000). In comparison, many fungi produce Cyt P450 (Lah et al., 2008; Kelly and Kelly, 2013) and only fungal laccases are used currently in biotechnological applications (Sharma et al., 2007; Harms et al., 2011). A system integrating suspended phase bacteria with activated fungi to transform conventional and emerging categories of pollutants can potentially take advantage of both organisms. Fungi can be activated by inducing the production of high levels of oxidase enzymes and at the same time stimulating enzyme activity (e.g., by providing H_2O_2 as cofactor for bacterial Cyt P450). By controlling the carbon source and electron acceptor regimens, it is possible to induct biocatalyst expression and activation (Price et al., 2013; Chubukov et al., 2014). As whole-cell catalyzed reactions are reportedly 10- to 100-fold slower than reactions catalyzed by free enzymes (Sotirova et al., 2008), cell membranes can be permeabilized to enhance extracellular laccase secretion and promote cellular influx of pollutants to membrane bound Cyt P450. In addition, gel encapsulated of fungi or bacterial cells can be implemented for improved performance. Alternatively, iterating between different redox conditions can promote the growth of different microbial populations to improve micropollutant biodegradation (Falås et al., 2016).

2.2 Exploiting Metabolic Diversity

New bioprocesses to treat wastewater can be developed by exploiting the diverse metabolic capabilities of microbes. Traditionally, only metabolic capabilities in relation to aerobic heterotrophy, aerobic nitrification and anoxic denitrification, and phosphate accumulation have been used in biological wastewater treatment systems. Exploiting alternative metabolic capabilities such as heterotrophic sulfate-reduction (Zhang and Wang, 2014), autotrophic sulfur-oxidizing denitrification (Hao et al., 2014; Pokorna and Zabranska, 2015), anaerobic methane-oxidation denitrification (Raghoebarsing et al., 2006), partial nitrite reduction to nitrous oxide for energy generation (Scherson et al., 2013) and electron shuttle redox biotransformation (Van der Zee and Cervantes, 2009) could improve micropollutant removal from wastewaters. These metabolic capabilities can also be exploited to remove inorganic micropollutants (e.g., heavy metals and radioactive elements) in which redox transformations can alter the solubility and precipitate the contaminants (Groudev et al., 1999; Gadd, 2010).

2.3 Exploiting Biocatalyst Diversity

Many enzymes are "promiscuous" biocatalysts capable of transforming a variety of substrates that share structural similarity with their primary substrate. These promiscuous or generalized enzymes: (i) are frequently not essential, (ii) maintain low metabolic fluxes, and (iii) require less regulation of enzyme activity to control metabolic flux in dynamic environments than do specialized enzymes (Nam et al., 2012). This non-specificity makes them promising biocatalysts for organic micropollutant degradation. Bacteria are constantly developing new catabolic pathways in order to either access sources of carbon, energy and nutrients or simply to detoxify new compounds. However, unraveling these degradation processes in nature is made difficult by the large number of chemicals, their occurrence at mostly low concentrations, and the number of unknown chemicals resulting from bacterial transformation and biodegradation. Genomics has revealed that many microorganisms have far greater potential to produce specialized enzymes and metabolites than was thought from classic bioactivity screens; however, realizing their degradation potential has been hampered by the fact that many specialized metabolite biosynthetic gene clusters are not expressed in laboratory cultures (Keller et al., 2005; Miller et al., 2010; Mora-Pale et al., 2014; Rutledge and Challis, 2015). Theoretically bacteria can utilize the full space of catabolic biochemical reaction types and initiate several pathways to degrade micropollutants (Kolvenbach et al., 2014). For example, multiple biodegradation mechanisms have been discovered for bisphenol A (BPA), an industrial chemical found in a variety of plastics and epoxies and a putative endocrine disrupting compound. Similarly, enzymatic biodegradation pathways may be discoverable for other compounds using biocatalyst and daughter products screening techniques such as metagenomics (Fernández-Arrojo et al., 2010), metaproteomics, and metabolomics (Villas-Bôas and Bruheim, 2007; Helbling et al., 2010, 2012).

3. IMPLEMENTATION

Current designs focus on removing conventional pollutants (organic and inorganic substances present) that are present at much larger concentrations than micropollutants to be metabolized as a source of carbon and energy. The low concentrations of organic micropollutants suggests that these substances are more likely to be cometabolized. Hence, system design will need to focus on ways of stimulating the production of enzymes that could degrade the micropollutants. Current wastewater treatment system designs contain the basic ingredients required to formulate new processes for efficiently degrading micropollutants using the strategies discussed in the sections above. Biological treatment systems contain a wide variety of active microorganisms, and employ different redox environments are used to stimulate a wider range of chemical transformations. Rapid degradation of micropollutants could be achieved by stimulating the production of enzymes (e.g., oxidoreductases and laccases) by using hard-to-degrade substrates or by employing other environmental stresses. Liquids from anaerobic digestate could serve as carbon source that promotes such activity. The performance of such systems could be further enhanced by employing operational strategies that avert a lowering of enzyme production resulting from adaptation of microbial consortium to the imposed stress. Two plant configurations could potentially be developed using these approaches: (i) main stream micropollutant removal (i.e., modification of existing bioprocesses) or (ii) add-on reactors for micropollutant removal (i.e., dedicated tanks to remove micropollutants in the effluent of treatment plant).

4. CONCLUSIONS

Numerous organic micropollutants are present in wastewater. To convert wastewater into high quality water will require their removal. Current wastewater treatment systems are not designed to degrade organic micropollutants. Physicochemical treatment systems are effective but expensive while biological approaches cheap but ineffective and inconsistent. The next stage in the evolution of biological wastewater treatment processes could be based on discovering and employing novel metabolic traits of unconventional microbes (e.g., sulfur or iron oxidizing bacteria), or inducing the synthesis of enzymes capable of degrading micropollutants. Implementation of these new bioprocesses will involve the use of different waste streams used as substrate and new process operation strategies such as dynamic substrates control.

REFERENCES

1. Bartelt-Hunt, S. L., Snow, D. D., Damon, T., Shockley, J., and Hoagland, K. (2009). The occurrence of illicit and therapeutic pharmaceuticals in wastewater effluent and surface waters in Nebraska. *Environ. Pollut.* 157, 786–791. doi: 10.1016/j.envpol.2008.11.025
2. Bolong, N., Ismail, A. F., Salim, M. R., and Matsuura, T. (2009). A review of the effects of emerging contaminants in wastewater and options for their removal. *Desalination* 239, 229–246. doi: 10.1016/j.desal.2008.03.020
3. Cañizares, P., Paz, R., Sáez, C., and Rodrigo, M. A. (2009). Costs of the electrochemical oxidation of wastewaters: a comparison with ozonation and Fenton oxidation processes. *J. Environ. Manag.* 90, 410–420. doi: 10.1016/j.jenvman.2007.10.010
4. Chong, M. N., Sharma, A. K., Burn, S., and Saint, C. P. (2012). Feasibility study on the application of advanced oxidation technologies for decentralised wastewater treatment. *J. Cleaner Prod.* 35, 230–238. doi: 10.1016/j.jclepro.2012.06.003
5. Chubukov, V., Gerosa, L., Kochanowski, K., and Sauer, U. (2014). Coordination of microbial metabolism. *Nat. Rev. Microbiol.* 12, 327–340. doi: 10.1038/nrmicro3238
6. Escher, B. I., Baumgartner, R., Koller, M., Treyer, K., Lienert, J., and McArdell, C. S. (2011). Environmental toxicology and risk assessment of pharmaceuticals from hospital wastewater. *Water Res.* 45, 75–92. doi: 10.1016/j.watres.2010.08.019
7. Falås, P., Longrée, P., la Cour Jansen, J., Siegrist, H., Hollender, J., and Joss, A. (2013). Micropollutant removal by attached and suspended growth in a hybrid biofilm-activated sludge process. *Water Res.* 47, 4498–4506. doi: 10.1016/j.watres.2013.05.010
8. Falås, P., Wick, A., Castronovo, S., Habermacher, J., Ternes, T. A., and Joss, A. (2016). Tracing the limits of organic micropollutant removal in biological wastewater treatment. *Water Res.* 95, 240–249. doi: 10.1016/j.watres.2016.03.009
9. Fernández-Arrojo, L., Guazzaroni, M. E., López-Cortés, N., Beloqui, A., and Ferrer, M. (2010). Metagenomic era for biocatalyst identification. *Curr. Opin. Biotechnol.* 21, 725–733. doi: 10.1016/j.copbio.2010.09.006
10. Gadd, G. M. (2010). Metals, minerals and microbes: Geomicrobiology and bioremediation. *Microbiology* 156, 609–643. doi: 10.1099/mic.0.037143-0
11. Galloway, T., Cipelli, R., Guralnick, J., Ferrucci, L., Bandinelli, S., Corsi, A. M., et al. (2010). Daily bisphenol a excretion and associations with sex hormone concentrations: results from the inchianti adult population study. *Environ. Health Perspect.* 118, 1603–1608. doi: 10.1289/ehp.1002367
12. Grassi, M., Belgiorno, V., and Lofrano, G. (2012). "Removal of emerging contaminants from water and wastewater by adsorption process," in *Emerging Compounds Removal from Wastewater*, ed Lofrano G. (New York, NY: Springer), p15–p37.
13. Groudev, S. N., Bratcova, S. G., and Komnitsas, K. (1999). Treatment of waters polluted with radioactive elements and heavy metals by means of a laboratory passive system. *Minerals Eng.* 12, 261–270. doi: 10.1016/S0892-6875(99)00004-7
14. Gulde, R., Helbling, D. E., Scheidegger, A., and Fenner, K. (2014). pH-dependent biotransformation of ionizable organic micropollutants in activated sludge. *Environ. Sci. Technol.* 48, 13760–13768. doi: 10.1021/es5037139
15. Guo, W., Ngo, H.-H., and Li, J. (2012). A mini-review on membrane fouling. *Bioresour. Technol.* 122, 27–34. doi: 10.1016/j.biortech.2012.04.089
16. Hao, T.-W., Xiang, P., Mackey, H. R., Chi, K., Lu, H., Chui, H., et al. (2014). A review of biological sulfate conversions in wastewater treatment. *Water Res.* 65, 1–21. doi: 10.1016/j.watres.2014.06.043
17. Harms, H., Schlosser, D., and Wick, L. Y. (2011). Untapped potential: exploiting fungi in bioremediation of hazardous chemicals. *Nat. Rev. Microbiol.* 9, 177–192. doi: 10. 1038/nrmicro2519
18. Helbling, D. E., Hollender, J., Kohler, H.-E., Singer, H., and Fenner, K. (2010). High-throughput identification of microbial transformation products of organic micropollutants. *Environ. Sci. Technol.* 44, 6621–6627. doi: 10.1021/es100970m

19. Helbling, D. E., Johnson, D. R., Honti, M., and Fenner, K. (2012). Micropollutant biotransformation kinetics associate with WWTP process parameters and microbial community characteristics. *Environ. Sci. Technol.* 46, 10579–10588. doi: 10.1021/es3019012
20. Hellinga, C., Schellen, A. A. J. C., Mulder, J. W., van Loosdrecht, M. C. M., and Heijnen, J. J. (1998). The sharon process: an innovative method for nitrogen removal from ammonium-rich waste water. *Water Sci. Technol.* 37, 135–142. doi: 10.1016/S0273-1223(98)00281-9
21. Joss, A., Zabczynski, S., Göbel, A., Hoffmann, B., Löffler, D., McArdell, C. S., et al. (2006). Biological degradation of pharmaceuticals in municipal wastewater treatment: proposing a classification scheme. *Water Res.* 40, 1686–1696. doi: 10.1016/j.watres.2006.02.014
22. Keller, N. P., Turner, G., and Bennett, J. W. (2005). Fungal secondary metabolism - From biochemistry to genomics. *Nat. Rev. Microbiol.* 3, 937–947. doi: 10.1038/nrmicro1286
23. Kelly, S. L., and Kelly, D. E. (2013). Microbial cytochromes P450: biodiversity and biotechnology. Where do cytochromes P450 come from, what do they do and what can they do for us? *Philos. Trans. R. Soc. B* 368, 20120476. doi: 10.1098/rstb.2012.0476
24. Kolvenbach, B. A., Helbling, D. E., Kohler, H.-P. E., and Corvini, P. F.-X. (2014). Emerging chemicals and the evolution of biodegradation capacities and pathways in bacteria. *Curr.Opin. Biotechnol.* 27, 8–14. doi: 10.1016/j.copbio.2013.08.017
25. Kumar, S. (2010). Engineering cytochrome P450 biocatalysts for biotechnology, medicine and bioremediation. *Expert Opin. Drug Metab. Toxicol.* 6, 115–131. doi: 10.1517/17425250903431040
26. Lackner, S., Gilbert, E. M., Vlaeminck, S. E., Joss, A., Horn, H., and van Loosdrecht, M. C. M. (2014). Full-scale partial nitritation/anammox experiences – an application survey. *Water Res.* 55, 292–303. doi: 10.1016/j.watres.2014.02.032
27. Lah, L., Kraševec, N., Trontelj, P., and Komel, R. (2008). High diversity and complex evolution of fungal cytochrome P450 reductase: cytochrome P450 systems. *Fungal Genet. Biol.* 45, 446–458. doi: 10.1016/j.fgb.2007.10.004
28. Lah, L., Podobnik, B., Novak, M., Korošec, B., Berne, S., Vogelsang, M., et al. (2011). The versatility of the fungal cytochrome P450 monooxygenase system is instrumental in xenobiotic detoxification. *Mol. Microbiol.* 81, 1374–1389. doi: 10.1111/j.1365-2958.2011.07772.x
29. Loganathan, B., Phillips, M., Mowery, H., and Jones-Lepp, T. L. (2009). Contamination profiles and mass loadings of macrolide antibiotics and illicit drugs from a small urban wastewater treatment plant. *Chemosphere* 75, 70–77. doi: 10.1016/j.chemosphere.2008.11.047
30. Luo, Y., Guo, W., Ngo, H. H., Nghiem, L. D., Hai, F. I., Zhang, J., et al. (2014). A review on the occurrence of micropollutants in the aquatic environment and their fate and removal during wastewater treatment. *Sci. Total Environ.* 473–474, 619–641. doi: 10.1016/j.scitotenv.2013.12.065
31. Majewsky, M., Gallé, T., Zwank, L., and Fischer, K. (2010). Influence of microbial activity on polar xenobiotic degradation in activated sludge systems. *Water Sci. Technol.* 62, 701–707. doi: 10.2166/wst.2010.925
32. McKie, R. (2012). Drug and water firms attack EU plan to reduce ethinylestradiol pollution. *BMJ* 344:4029. doi: 10.1136/bmj.e4029
33. Miller, L. D., Mosher, J. J., Venkateswaran, A., Yang, Z. K., Palumbo, A. V., Phelps, T. J., et al. (2010). Establishment and metabolic analysis of a model microbial community for understanding trophic and electron accepting interactions of subsurface anaerobic environments. *BMC Microbiol.* 10:149. doi: 10.1186/1471-2180-10-149
34. Mora-Pale, M., Sanchez-Rodriguez, S. P., Linhardt, R. J., Dordick, J. S., and Koffas, M. A. G. (2014). Biochemical strategies for enhancing the in vivo production of natural products with pharmaceutical potential. *Curr. Opin. Biotechnol.* 25, 86–94. doi: 10.1016/j. copbio.2013.09.009
35. Nam, H., Lewis, N. E., Lerman, J. A., Lee, D., Chang, R. L., Kim, D., et al. (2012). Network context and selection in the evolution to enzyme specificity. *Science* 337, 1101–1104. doi: 10.1126/science.1216861

36. Oulton, R. L., Kohn, T., and Cwiertny, D. M. (2010). Pharmaceuticals and personal care products in effluent matrices: a survey of transformation and removal during wastewater treatment and implications for wastewater management. *J. Environ. Monit.* 12, 1956–1978. doi: 10.1039/c0em00068j
37. Petrie, B., Barden, R., and Kasprzyk-Hordern, B. (2015). A review on emerging contaminants in wastewaters and the environment: current knowledge, understudied areas and recommendations for future monitoring. *Water Res.* 72, 3–27. doi: 10.1016/j.watres.2014.08.053
38. Petrie, B., McAdam, E. J., Lester, J. N., and Cartmell, E. (2014). Assessing potential modifications to the activated sludge process to improve simultaneous removal of a diverse range of micropollutants. *Water Res.* 62, 180–192. doi: 10.1016/j. watres.2014.05.036
39. Pokorna, D., and Zabranska, J. (2015). Sulfur-oxidizing bacteria in environmental technology. *Biotechnol. Adv.* 33, 1246–1259. doi: 10.1016/j.biotechadv.2015.02.007
40. Price, M. N., Deutschbauer, A. M., Skerker, J. M., Wetmore, K. M., Ruths, T., Mar, J. S., et al. (2013). Indirect and suboptimal control of gene expression is widespread in bacteria. *Mol. Sys. Biol.* 9, 660. doi: 10.1038/msb.2013.16
41. Racz, L., and Goel, R. (2010). Fate and removal of estrogens in municipal wastewater. *J. Environ. Monit.* 12, 58–70. doi: 10.1039/B917298J
42. Raghoebarsing, A. A., Pol, A., van de Pas-Schoonen, K. T., Smolders, A. J. P., Ettwig, K. F., Rijpstra, W. I. C., et al. (2006). A microbial consortium couples anaerobic methane oxidation to denitrification. *Nature* 440, 918–921. doi: 10.1038/nature04617
43. Richardson, S. D., and Kimura, S. Y. (2016). Water analysis: emerging contaminants and current issues. *Anal. Chem.* 88, 546–582. doi: 10.1021/acs.analchem.5b04493
44. Richardson, S. D., and Ternes, T. A. (2014). Water analysis: emerging contaminants and current issues. *Anal. Chem.* 86, 2813–2848. doi: 10.1021/ac500508t
45. Riva, S. (2006). Laccases: blue enzymes for green chemistry. *Trends Biotechnol.* 24, 219–226. doi: 10.1016/j.tibtech.2006.03.006
46. Rosal, R., Gonzalo, M. S., Boltes, K., Letón, P., Vaquero, J. J., and García-Calvo, E. (2009). Identification of intermediates and assessment of ecotoxicity in the oxidation products generated during the ozonation of clofibric acid. *J. Hazard. Mat.* 172, 1061–1068. doi: 10.1016/j.jhazmat.2009.07.110
47. Rossner, A., Snyder, S. A., and Knappe, D. R. U. (2009). Removal of emerging contaminants of concern by alternative adsorbents. *Water Res.* 43, 3787–3796. doi: 10.1016/j. watres.2009.06.009
48. Rutledge, P. J., and Challis, G. L. (2015). Discovery of microbial natural products by activation of silent biosynthetic gene clusters. *Nat. Rev. Microbiol.* 13, 509–523. doi: 10.1038/nrmicro3496
49. Scherson, Y. D., Wells, G. F., Woo, S.-G., Lee, J., Park, J., Cantwell, B. J., et al. (2013). Nitrogen removal with energy recovery through N_2O decomposition. *Energy Environ. Sci.* 6, 241–248. doi: 10.1039/C2EE22487A
50. Service, R. F. (2006). Desalination freshens up. *Science* 313, 1088–1090. doi: 10.1126/science.313.5790.1088
51. Sharma, P., Goel, R., and Capalash, N. (2007). Bacterial laccases. *World J. Microbiol. Biotechnol.* 23, 823–832. doi: 10.1007/s11274-006-9305-3
52. Sotirova, A. V., Spasova, D. I., Galabova, D. N., Karpenko, E., and Shulga, A. (2008). Rhamnolipid-biosurfactant permeabilizing effects on gram-positive and gram-negative bacterial strains. *Curr. Microbiol.* 56, 639–644. doi: 10.1007/s00284-008-9139-3
53. Spongberg, A. L., and Witter, J. D. (2008). Pharmaceutical compounds in the wastewater process stream in Northwest Ohio. *Sci. Total Environ.* 397, 148–157. doi: 10.1016/j.scitotenv.2008.02.042
54. Vandenberg, L. N., Ehrlich, S., Belcher, S. M., Ben-Jonathan, N., Dolinoy, D. C., Hugo, E. R., et al. (2013). Low dose effects of bisphenol A: an integrated review of *in vitro*, laboratory animal, and epidemiology studies. *Endocr. Disrupt.* 1, 25078. doi: 10.4161/endo.26490

55. van der Roest, H. F., de Bruin, L. M. M., Gademan, G., and Coelho, F. (2011). Towards sustainable waste water treatment with Dutch Nereda® technology. *Water Pract. Technol.* 6, 1–2. doi: 10.2166/wpt.2011.059
56. Van der Zee, F. P., and Cervantes, F. J. (2009). Impact and application of electron shuttles on the redox (bio)transformation of contaminants: a review. *Biotechnol. Adv.* 27, 256–277. doi: 10.1016/j.biotechadv.2009.01.004
57. van Dongen, U., Jetten, M. S., and van Loosdrecht, M. C. (2001). The SHARON-Anammox process for treatment of ammonium rich wastewater. *Water Sci. Technol.* 44, 153–160.
58. Villas-Bôas, S. G., and Bruheim, P. (2007). The potential of metabolomics tools in bioremediation studies. *OMICS A J. Integr. Biol.* 11, 305–313. doi: 10.1089/omi.2007.0005
59. Werck-Reichhart, D., and Feyereisen, R. (2000). Cytochromes P450: a success story. *Genome Biol.* 1:reviews3003. doi: 10.1186/gb-2000-1-6-reviews3003
60. Zhang, M., and Wang, H. (2014). Organic wastes as carbon sources to promote sulfate reducing bacterial activity for biological remediation of acid mine drainage. *Minerals Eng.* 69, 81–90. doi: 10.1016/j.mineng.2014.07.010

CHAPTER 20

Combination of Electrochemical Processes with Membrane Bioreactors for Wastewater Treatment and Fouling Control: A Review

Benny M. B. Ensano[1], Laura Borea[2], Vincenzo Naddeo[2], Vincenzo Belgiorno[2*], Mark D. G. de Luna[1,3] and Florencio C. Ballesteros Jr.[1,3]

[1]*Environmental Engineering Program, National Graduate School of Engineering, University of the Philippines, Quezon City, Philippines*

[2]*Sanitary Environmental Engineering Division, Department of Civil Engineering, University of Salerno, Fisciano, Italy*

[3]*Department of Chemical Engineering, University of the Philippines, Quezon City, Philippines*

This paper provides a critical review about the integration of electrochemical processes into membrane bioreactors (MBR) in order to understand the influence of these processes on wastewater treatment performance and membrane fouling control. The integration can be realized either in an internal or an external configuration. Electrically enhanced membrane bioreactors or electro membrane bioreactors (eMBRs) combine biodegradation, electrochemical and membrane filtration processes into one system providing higher effluent quality as compared to conventional

MBRs and activated sludge plants. Furthermore, electrochemical processes, such as electrocoagulation, electrophoresis, and electroosmosis, help to mitigate deposition of foulants into the membrane and enhance sludge dewaterability by controlling the morphological properties and mobility of the colloidal particles and bulk liquid. Intermittent application of minute electric field has proven to reduce energy consumption and operational cost as well as minimize the negative effect of direct current field on microbial activity which are some of the main concerns in eMBR technology. The present review discusses important design considerations of eMBR, its advantages as well as its applications to different types of wastewater. It also presents several challenges that need to be addressed for future development of this hybrid technology which include treatment of high strength industrial wastewater and removal of emerging contaminants, optimization study, cost benefit analysis and the possible combination with microbial electrolysis cell for biohydrogen production.

Keywords: electro membrane bioreactors (eMBRs), fouling precursors, bacterial activity, electrocoagulation, electroosmosis, electrophoresis, extracellular polymeric substances (EPS), soluble microbial products (SMP)

1. INTRODUCTION

Numerous water and wastewater treatment processes have been employed in order to minimize water pollution and to augment drinking water resources (de Luna et al., 2009; Naddeo et al., 2011; Cesaro et al., 2013; Secondes et al., 2014; Ballesteros et al., 2016). Among these techniques, conventional activated sludge (CAS) process is the most widely used method due to its effective mineralization of organic and inorganic compounds from various types of wastewater. However, CAS has been shown to have limited capability of degrading recalcitrant pollutants such as pharmaceuticals, flame retardants, persistent organic compounds, anionic surfactants, etc. (Bernhard et al., 2006; Singhal and Perez-Garcia, 2016). Besides, this treatment method also poses several disadvantages which include high footprint and reactor volume, the production of large amount of excess sludge, that requires further treatment and disposal, and the installation of the sedimentation tank for separation of the mixed liquor from the treated effluent. Furthermore, a final disinfection treatment (e.g., UV, ozone, chlorination, etc.) is required for the removal of pathogenic microorganisms.

A more promising alternative for wastewater remediation is the combination of the activated sludge treatment with membrane filtration process. This technology, also known as membrane bioreactor (MBR), differs from CAS since a membrane module is used instead of the secondary clarifier to separate activated sludge from the final effluent (Xing et al., 2000). MBR offers many advantages such as higher effluent quality, smaller footprint and reactor volume, lower energy consumption and less sludge production (Cicek, 2003; Yuniarto et al., 2008). As a result, MBR technology has experienced unprecedented growth in recent years. Its applications have extended widely from treatment of municipal wastewater (Weiss and Reemtsma, 2008; Ma et al., 2013; Zhou et al., 2015) to removal of toxic and hazardous compounds in landfill leachate (Ahmed and Lan, 2012), and industrial wastewaters (Artiga et al., 2005; Yigit et al., 2009; Palogos et al., 2014).

However, an inevitable fouling problem limits the application of this technology into full scale practical operation. The deposition of sludge particles and colloids onto and into the membrane leads to a severe decline of the permeate flux and affects stable operational performance (Kimura et al., 2005). There are several approaches that can reduce fouling in membranes (Naddeo et al., 2015a,b; Scannapieco et al., 2015) but the most common techniques used in MBR technology are the excess aeration and frequent membrane cleaning. However, these methods also result in an increase of energy demand and operating costs as well as reduction of membrane lifespan (Wei C. H. et al., 2011; Zhang et al., 2011; Shi et al., 2014).

Recent advances have demonstrated that the application of an electrical field to membrane bioreactors (electrically enhanced membrane bioreactors or electro membrane bioreactor, eMBR) can effectively reduce fouling (Chen et al., 2007; Akamatsu et al., 2010; Liu et al., 2012a). This method induces electrochemical mechanisms such as electrocoagulation, electroosmosis, and electrophoresis that help in degradation of pollutants and, at the same time, control the mobility, and deposition of foulants onto the membrane surface (Bani-Melhem and Elektorowicz, 2011). Electrocoagulation is the main mechanism influencing the removal of organic compounds with high fouling potential from wastewater. It involves the *in situ* generation of coagulants from the electrochemical dissolution of immersed sacrificial anodes (usually iron or aluminum) (Figure 1C). Depending on the pH of the solution, various metal species are produced during electrocoagulation which react with the pollutants leading to the destabilization and aggregation of suspended particles and the precipitation and adsorption of dissolved contaminants (Giwa et al., 2016). Moreover, the applied voltage also allows the negatively charged foulants such as activated sludge and secreted polymers to drift toward the oppositely charged electrode and away from the membrane via electrophoretic motion (Figure 1A) (Chen et al., 2007; Akamatsu et al., 2010, 2012). Consequently, electroosmotic force propels the removal of bound water from the microbial flocs electrical double layer which decreases the sludge specific resistance to filtration hence improving fouling control (Figure 1B) (Ibeid et al., 2013a).

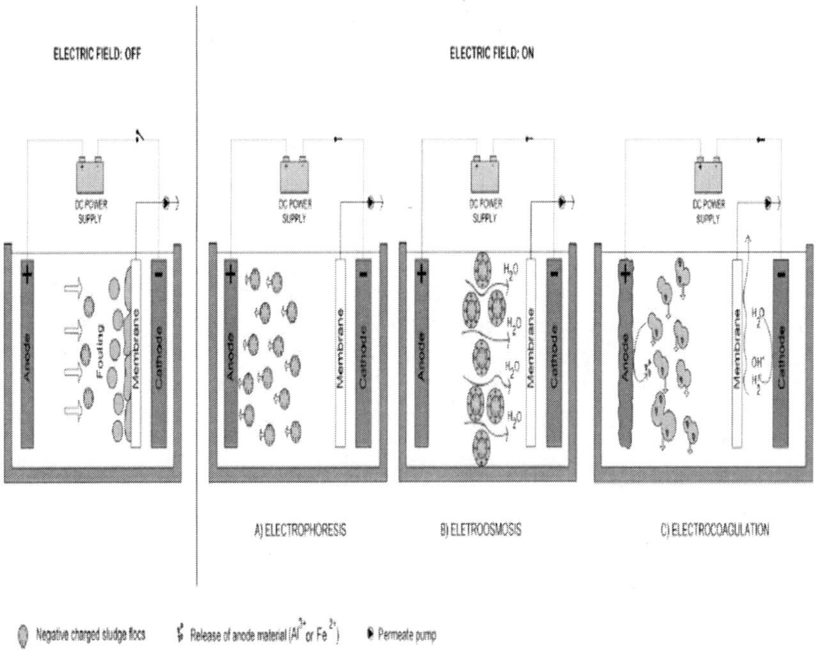

Figure 1. Electrochemical mechanisms occurring in the operation of the electro mbr (embr): (a) electrophoresis, (b) electroosmosis, (c) electrocoagulation.

The application of a direct current (DC) field to MBRs has proven to be more effective compared to other fouling control methods, however, it is also associated with high energy consumption and affect bacterial lifespan (Wei V. et al., 2011; Giwa et al., 2015; Zeyoudi et al., 2015). New and cheaper developments suggest that intermittent application of minute electric field also enhances membrane filterability and is similarly effective in controlling fouling as that with continuous DC field (Akamatsu et al., 2010, 2012; Liu et al., 2012a,b). It also minimizes the direct exposure of bacteria to the electric field, hence, reducing the negative effects to microbial community (Akamatsu et al., 2010; Bani-Melhem and Elektorowicz, 2010; Liu et al., 2012a; Hasan et al., 2014). Some studies also focused on extracting internal energy contained in wastewater (e.g., 17.8–

28.7 kJ/gCOD) (Heidrich et al., 2011) which can be used to offset the input of external DC field for fouling control. This can be achieved by combining bioelectrochemical systems (BES) with MBRs. Microbial fuel cells (MFCs) and microbial electrolysis cells (MECs) are two types of BES that use exoelectrogenic microbes to produce energy from wastewater by converting biodegradable organic matter directly into electricity, hydrogen, and other valuable products via biological and electrochemical processes (Zhang F. et al., 2014). They are innovative methods for simultaneous renewable energy production and wastewater treatment (Wang et al., 2015). Hybrid systems combining BES with MBR have also been recently investigated for cost-effective wastewater remediation and fouling control (Wang et al., 2012; Liu et al., 2013a, 2014; Li et al., 2014).

This paper critically reviews the combination of electrochemical processes with MBR. This is the first review paper that explores the advantages, the limitations, future challenges, and developments of this hybrid technology in wastewater treatment. The design considerations, applications to different types of wastewater and mechanisms of pollution removal and fouling control using electrically enhanced membrane bioreactor are also discussed.

2. ELECTRICALLY ENHANCED MEMBRANE BIOREACTOR (EMBR)

This section is divided into three parts: the first part highlights three of the most important considerations for the design of eMBR which are the main configurations, materials and operating conditions; the second part enumerates the numerous applications of this technology on different types of wastewater and the third discusses the effects of appending electricity on pollutant removal, fouling control, and bacterial activity in an eMBR technology.

3. DESIGN CONSIDERATIONS

3.1 Main Configurations

The main eMBR configurations applied in wastewater treatment can be generally classified into two: (1) External electrochemical unit (usually electrocoagulation unit) preceding a submerged membrane bioreactor (EC-SMBR) (Figure 2A) and (2) submerged membrane electro-bioreactor (SMEBR) (Figure 2B).

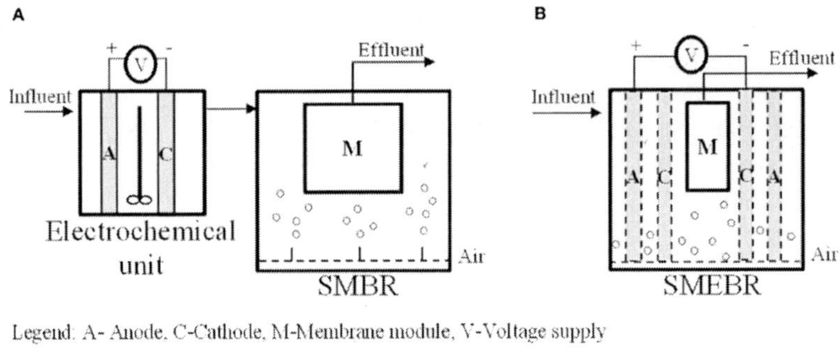

Figure 2. Main embr configurations applied to wastewater treatment: (a) external electrochemical unit (usually electrocoagulation unit) preceding a submerged membrane bioreactor (ec-smbr), (b) submerged membrane electro-bioreactor (smebr).

EC-SMBR was used by Kim et al. (2010), Bani-Melhem and Smith (2012), and Keerthi et al. (2013) to treat municipal wastewater, actual gray water, and tannery wastewater, respectively. Here, the influent is first treated via electrocoagulation process in a separate unit prior to biological degradation and membrane filtration in a conventional submerged membrane bioreactor (SMBR). This system was developed primarily to enhance the phosphorus removal from wastewater and prevent the applied DC field to have direct contact with the microbial community since it could have inhibitory effects on bacterial activity (Bani-Melhem and Smith, 2012). It was also proven to positively improve chemical oxygen demand (COD) and color removal from tannery wastewater (Keerthi et al., 2013).

In SMEBR, on the other hand, electrochemical processes are integrated with membrane bioreactor by the addition of sacrificial electrodes inside the SMBR. The wastewater first passes through the biological treatment zone for biodegradation, that is between the reactor wall and the anode, then across the electrical zone (i.e., between the electrodes) to undergo some electrochemical processes and further degradation, and eventually is filtered out through the membrane module to remove the solids (Hasan et al., 2014) (Figure 1). In some SMEBR studies, the cathode is designed as part of the membrane module [e.g., carbon cloth electrode fabricated with polyvinylidene fluoride (PVDF) membrane or copper wire cathode inside a PP non-woven sheet] or an electrically conductive low- cost membrane is used as the cathode itself [e.g., terylene cloth modified with polypyrrole (Ppy), polyester filter cloth modified with PPy coated graphene (GR/Ppy), or with PPy coated graphene oxide (GO/Ppy)] (Akamatsu et al., 2012; Liu et al., 2012a,b, 2013b; Zhang et al., 2015). Details about membrane and electrode materials are discussed in Section Materials. The resulting effluent quality of eMBR is observed to be better than the EC-SMBR (Tables 1, 2). Moreover, fouling alleviation is further enhanced, as reported in the Section Membrane Fouling Reduction, in this configuration due to the combined efforts of electrocoagulation, electroosmosis, and electrophoresis inside the SMEBR system. A novel design of SMEBR was proposed and developed by Bani-Melhem and Elektorowicz (2010) which consists of cylindrical perforated electrodes immersed inside the bioreactor with a membrane module placed at the center.

Table 1. Application of embr on real wastewater.

Type of wastewater	Wastewater characteristics	Configurations	Membrane materials	Electrode materials	Operational conditions	Results	References
Combined industrial wastewater	pH = 7.3 SS = 103 mg/L COD = 251 mg/L Total dissolved solids (TDS) = 1130 mg/L DO = 3 mg/L TN = 43 mg/L Alkalinity as $CaCO_3$ = 212 mg/L	MBR vs. SMEBR	Polyethersulfone (PES) Flat sheet UF membrane	Anode: Fe Cathode: Fe Electrode gap = 4–6 cm	Electrolysis Time = 105–150 min CD = 0–30 mA/cm^2 DC Exposure time = continuous pH = 6–8 SRT = 32 d Initial MLSS = 4500 mg/L	COD Removal Efficiencies: **MBR** = 60%, **SMEBR** = 85%; Optimum operational conditions: Electrolysis Time = 105 min, pH = 7 CD = 10 mA/cm^2 Electrode gap = 5 cm	Hosseinzadeh et al., 2015
Tannery wastewater	pH = 7.4, TS = 9920 mg/L, TDS = 5720 mg/L, TSS = 4200 mg/L, COD = 1600 mg/L, Color = Black	EC-SMBR vs. MBR	PVDF membranes	Anode: Al Cathode: SS Electrode gap = 3 cm	Operation time = 10–11 h CD = 15 mA/cm^2 DC Exposure time = continuous pH = 7.4 SRT = not specified (no sludge waste) Initial MLSS = 6 g/L	Removal Efficiencies: **EC-SMBR**: COD = 90.2%, Color = 92.75%, **MBR**: COD = 72.09%, Color = 75.82%	Keerthi et al., 2013
Gray water	0.59 mS/cm, pH = 7.5 T = 23.5 °C DO = 1.63 mg/L, TDS = 291 mg/L, Conductivity = 0.59 ms/cm, TSS = 78 mg/L, COD = 483 mg/L, NH_3-N = 11.3 mg/L, NO_3-N = 0.93 mg/L, PO_4-P = 0.53 mg/L	EC-SMBR vs SMBR	ZeeWeed-1 Hollow fiber UF membrane	Anode: Al Cathode: Al Electrode gap = 5.8 cm	Operation time = 24 d DC Power supply = 12 V DC Exposure time = 15 min ON/45 min OFF SRT (no sludge waste) Initial MLSS = 3500 mg/L (SMBR), 3040 mg/L (EC-SMBR)	Removal Efficiencies: **EC-SMBR**: COD = 88.9%, NH_3-N = 77.8%, PO_4-P = 94.3%, **SMBR**: COD = 86%, NH_3-N = 97.4% PO_4-P = 05%	Bani-Melhem and Smith, 2012
Municipal wastewater	BOD = 70.1 mg/L, COD = 177.5 mg/L, TN = 25.1 mg/L, TP = 3.1 mg/L	MBR vs. EC-MBR	PVDF vs. PES Flat sheet membranes	Anode:Al Cathode: Al	Operation time = 5 months DC Power supply = 4.9–5.4 A (1.4–2.0 V) DC Exposure time = continuous HRT = 6.7–7 h SRT = 42–69 days MLSS = 6770–12,150 mg/L	Removal Efficiency: **MBR**: PO_4-P = 59.0 %, Total P = 41.0% **EC-MBR**: PO_4-P = 77.2 %, Total P = 79.9 %	Kim et al., 2010
Municipal wastewater	**SMEBR** COD = 160–700 mg/L PO_4-P = 2–11 mg/L NH_3-N = 19–70 mg/L **MBR** COD = 160–950 mg/L PO_4-P = 2–7 mg/L NH_3-N = 26–68 mg/L	SMEBR vs. MBR	Microza Hollow fiber MF membrane	Anode:Al Cathode: Al	Operation Time = 7 weeks CD = 12 A/m^2 DC Exposure time = Intermittent SRT = 10 d HRT = 11 h	Removal Efficiency: **SMEBR**: PO_4-P = 99%, NH_3-N = 99%, and COD = 92% **MBR**: PO_4-P = 59%, NH_3-N = 97%, and COD = 67%	Hasan et al., 2012
Municipal wastewater	COD = 316 ± 148 mg/L PO_4-P = 4.1 ± 1.6 mg/L NH_3-N = 43 ± 11 mg/L NO_3-N = 0.4 ± 0.1 mg/L TSS = 120 ± 10 mg/L T = 15.4 ± 2.5 °C	SMEBR	PVDF Hollow Fiber MF membrane	Anode: Al Cathode: SS	Operation time = 7 weeks CD = 12 A/m^2 DC Exposure time = 5 min ON/10 min OFF SRT = 10 d HRT = 11 h	Removal efficiencies: NH_3-N = 99%, PO_4-P = 99%, COD = 92% Specific energy consumption = 1.1–1.6 kW h/m^3	Hasan et al., 2014

Municipal wastewater	COD = 389–1228 mg/L NH_4-N = 37.4–154 mg/L NO_3-N = 0.29–0.86 mg/L TN = 69.7–179.37 mg/L PO_4-P = 2.31–8.80 mg/L TP = 2.90–12.60 mg/L Ni = 0.49–1.41 mg/L Fe = 0.29–3.74 mg/L Cr = 0.03–0.32 mg/L	SMEBR	Combination of acrylonitrile butadiene styrene (ABS), polypropylene, polyethylene terephthalate and chlorinated polyethylene MF membranes	Anode: Al Cathode: SS Electrode gap = 5 cm	Operation time = 60 days CD = 15 A/m^2 DC Exposure time = 5 min ON/15 min OFF HRT = 13.5 h SRT = 10 d DO = 1–2 mg/L (anoxic zone), 5–8 mg/L (filtration zone)	Removal efficiencies: COD = 99.9% TP = 99% TN = 91% Ni = 79% Fe = 89% Cr = 80%	Giwa and Hasan, 2015

Between the two configurations, SMEBR is increasingly becoming more popular and has been utilized in many studies (Bani-Melhem and Elektorowicz, 2010; Ibeid et al., 2013a; Hasan et al., 2014). Since it allows all the three fundamental operations (i.e., biodegradation, electrochemical processes, and membrane filtration) to take place in a single reactor vessel (smaller footprint) and it can be adapted as the most suitable design for large scale applications. Its compact characteristics and high treatment efficiency make SMEBR an economically attractive technology for wastewater treatment.

Table 2. Application of embr on synthetic wastewater.

Type of wastewater	Wastewater characteristics	Configurations	Membrane materials	Electrode materials	Operational conditions	Results	References
Synthetic wastewater (Composition (mg/L): Glucose = 240, Yeast extract = 160, $MnSO_4 \cdot H_2O$ = 4.58, $FeSO_4 \cdot 7H_2O$ = 3.40, KCl = 10.50, K_2HPO_4 = 60, $NaHCO_3$ = 550, $CaCl_2$ = 5.50, NH_4Cl = 125, $MgSO_4 \cdot 7H_2O$ = 75)	COD~550 mg/L NH_4-N~30 mg/L PO_4-P~12.5 mg/L	SMEBR	ZeeWeed-1 Hollow fiber UF membrane	Anode: Al Cathode: Al Electrode gap = 1.1 cm	Operation time = 21 days Current = 0.05 A Exposure time = 2 min ON/2 min OFF HRT = 13.8 h SRT = 21 d DO (aerobic zone) = 5.0–5.5 mg/L, DO (anoxic zone) = 0.1 mg/L MLSS = 10 g/L	Removal efficiencies: **Control MBR**: COD = 94.3%, TN = 76.7%, PO_4-P = 20.8%, NH_4-N = 100% **SMEBR**: COD = 94.3%, TN = 77% and PO_4-P = 86.6%, NH_4-N = 100%	Wei et al., 2009
Synthetic wastewater (Composition (mg/L): glucose = 310, peptone = 252, yeast extract = 300, $(NH_4)2SO_4$ = 200, KH_2PO_4 = 37, $MgSO_4 \cdot 7H_2O$ = 40, $MnSO_4 \cdot H_2O$ = 4.5, $FeCl_3 \cdot 6H_2O$ = 0.4, $CaCl_2 \cdot 2H_2O$ = 4, KCl = 25, and $NaHCO_3$ = 25)	COD = 334 ± (23) mg/L NH_3-N = 30.8 ± (2.1) mg/L NO_3-N = Not detected PO_4-P = 27.2 ± (1.9) mg/L	Stage 1: SMBR vs. Stage 2: SMEBR	ZeeWeed-1 Hollow fiber UF membrane	Anode: Fe Cathode: Fe Electrode gap = 5.5 cm	Operation time = 26 d (Stage I), 27 d (Stage II) Electric voltage = 1 V/cm (stage II) Exposure time: 15 min ON/45 min OFF (stage II) SRT = 268 d DO = 5–8 mg/L MLSS = 2100–5000 mg/L	COD removal efficiency: Stage I (**SMBR**) = 75–90 %, Stage II (**SMEBR**) = 85–95%	Bani-Melhem and Elektorowicz, 2010

Type of wastewater	Wastewater characteristics	Configurations	Membrane materials	Electrode materials	Operational conditions	Results	References
Synthetic wastewater (Composition (mg/L): glucose = 310, peptone = 252, yeast extract = 300, $(NH_4)_2SO_4$ = 200, KH_2PO_4 = 37, $MgSO_4 \cdot 7H_2O$ = 40, $MnSO_4 \cdot H_2O$ = 4.5, $FeCl_3 \cdot 6H_2O$ = 0.4, $CaCl_2 \cdot 2H_2O$ = 4, KCl = 25, and $NaHCO_3$ = 25)	COD = 334 ± (23) mg/L NH_3-N = 30.8 ± (2.1) mg/L NO_3-N = Not detected PO_4-P = 27.2 ± (1.9) mg/L	Stage 1: SMBR vs. Stage 2. SMEBR	ZeeWeed-1 Hollow fiber UF membrane	Anode: Fe Cathode: Fe Electrode gap = 5.5 cm	Operation time = 26 d (Stage I), 27 d (Stage II) Electric voltage = 1 V/cm (stage II) Exposure time: 15 min ON/45 min OFF (stage II) SRT = 288 d DO = 5-8 mg/L MLSS = 2100-5000 mg/L	COD removal efficiency: Stage I (**SMBR**) = 75-90 %, Stage II (**SMEBR**) = 85-96%	Bani-Melhem and Elektorowicz, 2010
Synthetic wastewater (Composition (mg/L): glucose = 310, peptone = 252, yeast extract = 300, $(NH_4)_2SO_4$ = 200, KH_2PO_4 = 37, $MgSO_4 \cdot 7H_2O$ = 40, $MnSO_4 \cdot H_2O$ = 4.5, $FeCl_3 \cdot 6H_2O$ = 0.4, $CaCl_2 \cdot 2H_2O$ = 4, KCl = 25, and $NaHCO_3$ = 25)	COD = 334 ± (23) mg/L NH_3-N = 30.8 ± (2.1) mg/L NO_3-N = Not detected PO_4-P = 27.2 ± (1.9) mg/L	Stage 1: SMBR vs. Stage 2. SMEBR	ZeeWeed-1 Hollow fiber UF membrane	Anode: Fe Cathode: Fe Electrode gap = 5.5 cm	Operation time = 26 d (Stage I), 27 d (Stage II) Electric Field = 1 V/cm (stage II) DC Exposure time: 15 min ON/45 min OFF (stage II) SRT = 288 d DO = 5-8 mg/L MLSS = 2100-5000 mg/L	Removal Efficiency: Stage I (**SMBR**): COD = 75-90 %, NH_3-N = 97 %, PO_4-P = 75-96 %, stage II (**SMEBR**): COD = 85-96%, NH_3-N = 70%, PO_4-P ~ 98%	Bani-Melhem and Elektorowicz, 2011
Synthetic wastewater (Composition: Mixture of organic compounds, phosphorus, nitrate and soluble salts)	COD = 700–1000 mg/L NH_3-N = 16–19 mg/L PO_4-P = 15-22 mg/L	SMEBR	ZeeWeed-1 Hollow fiber UF membrane V = 22.5 L	Anode: Al Cathode: SS	Operation time = 27 weeks CD < 20 A/m^2 DC Exposure Mode = 5 min ON/20 min OFF SRT = 15 d HRT = 12 h MLSS = 2482 mg/L DO = 3-8 mg/L	Removal Efficiency: NH_3-N = 99.9%, PO_4-P = 99.7%, and COD = 97.8%	Elektorowicz et al., 2014
Synthetic wastewater (Composition (mg/L): $C_6H_{12}O_6$ = 175, starch = 139, peptone = 19, urea = 32, NH_4Cl = 153, K_2HPO_4 = 22, $MgSO_4 \cdot 7H_2O$ = 66, $CaCl_2 \cdot 2H_2O$ = 23, $NaHCO_3$ = 111, $MnSO_4$ = 2.6 and $FeSO_4$ = 0.4)	COD = 335 ± 33.1 mg/L NH_4-N = 28 ± 1.2 mg/L TP = 4.91 ± 0.4 mg/L	SMEBR vs SMBR	Ceramic membrane Flat Sheet	Anode: SS (Fe-MBR) vs. Ti (Ti-MBR) Cathode: Ti	Operation time = 450 h Current = 0.1 A DC Exposure Mode = 2 min ON/10 min OFF SRT = 15 d HRT = 12 h MLSS = 2482 mg/L DO = 2 mg/L	Removal efficiencies: **SMBR**: COD = 94.0 ± 0.8%, TP = 40.2 ± 11.6%, NH_4-N = 88.9 ± 1.3% **Fe-MBR:** COD = 95.1 ± 1.3%, TP = 91.7 ± 3.4%, NH_4-N = 81.9 ± 1.1% **Ti-MBR:** COD = 93.7 ± 0.1%, TP = 42.4 ± 13.4%, NH_4-N = 52.4 ± 3.1%	Zhang et al., 2015
Synthetic wastewater (Composition (mg/L): molasses = 600, $(NH_4)_2SO_4$ = 10, CH_4N_2O = 70, KH_2PO_4 = 35, $MgSO_4 7H_2O$ = 10, $MnSO_4 H_2O$ = 5, KCl = 20, and $NaHCO_3$ = 20)	COD = 350 ± (27) mg/L NH_3-N = 30 ± (4.1) mg/L NO_3-N = Not detected PO_4-P = 20 ± (3.5) mg/L	SMEBR	PVDF Hollow fiber	Anode: SS Cathode: SS	Operation time = 14 d CD = 5, 12.5, 20, 23 A/m^2 DC Exposure time = 480 s OFF/120 s ON, 200 s OFF/400 s ON, 496 s OFF/104 s ON, 415 s OFF/185 s OFF, Continuously ON HRT = 9 h SRT = 60 d MLSS = 6000-8000 mg/L	Optimum Parameters: CD = 12.5 A/m^2, exposure mode = 415 s OFF–185 s ON, Removal efficiencies: COD~ 95%, PO_4-P~85%, NH_3-N >80%	Tafti et al., 2015

4. MATERIALS

Membrane and electrode materials are equally important in designing eMBR as both can significantly affect the reactor's performance, pollutant removal, and fouling control. The membranes are responsible for the retention of microbes in the reactor which improves biological treatment and for further purification of the resulting effluent via filtration (Giwa et al., 2016). Among the numerous papers available in literature for eMBR, only few studies have been made comparing the different performances of membrane materials and these studies only focused on the effects of the membrane's pore size on fouling.Kim et al. (2010) used two plate type membranes: PVDF (mean pore size: 0.08 μm) and polyethersulfone (PES) (mean pore size: 0.2 μm) in an EC-SMBR system and observed that although both membranes maintained a stable transmembrane pressure (TMP) for a period of 5 months without chemical cleaning, PVDF has better performance, as shown in its lower average TMP (PVDF = 10 kPa vs. PES = 16 kPa) which was attributed to its smaller pore size. Apparently, membranes with large pore sizes are more susceptible to fouling as more smaller particles can migrate, attach and accumulate into the interior surfaces of the pores resulting in pore narrowing and pore blocking and eventually to irreversible fouling (Kim et al., 2010). The same conclusion was reached by Akamatsu et al. (2010) when they used MF-1 (0.1 μm) and MF-2 membranes (0.22 μm) in eMBR, both membranes were made up of a mixture of cellulose acetate and cellulose nitrate.

The high cost of membrane materials limits eMBR in large scale application. This brought the development of low cost alternatives such as non-wovens, meshes, and filter cloths (Liu et al., 2012a). Though cost efficient, these alternative filters have relatively large pore size and inadequate fouling control. Such constraint has been addressed by a group of researchers modifying low cost filters to improve its effectivity.Liu et al. (2012b) utilized polypyrrole (PPy) modified polyester filter cloth as both the cathode and filtration material in eMBR. Membrane modification was done by liquid and vapor phase polymerization using different concentrations of $FeCl_3{\cdot}6H_2O$ (10, 20, and 40%). Results showed that vapor phased PPy modification with 20% iron oxidants has substantially increased the membrane's electrical conductivity (lower specific resistances) and a DC field as low as 0.2 V/cm is sufficient enough to drive electrostatic repulsion between negatively charged foulants and membrane. This development also paved the way of addressing the issue on the effects of direct application of electric field on bacterial activity which is discussed in Section Microbial Activity. In another study, vapor modified PPy membranes (VPM) were coated by graphene (Gr/PPy) and graphene oxides (GO/PPy) to further enhance their conductivity (Liu et al., 2013b). Graphene and its oxides have unique electronic properties which can be enhanced by coating with polypyrrole (PPy). The electric resistance of GO/PPy modified membrane is 0.71 kΩ/cm while for Gr/PPy is 0.68 kΩ/cm. These values are significantly lower compared to 2.03 kΩ/cm for PPy modified filter cloth (Liu et al., 2013b). In terms of anti-fouling property, membrane with higher conductivity (i.e., lower electric resistance) shows improvement in permeate flux.Liu et al. (2013b) tested the antifouling performance of Gr/PPy and compared it with PPy modified filter cloth under 1.0 V/cm electric field. VPM modified Gr/PPy membrane can suppress the fouling more effectively than PPy by increasing the permeate flux and cumulative permeate volume by 20% (Table 3). PPy modified filter only increases 10% of cumulative permeate volume (Table 3).

Table 3. Influence of the applied electric field on membrane fouling.

Applied electric field	Exposure mode	Membrane material	Effects on membrane fouling	References
0–20 V/cm	Continuous ON	Polypropylene Hollow fibrous membrane assembly (inner diameter = 0.32 mm)	↑Electric field = ↑ Flux, ↑ electrophoresis rate, ↑electroosmosis Optimum electric field = 20 V/cm	Chen et al., 2007
1 V/cm	Continuous ON	VPM Gr/Ppy modified polyester VPM Ppy modified polyester	**VPM Gr/Ppy modified polyester:** ↑ permeate flux by ~20% ↑ cumulative permeate volume by ~20% **VPM Ppy modified polyester:** ↑ cumulative permeate volume by ~10%	Liu et al., 2013b
0, 4, and 6 V/cm	Continuous ON, Intermittent: 90 s ON/90 s OFF, 60 s ON/90 s OFF, 30 s ON/90 s OFF	Mixture of Cellulose acetate and cellulose nitrate (MF-1 pore size = 0.1 μm, MF-2 pore size = 0.22 μm)	**Continuous:** ↑ electric field = ↑ Flux, ↑ electrostatic repulsion **Intermittent:** permeate flux always recovers when 6 V/cm is applied for 90 s due to electrophoretic force (optimum application = 6 V/cm, 90 s ON/90 s OFF)	Akamatsu et al., 2010
5, 12.5, 20, and 23 A/m^2	Continuous ON, Intermittent: 200 s ON/400 s OFF, 415 s ON/185 s OFF, 480 s ON/120 s OFF, 496 s ON/104 s OFF	Hollow fiber PVDF membrane (pore size = 0.1 μm)	↑Current Density = ↓ sludge conductivity, ↑ settleability, ↓ SVI, ↓ SMP, ↑Flocs size, ↑ electrophoresis,	Tafti et al., 2015
0, 15 A/m^2	Intermittent: 5 min ON/20 min OFF	Ultrafiltration Zeeweed-1 (pore size = 0.04 μm)	**SRT = 20 d** ↓ SRF (MBR = 24 × 1012 m/kg, SEMBR = 3.2 × 1012 m/kg) ↓SMP (MBR = 75–135 mg/L, SEMBR = 40–60 mg/L) ↓bound water (MBR = 1.2 gH_2O/g TS, SEMBR = 1.2 gH_2O/g TS) ↓VSS/MLSS (MBR = 0.85, SEMBR = 0.52) ↓PSD (MBR = 200 μm, SEMBR = 400 μm) ↓Zeta potential (MBR = −30 to −35 mV, SEMBR = −20 to −25 mV) **SRT = 150 d** ↓ SRF (MBR = 688 × 1012 m/kg, SEMBR = 3.7 × 1012 m/kg) ↓SMP (MBR = 130–195 mg/L, SEMBR = 75–85 mg/L) ↓bound water(MBR = 20 gH_2O/g TS, SEMBR = 6.8 gH_2O/g TS) ↓VSS/MLSS (MBR = 0.85, SEMBR = 0.59) ↓PSD (MBR = 200 μm, SEMBR = 400 μm) ↓Zeta potential (MBR = −30 to −35 mV, SEMBR = −20 to −25 mV)	Ibeid et al., 2015
15 A/m^2	Intermittent: 5 min ON/20 min OFF	Hollow fiber PVDF membrane (pore size = 0.1 μm)	↑ TMP from 2 to 8 kPa ↓ Fouling rate by 3×: SEMBR = 0.13 kPa/d vs. MBR = 0.37 kPa/d	Ibeid et al., 2013a
30 and 50 V/cm	Intermittent: 4 min ON/4 min OFF	PVDF Membrane-Carbon Cloth assembly (pore size = 0.1 μm)	The flux always recover completely to the initial flux level when 50 V/cm is applied. For 30 V/cm, recovery is not complete. ↑Electric field = ↑repulsive force acting on the foulant substances at the membrane surface	Akamatsu et al., 2012

0.036 and 0.073 V/cm	Intermittent (5 min ON/1 min OFF)	Non-woven PP membrane (pore size = 5 μm) Terylene Filter cloth (pore size = 22 μm)	**Non-woven PP membrane (0.036 V/cm):** ↓ Irreversible Resistance by 20x ↓ flocs size ↓ EPS concentration by 6–7 mg /gMLVSS **Terylene Filter cloth (0.073 V/cm):** 2x voltage drop is required due to larger pore size and thickness	Liu et al., 2012a
0.2 V/cm	Intermittent: 5 min ON/1 min OFF	Vapor Phased Polymerized (VPM) Ppy membrane	↑ Flux by 2x ↑ Effluent volume by 30% after 4 h of filtration	Liu et al., 2012b
0.2 V/cm	Intermittent: 5 min ON/1 min OFF	Ppy/SDBS modified polyester	↓ Irreversible resistances by 5x ↓ Total resistances by 2x ↓ Deposited sludge in cake layer by 22%	Liu et al., 2013c

↑, *increase of variable value;* ↓, *decrease of variable value; x, times.*

Above aforementioned studies have proven that using electrically conductive membranes as the cathode itself was possible and very effective in fouling suppression. Some low-cost membranes used in eMBR and their performances with applied electric field are included in Table 3.

Utilizing better configuration of electrode and membrane materials helps minimize electric energy consumption in fouling reduction.Akamatsu et al. (2012) fabricated a low cost membrane-carbon cloth assembly made of PVDF flat-sheet microfiltration (MF) and plain-woven carbon cloth as the membrane and electrodes, respectively. The study successfully demonstrated that upon application of electricity, either continuously or intermittently, the membrane flux always recovers to the initial flux level (Table 3).Liu et al. (2012a) placed copper wires inside the flat sheet membrane module and two stainless mesh anodes outside the module. This configuration ensures constant electrophoretic forces against the deposition of foulant on the membrane (Table 3). This suggests that inexpensive electrode materials can also be used in eMBR instead of high cost membranes (e.g., PVDF, PES, titanium) which can be a valuable consideration for full scale use.

Anode material is one of the most important parameters in studying eMBR since it significantly affects the current density which is one of the typical performance indicator in any electrochemical system. The two most commonly used anodes in eMBR are aluminum (Al) and iron (Fe) electrodes. Both of them are proven to be useful for pollution removal (Wei et al., 2009) as well as for membrane fouling reduction (Zhang J. et al., 2014). Most importantly, they also serve as micronutrient necessary for microbial growth (Bani-Melhem and Elektorowicz, 2010). However, higher concentrations of dissolved anode materials may create inhibitory effects on bacterial metabolism, hence, electrode impact should be tested according to the applied DC field prior to the eMBR design. There are other anode materials used in eMBR study but this paper focused only on these two sacrificial anodes as they have more benefits than other tested materials (Zhang et al., 2015). The mechanisms of pollutant removal and fouling suppression using these aluminum and iron electrodes are discussed in the succeeding Sections Pollutants Removal and Membrane Fouling Reduction. Aluminum electrodes are preferred compared to iron electrodes due to the higher surface area of aluminum hydroxides, generated from electrocoagulation process, which increases the adsorption capacity (Emamjomeh and Sivakumar, 2009) of soluble compounds and the entrapment of particulate materials.

5. OPERATING CONDITIONS

Operating parameters such as aeration intensity, hydraulic retention time (HRT), sludge retention time (SRT), mixed liquor suspended solids (MLSS), current density (CD), and direct current (DC) exposure time play a vital role in determining both treatment performance and membrane fouling in eMBR. In conventional SMBR, surplus aeration is applied to aid in the scouring of flocs that have adhered on the membrane surfaces for fouling control (Rahimi et al., 2011). In eMBR, excess amount of oxygen is not needed since deposition of foulants onto the membrane surface is controlled by the applied DC field. Studies have shown that the running costs of eMBRs are much lower than MBRs using excess aeration for fouling control (Akamatsu et al., 2010). However, an optimum aeration condition in eMBR is not yet fully studied as it varies accordingly on the design and configuration of the bioreactor. In EC-SMBR, for example, 4.2 L/min of compressed air was considered adequate to supply oxygen for microbial incubation, to provide good mixing of the sludge suspension and to create a shear stress for effective scouring of membrane surfaces (Bani-Melhem and Smith, 2012).Zhang et al. (2015), on the other hand, applied an aeration intensity equal to 1 m^3/h to maintain 2 ppm dissolved oxygen concentration in the SMEBR. Large aeration intensity, greater than 800 L/h, was observed to be harmful for microbial growth and may cause break down of floc's formation which promotes the release of colloids and solutes resulting in severe pore blocking (Meng et al., 2008).

The effects of HRT and sludge properties (SRT and MLSS) in wastewater treatment using eMBR were explored by Giwa and Hasan (2015). In their experiments, reduction of effluent characteristics [COD, total nitrogen (TN), and total phosphorus (TP)] were realized at increasing HRT due to more exposure time of reactor content to biodegradation and electrocoagulation. HRT was varied by increasing the influent flow rate from 20 to 50 L/d. COD and TP concentrations in the effluent were lower at shorter SRT due to the decline in the MLSS concentration, while high TN removal was observed when SRT was increased from 5 to 20 days due to the provision of ample time for the growth of nitrifying bacteria responsible for the biological nitrification processes (Giwa and Hasan, 2015). The values of HRT in some studies range from 6 to 14 h (Wei et al., 2009; Kim et al., 2010; Elektorowicz et al., 2014; Hasan et al., 2014) while, regarding SRT, it can last up to 268 days (Bani-Melhem and Elektorowicz, 2011) (Tables 1, 2). Longer SRT is one of the many advantages of eMBR over CAS process. Sludge wasting can be minimized, hence, eMBR has lower sludge production.

Meanwhile, the effect of MLSS on fouling control varies in some studies.Giwa et al. (2015) proved that increasing MLSS (1455–4111 mg/L) causes more severe fouling due to the increase in sludge viscosity leading to higher hydrodynamic force required to propel sludge circulation. Bani-Melhem and Elektorowicz (2011), on the other hand, showed that MLSS concentration within the range of 2100–5000 mg/L has less effect on membrane fouling. To date, there is no study yet on the optimum range of MLSS in eMBR that would give excellent pollutant removal and fouling minimization.

Current density (CD = current [A]/anode electrode area [m^2]) is a crucial factor which affects the performance of the electrochemical treatment in eMBR. It controls the dosing rate of metal ions as well as the gas bubble density that are released in the solution via redox reactions of electrodes (Ibeid et al., 2013b). The higher the current density, the more coagulants are formed in a shorter period of time and the stronger electrokinetic forces are induced between foulants, bulk liquid, and membrane, hence, improving pollutant removal and fouling abatement. Excessive CD also results in waste of energy in the form of heat dissipation inside the reactor (Zhang et al., 2015). At the same time, DC exposure time is also important since both the intensity and application time can significantly influence bacterial community in the eMBR system (Giwa and Hasan, 2015). Optimization of these parameters is, therefore, a must.Tafti et al. (2015) studied the impact of CD ranging from 5 to 23 A/m^2 in combination with different DC exposure modes as reported in Table 2 (480 s OFF/120 s ON, 200 s OFF/400 s ON, 496 s OFF/104 s ON, 415 s OFF/185 s OFF and continuously ON) on activated sludge properties and pollutant removal efficiencies using synthetic municipal wastewater. Results showed that at higher CD (>20 A/m^2), eMBR exhibits higher COD, phosphate and ammonia removals as compared to conventional SMBR and that the microbial flocs were able to tolerate the electrical shocks at these values. The optimum conditions observed by Tafti et al. (2015) were CD = 12.5 A/m^2 and exposure mode equal to 415 s OFF/185 s ON. Other papers employed CD intensities ranging from 10 to 12 A/m^2 in their eMBR studies and obtained similar results (Hasan et al., 2012, 2014; Hosseinzadeh et al., 2015) (Table 1). Moreover, the amount of DC exposure time used in eMBR primarily depends on the current density and sludge properties (i.e., MLSS). At CD equals to 25 A/m^2 and MLSS below 10,000 mg/L, shorter time-OFF was needed in each electrical cycle while at higher MLSS values, longer time-OFF is required (Ibeid et al., 2013b). The effects of electricity on pollutant removal, fouling control and microbial activity are further discussed in Section Effects of the Electric Field on eMBR Performance.

As mentioned, all of these parameters influence eMBR performance by directly affecting the sludge properties. Therefore, it is important to note that understanding the relationship between these parameters and identifying their appropriate combination, which will depend on the type of wastewater (low, medium, high-strength wastewater) being treated and configuration used, result in an enhancement of eMBR performance.

6. APPLICATION WITH DIFFERENT INFLUENT WASTEWATER

Due to numerous advantages that the technology provides, eMBR has been used in the treatment of different types of influent wastewater. It has been employed to remove organic substances (COD) and nutrients (nitrogen and phosphorous) as well as color from combined industrial wastewater (Hosseinzadeh et al., 2015), tannery wastewater (Keerthi et al., 2013), and municipal or domestic wastewater (Kim et al., 2010; Bani-Melhem and Smith, 2012; Hasan et al., 2012, 2014). Tables 1, 2 provide a comprehensive summary of the applications of electrically induced membrane bioreactor on real and synthetic wastewater, respectively.

Most investigations of this hybrid reactor were still limited to lab scale reactors fed with synthetic or low-to-medium strength wastewater and only few focuseson pilot scale application. The challenge, therefore, centers on scale-up of this system and its application to wastewater with high pollutant loading (e.g., landfill leachate, food processing wastewater, etc.). Moreover, eMBR could also provide a suitable environment for the biodegradation of emerging contaminants (i.e., endocrine disrupting substances, pharmaceutical products, flame retardants, and personal care products) from water and wastewater streams due to its high solid retention and use of diverse microbial culture.

7. EFFECTS OF THE ELECTRIC FIELD ON EMBR PERFORMANCE

7.1 Pollutants Removal

Previous studies have reported high satisfactory results in eMBR performance. In fact, removal efficiencies of organic and inorganic compounds in wastewater using eMBR exceeded the performance of conventional MBR systems (Wei et al., 2009; Bani-Melhem and Elektorowicz, 2010, 2011; Kim et al., 2010; Bani-Melhem and Smith, 2012; Hasan et al., 2012; Zhang et al., 2015). Both MBR and eMBR systems exhibit excellent removal of physical impurities i.e., color (>91%), turbidity (>95%), and suspended solids (100%) due to the effective solid retention by the membrane module (Bani-Melhem and Smith, 2012). However, in terms of COD, eMBR gives higher removal efficiency (85–98%) compared to MBR (72–95%) (Tables 1, 2) which is attributed to the effects of electrochemical processes (i.e., electrocoagulation, electroosmosis, electrophoresis) in addition to biodegradation and membrane filtration (Hosseinzadeh et al., 2015). During electrocoagulation (Figure 1C), various monomeric and polymeric metal complexes are generated from the electrochemical dissolution (electro-oxidation) of sacrificial metal electrodes (aluminum or iron) (Wei et al., 2009).

In the case of aluminum anode, Al^{3+} ions react with OH^-ions to form amorphous $Al(OH)_{3(s)}$ while for iron, gelatinous iron hydroxide $Fe(OH)_{2(s)}$ and/or $Fe(OH)_{3(s)}$ is formed in the aqueous stream due to the reaction between Fe^{2+} and Fe^{3+} with OH^- (Equations 1–8). These metal complexes react with contaminants in the wastewater to form flocs which can both destabilize and aggregate colloidal particles or precipitate and adsorb dissolved compounds (Elabbas et al., 2016). Oxidation and reduction of some pollutants may also be possible which lead to its deposition at the anode and cathode, respectively (Giwa et al., 2016).

For Aluminum:

$$\text{At the anode: } Al(s) \rightarrow Al3^{+} + 3e- \quad (1)$$

$$\text{At the cathode: } 2H_2O + 2e^{-} \rightarrow H_2(g) + 2OH\text{-} \quad (2)$$

$$\text{In the solution: } Al^{3+} + 3OH- \rightarrow Al(OH)_{3(s)} \quad (3)$$

For Iron:

$$\text{At the anode: } Fe_{(s)} \rightarrow Fe^{2+} + 2e^{-} \quad (4)$$

$$\text{At the cathode: } 2H_2O + 2e^{-} \rightarrow H_{2(g)} + 2OH^{-} \quad (5)$$

$$\text{In the solution: } Fe^{2+} + 2OH^{-} \rightarrow Fe(OH)_{2(s)} \quad (6)$$

Likewise, eMBR performs better in phosphorus removal than SMBR. EC-SMBR can remove as high as 94.3% PO3−4PO43−P (compared to 65% in SMBR) (Bani-Melhem and Smith, 2012) while SMEBR can remove up to 99% (MBR = 59% removal) (Hasan et al., 2012) from both raw (Table 1) and synthetic wastewaters (Table 2). This is because generated Al and Fe coagulants facilitate adsorption of soluble phosphorus in the bioreactor (Kim et al., 2010; Bani-Melhem and Smith, 2012). Biodegradation alone does not adequately eliminate PO3−4PO43−P from aqueous solution hence the low removal efficiency in conventional MBR (as low as 20.8%) (Wei et al., 2009). Using aluminum anode, cationic polymers such as Al^{3+} and $[Al_6(OH_{15})]^{3+}$ react with phosphate ions to precipitate $AlPO_{4(s)}$ and $Al_6(OH_{15})PO_{4(s)}$, respectively, according to Equations (7), (8) (Bani-Melhem and Smith, 2012). In the case of sacrificial Fe anode, the iron hydroxides that are formed adsorb the soluble phosphorus and the excess Fe^{3+} combines with phosphorus ions to form $FePO_{4(s)}$ precipitates (Equation 9) (Bani-Melhem and Elektorowicz, 2011).

$$Al^{3+} + PO_4^{3-} \rightarrow AlPO_{4(s)} \quad (7)$$

$$Al_6(OH)_{15}^{3+} + PO_4^{3-} \rightarrow [Al_6(OH)_{15}]PO_{4(s)} \quad (8)$$

$$Fe^{3+} + PO_4^{3-} \rightarrow FePO_{4(s)} \quad (9)$$

Depending on their net surface charge, the pollutants can also migrate and deposit on oppositely charged electrodes via electrophoretic motion upon DC application (Giwa et al., 2015). For example, the deposition of anions on the anode as represented by Equation (10) where M^{+} and A^{-} correspond to metal ions and anion, respectively. In addition, electroosmosis mechanism (Figure 1B) also occurs within the system which enhances dewatering of the smaller floc particles and decreasing specific resistance to filtration (Ibeid et al., 2013a). Heavier flocs that are not adsorbed on the electrode surface settled at the bottom of the reactor as sludge due to gravity and to the gas produced at the electrodes (Giwa et al., 2015).

$$M^{+} + A^{-} \rightarrow MA + (MA)^{+} \text{ or } (MA)^{-} \quad (10)$$

Treatment efficiencies of ammonia nitrogen (NH_3-N) were consistently excellent in conventional MBR (~89% and up) but vary in electro-enhanced MBR (70–100%) (Tables 1, 2). There are two factors being considered for such results: (1) the application of high DC field and (2) excessive metal ion concentration (Bani-Melhem and Elektorowicz, 2011; Tafti et al., 2015). Nitrifying bacteria, which are responsible for the conversion of ammonia nitrogen (NH_3-N) to nitrate nitrogen (NO_3-N), are very much sensitive to high voltage. Direct exposure to applied DC >2.5 A/m^2 was proved to have inhibitory effects in the metabolism of nitrifying bacteria (Li et al., 2001). Furthermore, accumulation of metal ion complexes as a result of electro-dissolution of sacrificial anodes is also detrimental to nitrifying bacteria. Iron complexes, in particular, form a barrier that hinders the transfer of enzymes and nutrients through the microbial cell membrane (Bani-Melhem and Elektorowicz, 2011). This is the reason why ammonia removal using iron and stainless steel anodes are relatively lower (70 and 81.9%, respectively) (Bani-Melhem and Elektorowicz, 2011; Zhang et al., 2015) as compared when using aluminum anode (98–100%) (Wei et al., 2009; Hasan et al., 2012; Giwa et al., 2016). Optimization of electrolysis conditions should, therefore, be done so as not to impede biological treatment.

8. MEMBRANE FOULING REDUCTION

Investigation on the influence of appending electricity on fouling control in eMBR was pioneered by Chen et al. (2007). In this study, six different DC field strengths ranging from 0 to 20 V/cm were applied and the results showed that better filtrate flux recovery was obtained at higher DC intensity due to the increase in particle electrophoresis. In SMEBR, the movement of the suspended particles and the water during membrane filtration can be controlled by the applied electrical field. Major foulants in wastewater usually possess negative surface charge and upon DC application, electrophoresis force counteracts permeation-induced deposition of these foulants away from the membrane surface and toward the electrode of opposite charge (anode) (Figure 1A) (Wei et al., 2009). On the other hand, the bulk liquid which is commonly positively charged is drifted toward the cathode then flows to the membrane facilitating the filtration, a process called electroosmosis (Figure 1B). Electroosmosis, thus, removes bound water from the microbial flocs enhancing sludge dewaterability and decreasing the specific resistance to filtration (Ibeid et al., 2013a). Specific cake resistance to filtration and modified fouling index were observed to decrease by 82 and 81%, respectively, when 12 A/m^2 current density was applied to eMBR in the treatment of raw municipal wastewater (Hasan et al., 2014).

Application of electric field for membrane fouling suppression does not only control the movement of charged particles and bulk liquid, it also modifies the sludge properties through the electrocoagulation process (Figure 1C) (Hasan et al., 2014). The use of sacrificial anodes releases metal ions which precipitate or adsorb negatively charged foulants [i.e., extracellular polymeric substances (EPS), soluble microbial products (SMP)] and facilitates formation of larger flocs which drives the particle back-transport from the membrane surface to the bulk solution (Ibeid et al., 2013a). In a study conducted by Zhang et al. (2015), two types of anode materials were used: stainless steel (Fe-MBR) and titanium (Ti-MBR). Between the two, Fe-MBR showed better fouling mitigation performance than the Ti-MBR due to the combined action of electric field force induced backwash (electrophoresis) and the formation of iron coagulants which reduced the SMP concentration (electrocoagulation). Titanium mesh anodes used in Ti-MBR were inactive metal hence it did not release any coagulant ions and fouling control relied mostly on the magnitude of electric field applied (Zhang et al., 2015). Moreover,Kim et al. (2010) proved that electrocoagulation does not only effectively removed phosphorus by precipitating $AlPO_{4(s)}$ crystals but these precipitates are also capable of adsorbing EPS.Hosseinzadeh et al. (2015) attributed the lower tansmembrane pressure (TMP) and sludge volume index (SVI) to electrocoagulation in SMEBR. In EC-SMBR, membrane fouling was reduced by 13% due to the electrocoagulation pre-treatment (Bani-Melhem and Smith, 2012).

Nevertheless, it should be noted that energy consumption and operational cost will obviously be affected by high electricity requirements and, therefore, must be reduced. Some researchers studied the use of intermittent application of minute electric field for effective fouling control and possible energy savings. Table 3 summarizes the different studies that investigates effects of continuous and intermittent application of electric field on membrane fouling.Akamatsu et al. (2010) developed a novel fouling suppression system in MBRs which controls the motion of activated sludge by applying electric current only when the permeate flux has drastically declined due to membrane fouling. By switching on and off every 90 s, electric field strength of 6 V/cm facilitates detachment of foulant activated sludge from the surface of the membrane thus significantly improved the average permeate flux (Figure 1A). Upon cost estimation, about half of the energy requirements were saved using this fouling suppression system. In another experiment, Akamatsu et al. (2012) investigated the performance of MBR on fouling suppression by applying 50 and 33 V/cm electric field at a 4 min ON/4 min OFF cycle directly on the membrane cathode made of carbon cloth. The results showed that the flux always recovered to the initial flux level when the electric field was ON and at voltage gradient applied equal to 50 V/cm. This proved that higher electric field intensity means stronger repulsive force between the foulants and the membrane surfaces enhancing the fouling control (Akamatsu et al., 2012). A minute electric field, 0.2 V (0.036 V/cm) and 0.4 V (0.073 V/cm) were applied by Liu et al. (2012a) using a low cost, micro-porous PP non-woven sheet and a terylene filter cloth. For both membranes, application of either 0.2 or 0.4 V electric field causes increase in permeate flux, enhancement of microbial growth and activity, decrease of filtration resistance and reduction of sludge EPS content which all contributed to fouling control.

9. MICROBIAL ACTIVITY

Like any other biological wastewater treatment processes, eMBR relies mostly on the mixed microorganisms present in the system. The ability of bacterial community to effectively degrade organic matter (COD) and nutrients (N and P) from raw wastewater makes biological process the most widely used treatment method. Therefore, integrating electrochemical technologies with biological treatment should take into consideration the effects of electricity on bacterial viability.

Microbial analyses have revealed that a stressor such as strong electric current provides negative impact on the microbial properties (i.e., metabolism, physiology, shape, and mobility) (Wei V. et al., 2011). Increasing intensity of electric field (DC > 2.5 A/m^2) decreases nitrification rate by 20% (Li et al., 2001). As a result, low ammonia removal is observed due to the inhibition of nitrobacteria and denitrifying bacteria by large voltage (Bani-Melhem and Elektorowicz, 2011; Bani-Melhem and Smith, 2012; Hasan et al., 2012). Moreover, at higher voltage, breakage of bacterial cells and production of SMP are observed to occur which increases fouling (Zhang et al., 2015). Wei V. et al. (2011) proved that increasing DC intensity (6.2, 12.3, and 24.7 A/m^2) also increases pH (from nearly neutral to around pH 10) of electrolytic biomass fluid causing 10, 15, and 29% death percentage, respectively.

It should be noted that beneficial effects can also be achieved in applying minute electric field in eMBR system. Low voltage (0.036–0.073 V/cm) promotes the growth of filamentous bacteria which are proven to improve sludge filterability (Liu et al., 2012a). Intermittent application prevents prolonged direct contact between electric field and microbial community hence minimizing the negative effects of high DC field. To date, there are no studies yet on the optimal electric field suitable for enhancing bacterial activity in an eMBR application.

10. CONCLUSIONS

Membrane fouling is a major limitation in MBR operation that should be resolved. Among the different fouling control techniques, integration of electrochemical processes into MBR seems to be a promising method. Electrochemical phenomena such as electrocoagulation, electroosmosis and electrophoresis are the main mechanisms of membrane fouling control in eMBR. It also enhances pollutant removal. However, direct application of electric field to MBRs incurs high energy consumption and has negative effects on bacterial viability which is responsible for the biodegradation of the organic pollutants. Optimum configuration of electric field helps to reduce energy consumption while simultaneously reducing inadvertent interferences with eMBR performance. This can be done by using better configuration of electrode and membrane materials and intermittent application of low or moderate electric field.

The combination of electrochemical processes with MBR processes is still in its infancy stage, with a small number of scientific papers published during the last decade. This review focused only on those publications with international significance. The studies on eMBR are limited to bench or pilot scale reactor fed with real or synthetic wastewater. Large scale applications are not reported. Optimization study, modeling and cost benefit analysis between MBR and eMBR are, therefore, useful for scaling up the system. Moreover, treatment of high strength industrial wastewater and the removal of emerging contaminants through eMBRs have not been yet highly explored. There are numerous already published papers regarding the effective removal of these contaminants by electrochemical processes or MBR, but not by the combined process. Moreover, the potential of this combined system as an efficient technology for biohydrogen production, due to hydrogen ions reduction at the cathode side, has not been yet investigated. Therefore, further studied could be focused on the combination of MECs with MBR through the operation of the electro MBR at anaerobic conditions for the production of renewable energy.

ACKNOWLEDGMENTS

The authors gratefully thank the Department of Science and Technology-Engineering Research Development and Technology (ERDT-DOST), the University of the Philippines—Diliman, and the Sanitary Environmental Engineering Division (SEED) of the University of Salerno in Italy for the Ph.D. Sandwich program and research funds. Research activities were partially funded by the FARB project of the University of Salerno.

REFERENCES

1. Ahmed, F. N., and Lan, C. Q. (2012). Treatment of landfill leachate using membrane bioreactors: a review. *Desalination* 287, 41–54. doi: 10.1016/j.desal.2011.12.012
2. Akamatsu, K., Lu, W., Sugawara, T., and Nakao, S. (2010). Development of a novel fouling suppression system in membrane bioreactors using an intermittent electric field. *Water Res.* 44, 825–830. doi: 10.1016/j.watres.2009.10.026
3. Akamatsu, K., Yoshida, Y., Suzaki, T., Sakai, Y., Nagamoto, H., and Nakao, S. (2012). Development of a membrane – carbon cloth assembly for submerged membrane bioreactors to apply an intermittent electric field for fouling suppression. *Sep. Purif. Technol.* 88, 202–207. doi: 10.1016/j.seppur.2011.12.031
4. Artiga, P., Ficara, E., Malpei, F., Garrido, J. M., and Méndez, R. (2005). Treatment of two industrial wastewaters in a submerged membrane bioreactor. *Desalination* 179, 161–169. doi: 10.1016/j.desal.2004.11.064
5. Ballesteros, F. Jr., Vuong, T. H., Secondes, M. F., and Tuan, P. D. (2016). Removal efficiencies of constructed wetland and efficacy of plant on treating benzene. *Sustain. Environ. Res.* 26, 93–96. doi: 10.1016/j.serj.2015.10.002
6. Bani-Melhem, K., and Elektorowicz, M. (2010). Development of a novel submerged membrane electro-bioreactor (SMEBR): performance for fouling reduction. *Environ. Sci. Technol.* 44, 3298–3304. doi: 10.1021/es902145g
7. Bani-Melhem, K., and Elektorowicz, M. (2011). Performance of the submerged membrane electro-bioreactor (SMEBR) with iron electrodes for wastewater treatment and fouling reduction. *J. Membr. Sci.* 379, 434–439. doi: 10.1016/j.memsci.2011.06.017
8. Bani-Melhem, K., and Smith, E. (2012). Grey water treatment by a continuous process of an electrocoagulation unit and a submerged membrane bioreactor system. *Chem. Eng. J.* 198–199, 201–210. doi: 10.1016/j.cej.2012.05.065
9. Bernhard, M., Müller, J., and Knepper, T. P. (2006). Biodegradation of persistent polar pollutants in wastewater: comparison of an optimised lab-scale membrane bioreactor and activated sludge treatment. *Water Res.* 40, 3419–3428. doi: 10.1016/j.watres.2006.07.011
10. Cesaro, A., Naddeo, V., and Belgiorno, V. (2013). Wastewater treatment by combination of advanced oxidation processes and conventional biological systems. *J. Bioremed. Biodeg.* 4:208. doi: 10.4172/2155-6199.1000208
11. Chen, J., Yang, C., Zhou, J., and Wang, X. (2007). Study of the influence of the electric field on membrane flux of a new type of membrane bioreactor. *Chem. Eng. J.* 128, 177–180. doi: 10.1016/j.cej.2006.10.010
12. Cicek, N. (2003). A review of membrane bioreactor and their potential application in the treatment of agricultural wastewater. *Can. Biosyst. Eng.* 45, 637–649.
13. de Luna, M. D. G., and Warmadewanthi Liu, J. C. (2009). Combined treatment of polishing wastewater and fluoride-containing wastewater from a semiconductor manufacturer. *Colloids Surf. A* 347, 64–68. doi: 10.1016/j.colsurfa.2008.12.006
14. Elabbas, S., Ouazzani, N., Mandi, L., Berrekhis, F., Perdicakis, M., Pontvianne, S., et al. (2016). Treatment of highly concentrated tannery wastewater using electrocoagulation: influence of the quality of aluminium used for the electrode. *J. Hazard. Mater.* 319, 69–77. doi: 10.1016/j.jhazmat.2015.12.067
15. Elektorowicz, M., Arian, Z., and Ibeid, S. (2014). "Submerged membrane electro-bioreactor for water recovery," in *Membranes and Membrane Processes in Environmental Protection*, Vol. 119, eds M. Bodzek and J. Pelczar (Monographs of the Environmental Engineering Committee, Polish Academy of Sciences), 93–98.
16. Emamjomeh, M. M., and Sivakumar, M. (2009). Review of pollutants removed by electrocoagulation and electrocoagulation/flotation processes. *J. Environ. Manage.* 90, 1663–1679. doi: 10.1016/j.jenvman.2008.12.011

17. Giwa, A., Ahmed, I., and Hasan, S. W. (2015). Enhanced sludge properties and distribution study of sludge components in electrically-enhanced membrane bioreactor. *J. Environ. Manage.* 159, 78–85. doi: 10.1016/j.jenvman.2015.05.035
18. Giwa, A., Daer, S., Ahmed, I., Marpu, P. R., and Hasan, S. W. (2016). Experimental investigation and artificial neural networks ANNs modeling of electrically-enhanced membrane bioreactor for wastewater treatment. *J. Water Process Eng.* 11, 88–97. doi: 10.1016/j.jwpe.2016.03.011
19. Giwa, A., and Hasan, S. W. (2015). Theoretical investigation of the influence of operating conditions on the treatment performance of an electrically-induced membrane bioreactor. *J. Water Process Eng.* 6, 72–82. doi: 10.1016/j.jwpe.2015.03.004
20. Hasan, S. W., Elektorowicz, M., and Oleszkiewicz, J. A. (2012). Correlations between trans-membrane pressure (TMP) and sludge properties in submerged membrane electro-bioreactor (SMEBR) and conventional membrane bioreactor (MBR). *Bioresour. Technol.* 120, 199–205. doi: 10.1016/j.biortech.2012.06.043
21. Hasan, S. W., Elektorowicz, M., and Oleszkiewicz, J. A. (2014). Start-up period investigation of pilot-scale submerged membrane electro-bioreactor (SMEBR) treating raw municipal wastewater. *Chemosphere* 97, 71–77. doi: 10.1016/j.chemosphere.2013.11.009
22. Heidrich, E. S., Curtis, T. P., and Dolfing, J. (2011). Determination of the internal chemical energy of wastewater. *Environ. Sci. Technol.* 45, 827–832. doi: 10.1021/es103058w
23. Hosseinzadeh, M., Bidhendi, G. N., Torabian, A., Mehrdadi, N., and Pourabdullah, M. (2015). A new flat sheet membrane bioreactor hybrid system for advanced treatment of effluent, reverse osmosis pretreatment and fouling mitigation. *Bioresour. Technol.* 192, 177–184. doi: 10.1016/j.biortech.2015.05.066
24. Ibeid, S., Elektorowicz, M., and Oleszkiewicz, J. A. (2013a). Novel electrokinetic approach reduces membrane fouling. *Water Res.* 47, 6358–6366. doi: 10.1016/j.watres.2013.08.007
25. Ibeid, S., Elektorowicz, M., and Oleszkiewicz, J. A. (2013b). Modification of activated sludge properties caused by application of continuous and intermittent current. *Water Res.* 47, 903–910. doi: 10.1016/j.watres.2012.11.020
26. Ibeid, S., Elektorowicz, M., and Oleszkiewicz, J. A. (2015). Electro-conditioning of activated sludge in a membrane electro-bioreactor for improved dewatering and reduced membrane fouling. *J. Membr. Sci.* 494, 136–142. doi: 10.1016/j.memsci.2015.07.051
27. Keerthi, S. V, Mahalakshmi, M., and Balasubramanian, N. (2013). Development of hybrid membrane bioreactor for tannery effluent treatment. *Desalination* 309, 231–236. doi: 10.1016/j.desal.2012.10.014
28. Kim, H. G., Jang, H. N., Kim, H. M., Lee, D. S., and Chung, T. H. (2010). Effect of an electro phosphorous removal process on phosphorous removal and membrane permeability in a pilot-scale MBR. *Desalination* 250, 629–633. doi: 10.1016/j.desal.2009.09.038
29. Kimura, K., Yamato, N., Yamamura, H., and Watanabe, Y. (2005). Membrane fouling in pilot-scale membrane bioreactors (MBRs) treating municipal wastewater. *Environ. Sci. Technol.* 39, 6293–6299. doi: 10.1021/es0502425
30. Li, X. G., Cao, H. B., Wu, J. C., and Yu, K. T. (2001). Inhibition of the metabolism of nitrifying bacteria by direct electric current. *Biotechnol. Lett.* 23, 705–709. doi: 10.1023/A:1010346501857
31. Li, Y., Liu, L., Liu, J., Yang, F., and Ren, N. (2014). PPy/AQS (9, 10-anthraquinone-2-sulfonic acid) and PPy/ARS (Alizarin Red's) modified stainless steel mesh as cathode membrane in an integrated MBR/MFC system. *Desalination* 349, 94–101. doi: 10.1016/j.desal.2014.06.027
32. Liu, J., Liu, L., Gao, B., and Yang, F. (2013a). Integration of bio-electrochemical cell in membrane bioreactor for membrane cathode fouling reduction through electricity generation. *J. Membr. Sci.* 430, 196–202. doi: 10.1016/j.memsci.2012.11.046
33. Liu, J., Liu, L., Gao, B., Yang, F., Crittenden, J., and Ren, N. (2014). Integration of microbial fuel cell with independent membrane cathode bioreactor for power generation, membrane fouling mitigation and wastewater treatment. *Int. J. Hydrog. Energy* 39, 17865–17872. doi: 10.1016/j.ijhydene.2014.08.123

34. Liu, L., Liu, J., Bo, G., Yang, F., Crittenden, J., and Chen, Y. (2013c). Conductive and hydrophilic polypyrrole modified membrane cathodes and fouling reduction in MBR. *J. Membr. Sci.* 429, 252–258. doi: 10.1016/j.memsci.2012.11.066
35. Liu, L., Liu, J., Gao, B., and Yang, F. (2012a). Minute electric field reduced membrane fouling and improved performance of membrane bioreactor. *Sep. Purif. Technol.* 86, 106–112. doi: 10.1016/j.seppur.2011.10.030
36. Liu, L., Liu, J., Gao, B., Yang, F., and Chellam, S. (2012b). Fouling reductions in a membrane bioreactor using an intermittent electric field and cathodic membrane modified by vapor phase polymerized pyrrole. *J. Membr. Sci.* 394–395, 202–208. doi: 10.1016/j.memsci.2011.12.042
37. Liu, L., Zhao, F., Liu, J., and Yang, F. (2013b). Preparation of highly conductive cathodic membrane with graphene (oxide)/PPy and the membrane antifouling property in filtrating yeast suspensions in EMBR. *J. Membr. Sci.* 437, 99–107. doi: 10.1016/j.memsci.2013.02.045
38. Ma, D., Gao, B., Hou, D., Wang, Y., Yue, Q., and Li, Q. (2013). Evaluation of a submerged membrane bioreactor (SMBR) coupled with chlorine disinfection for municipal wastewater treatment and reuse. *Desalination* 313, 134–139. doi: 10.1016/j.desal.2012.12.016
39. Meng, F., Yang, F., Shi, B., and Zhang, H. (2008). A comprehensive study on membrane fouling in submerged membrane bioreactors operated under different aeration intensities. *Sep. Purif. Technol.* 59, 91–100. doi: 10.1016/j.seppur.2007.05.040
40. Naddeo, V., Belgiorno, V., Borea, L., Secondes, M. F. N., and Ballesteros, F. Jr. (2015a). Control of fouling formation in membrane ultrafiltration by ultrasound irradiation. *Environ. Technol.* 36, 1299–1307. doi: 10.1080/09593330.2014.985731
41. Naddeo, V., Borea, L., and Belgiorno, V. (2015b). Sonochemical control of fouling formation in membrane ultrafiltration of wastewater: Effect of ultrasonic frequency. *J. Water Process Eng.* 8, e92–e97. doi: 10.1016/j.jwpe.2014.12.005
42. Naddeo, V., Rizzo, L., and Belgiorno, V. (2011). *Water, Wastewater and Soil Treatment by Advanced Oxidation Processes*. Fisciano, SA. Available online at: www.Lulu.com
43. Palogos, I., Babatsouli, P., Costa, C., and Kalogerakis, N. (2014). Computer simulation of a submerged membrane bioreactor treating high COD industrial wastewater. *Front. Environ. Sci.* 2:30. doi: 10.3389/fenvs.2014.00030
44. Rahimi, Y., Torabian, A., Mehrdadi, N., Habibi-rezaie, M., Pezeshk, H., and Nabi-bidhendi, G. R. (2011). Optimizing aeration rates for minimizing membrane fouling and its effect on sludge characteristics in a moving bed membrane bioreactor. *J. Hazard. Mater.* 186, 1097–1102. doi: 10.1016/j.jhazmat.2010.11.117
45. Scannapieco, D., Naddeo, V., and Belgiorno, V. (2015). Control of fouling in MBRs through nanospheres addition. *Desalin. Water Treat.* 55, 702–711. doi: 10.1080/19443994.2014.942382
46. Secondes, M. F. N., Naddeo, V., Belgiorno, V., and Ballesteros, F. Jr. (2014). Removal of emerging contaminants by simultaneous application of membrane ultrafiltration, activated carbon adsorption, and ultrasound irradiation. *J. Hazard. Mater.* 264, 342–349. doi: 10.1016/j.jhazmat.2013.11.039
47. Shi, X., Tal, G., Hankins, N. P., and Gitis, V. (2014). Fouling and cleaning of ultrafiltration membranes: a review. *J. Water Process Eng.* 1, 121–138. doi: 10.1016/j.jwpe.2014.04.003
48. Singhal, N., and Perez-Garcia, O. (2016). Degrading organic micropollutants: the next challenge in the evolution of biological wastewater treatment processes. *Front. Environ. Sci.* 4:36. doi: 10.3389/fenvs.2016.00036
49. Tafti, A. D., Seyyed Mirzaii, S. M., Andalibi, M. R., and Vossoughi, M. (2015). Optimized coupling of an intermittent DC electric field with a membrane bioreactor for enhanced effluent quality and hindered membrane fouling. *Sep. Purif. Technol.* 152, 7–13. doi: 10.1016/j.seppur.2015.07.004
50. Wang, H., Luo, H., Fallgren, P. H., Jin, S., and Ren, Z. J. (2015). Bioelectrochemical system platform for sustainable environmental remediation and energy generation. *Biotechnol. Adv.* 33, 317–334. doi: 10.1016/j.biotechadv.2015.04.003

51. Wang, Y. P., Liu, X. W., Li, W. W., Li, F., Wang, Y. K., Sheng, G. P., et al. (2012). A microbial fuel cell-membrane bioreactor integrated system for cost-effective wastewater treatment. *Appl. Energy* 98, 230–235. doi: 10.1016/j.apenergy.2012.03.029
52. Wei, C. H., Huang, X., Ben Aim, R., Yamamoto, K., and Amy, G. (2011). Critical flux and chemical cleaning-in-place during the long-term operation of a pilot-scale submerged membrane bioreactor for municipal wastewater treatment. *Water Res.* 45, 863–871. doi: 10.1016/j.watres.2010.09.021
53. Wei, V., Elektorowicz, M., and Oleszkiewicz, J. A. (2011). Influence of electric current on bacterial viability in wastewater treatment. *Water Res.* 45, 5058–5062. doi: 10. 1016/j.watres.2011.07.011
54. Wei, V., Oleszkiewicz, J. A., and Elektorowicz, M. (2009). Nutrient removal in an electrically enhanced membrane bioreactor. *Water Sci. Technol.* 60, 3159–3163. doi: 10.2166/wst.2009.625
55. Weiss, S., and Reemtsma, T. (2008). Membrane bioreactors for municipal wastewater treatment - A viable option to reduce the amount of polar pollutants discharged into surface waters? *Water Res.* 42, 3837–3847. doi: 10.1016/j.watres.2008.05.019
56. Xing, C., Tardieu, E., Qian, Y., and Wen, X. (2000). Ultrafiltration membrane bioreactor for urban wastewater reclamation. *J. Membr. Sci.* 177, 73–82. doi: 10.1016/S0376-7388(00)00452-X
57. Yigit, N. O., Uzal, N., Koseoglu, H., Harman, I., Yukseler, H., Yetis, U., et al. (2009). Treatment of a denim producing textile industry wastewater using pilot-scale membrane bioreactor. *Desalination* 240, 143–150. doi: 10.1016/j.desal.2007.11.071
58. Yuniarto, A., Ujang, Z., and Noor, Z. Z. (2008). Performance of bio-fouling reducers in aerobic submerged membrane bioreactor for palm oil mill effluent treatment performance of bio-fouling reducers in aerobic submerged membrane bioreactor for palm oil mill effluent treatment. *Management* 49, 555–566.
59. Zeyoudi, M., Altenaiji, E., Ozer, L. Y., Ahmed, I., Yousef, A. F., and Hasan, S. W. (2015). Impact of continuous and intermittent supply of electric field on the function and microbial community of wastewater treatment electro-bioreactors. *Electrochim. Acta* 181, 271–279. doi: 10.1016/j.electacta.2015.04.095
60. Zhang, F., Li, J., and He, Z. (2014). A new method for nutrients removal and recovery from wastewater using a bioelectrochemical system. *Bioresour. Technol.* 166, 630–634. doi: 10.1016/j.biortech.2014.05.105
61. Zhang, J., Satti, A., Chen, X., Xiao, K., Sun, J., Yan, X., et al. (2015). Low-voltage electric field applied into MBR for fouling suppression: performance and mechanisms. *Chem. Eng. J.* 273, 223–230. doi: 10.1016/j.cej.2015.03.044
62. Zhang, J., Xiao, K., Liang, P., Waite, T. D., and Huang, X. (2014). Electrically released iron for fouling control in membrane bioreactors: a double-edged sword? *Desalination* 347, 10–14. doi: 10.1016/j.desal.2014.05.018
63. Zhang, K., Wei, P., Yao, M., Field, R. W., and Cui, Z. (2011). Effect of the bubbling regimes on the performance and energy cost of flat sheet MBRs. *Desalination* 283, 221–226. doi: 10.1016/j.desal.2011.04.023
64. Zhou, L., Xia, S., and Alvarez-Cohen, L. (2015). Structure and distribution of inorganic components in the cake layer of a membrane bioreactor treating municipal wastewater. *Bioresour. Technol.* 196, 586–591. doi: 10.1016/j.biortech.2015.08.026

INDEX

L

M

N

O

P

R

S